T0136982

Advances in Intelligent Systems and Computing

Volume 958

The series "Advances in Intelligent Systems and Computing" contains publications on theory, applications, and design methods of Intelligent Systems and Intelligent Computing. Virtually all disciplines such as engineering, natural sciences, computer and information science, ICT, economics, business, e-commerce, environment, healthcare, life science are covered. The list of topics spans all the areas of modern intelligent systems and computing such as: computational intelligence, soft computing including neural networks, fuzzy systems, evolutionary computing and the fusion of these paradigms, social intelligence, ambient intelligence, computational neuroscience, artificial life, virtual worlds and society, cognitive science and systems, Perception and Vision, DNA and immune based systems, self-organizing and adaptive systems, e-Learning and teaching, human-centered and human-centric computing, recommender systems, intelligent control, robotics and mechatronics including human-machine teaming, knowledge-based paradigms, learning paradigms, machine ethics, intelligent data analysis, knowledge management, intelligent agents, intelligent decision making and support, intelligent network security, trust management, interactive entertainment, Web intelligence and multimedia.

The publications within "Advances in Intelligent Systems and Computing" are primarily proceedings of important conferences, symposia and congresses. They cover significant recent developments in the field, both of a foundational and applicable character. An important characteristic feature of the series is the short publication time and world-wide distribution. This permits a rapid and broad dissemination of research results.

** Indexing: The books of this series are submitted to ISI Proceedings, EI-Compendex, DBLP, SCOPUS, Google Scholar and Springerlink **

More information about this series at http://www.springer.com/series/11156

Daniel N. Cassenti
Editor

Advances in Human Factors and Simulation

Proceedings of the AHFE 2019 International
Conference on Human Factors
and Simulation, July 24–28, 2019,
Washington D.C., USA

 Springer

Editor
Daniel N. Cassenti
RDRL-HRF-D
U.S. Army Research Laboratory
Aberdeen Proving Ground, MD, USA

ISSN 2194-5357   ISSN 2194-5365   (electronic)
Advances in Intelligent Systems and Computing
ISBN 978-3-030-20147-0   ISBN 978-3-030-20148-7   (eBook)
https://doi.org/10.1007/978-3-030-20148-7

This Springer imprint is published by the registered company Springer Nature Switzerland AG
The registered company address is: Gewerbestrasse 11, 6330 Cham, Switzerland

Advances in Human Factors
and Ergonomics 2019

AHFE 2019 Series Editors

Tareq Ahram, Florida, USA
Waldemar Karwowski, Florida, USA

10th International Conference on Applied Human Factors and Ergonomics and the
Affiliated Conferences

Proceedings of the AHFE 2019 International Conference on Human Factors and
Simulation, held on July 24–28, 2019, in Washington D.C., USA

Advances in Affective and Pleasurable Design	Shuichi Fukuda
Advances in Neuroergonomics and Cognitive Engineering	Hasan Ayaz
Advances in Design for Inclusion	Giuseppe Di Bucchianico
Advances in Ergonomics in Design	Francisco Rebelo and Marcelo M. Soares
Advances in Human Error, Reliability, Resilience, and Performance	Ronald L. Boring
Advances in Human Factors and Ergonomics in Healthcare and Medical Devices	Nancy J. Lightner and Jay Kalra
Advances in Human Factors and Simulation	Daniel N. Cassenti
Advances in Human Factors and Systems Interaction	Isabel L. Nunes
Advances in Human Factors in Cybersecurity	Tareq Ahram and Waldemar Karwowski
Advances in Human Factors, Business Management and Leadership	Jussi Ilari Kantola and Salman Nazir
Advances in Human Factors in Robots and Unmanned Systems	Jessie Chen
Advances in Human Factors in Training, Education, and Learning Sciences	Waldemar Karwowski, Tareq Ahram and Salman Nazir
Advances in Human Factors of Transportation	Neville Stanton

(continued)

(continued)

Advances in Artificial Intelligence, Software and Systems Engineering	Tareq Ahram
Advances in Human Factors in Architecture, Sustainable Urban Planning and Infrastructure	Jerzy Charytonowicz and Christianne Falcão
Advances in Physical Ergonomics and Human Factors	Ravindra S. Goonetilleke and Waldemar Karwowski
Advances in Interdisciplinary Practice in Industrial Design	Cliff Sungsoo Shin
Advances in Safety Management and Human Factors	Pedro M. Arezes
Advances in Social and Occupational Ergonomics	Richard H. M. Goossens and Atsuo Murata
Advances in Manufacturing, Production Management and Process Control	Waldemar Karwowski, Stefan Trzcielinski and Beata Mrugalska
Advances in Usability and User Experience	Tareq Ahram and Christianne Falcão
Advances in Human Factors in Wearable Technologies and Game Design	Tareq Ahram
Advances in Human Factors in Communication of Design	Amic G. Ho
Advances in Additive Manufacturing, Modeling Systems and 3D Prototyping	Massimo Di Nicolantonio, Emilio Rossi and Thomas Alexander

Preface

This volume is a compilation of cutting-edge research regarding how simulation and modeling support human factors. The compilation of the chapters is the result of efforts by the 9th International Conference on Applied Human Factors and Ergonomics (AHFE), which provides the organization for several affiliated conferences. Specifically, the chapters herein represent the 3rd International Conference on Human Factors and Simulation and the 7th International Conference on Digital Human Modeling and Applied Optimization.

Simulation is a technology that supports an approximation of real-world scenes and scenarios for a user. For example, a cockpit simulator represents the configuration of the inside of a cockpit and presents a sensory and motor experience to mimic flight. Simulations advance research by providing similar experiences to those scenarios that would otherwise be impractical to carry out in the real world for such reasons as monetary cost or safety concerns. Simulations can support numerous goals including training or practice on established skills.

Modeling is a somewhat different tool than simulation, though the two are often used interchangeably as they both imply estimation of real-world scenes or scenarios that bypass practical concerns. The difference in the context of this book is that modeling is not intended to provide a user with an experience, but rather to represent anything pertinent about the real world in computational algorithms, possibly including people and their psychological processing. Modeling may answer questions about large-scale scenarios that would be difficult to address otherwise, such as the effects of economic interventions, or smaller-scale scenarios such as the cognitive processing required to perform a task that is otherwise undetectable by measurement devices.

The goal of the research herein is to bring awareness and attention to advances that human factors specialists may make in their field to address the design of programs of research, systems, policies, and devices. This book provides a plethora of avenues for human factors research that may be helped by simulation and modeling.

This book is divided into the following sections:

Section 1   Advances in computational social sciences and simulated agents
Section 2   Virtual and augmented reality
Section 3   Human factors and simulation in transportation
Section 4   Physical, mental, and social effects in simulation
Section 5   Application for occupational safety

All papers in this book were either reviewed or contributed by the members of the editorial board. For this, I would like to recognize Board Members listed below:

Human Factors and Simulation

Hanon Alnizami, USA
Jasbir Arora, USA
Rajan Bhatt, USA
Patrick Craven, USA
Brianna Eiter, USA
Brian Gore, USA
Javier Irizarry, USA
Tiffany Jastrzembski, USA
Catherine Neubauer, USA
Debra Patton, USA
Brandon Perelman, USA
Teresita Sotomayor, USA
Simon Su, USA
Ming Sun, USA
Julia Wright, USA
Zining Yang, USA

This book covers diverse topics in simulation and modeling. I hope this book is informative and helpful for the researchers and practitioners in developing better products, services, and systems.

July 2019   Daniel N. Cassenti

Contents

Physical, Mental and Social Effects in Simulation

Application for Occupational Safety

Advances in Computational Social Sciences and Simulated Agents

Simulated Marksmanship Performance Methodology: Assessing Lethality, Mobility and Stability Across the Preparation, Execution and Recovery Stages of a Military Field Training Exercise

Stephanie A. T. Brown[(✉)], Jose Villa, Erika K. Hussey,
John W. Ramsay, and K. Blake Mitchell

Combat Capabilities Development Command, Soldier Center,
Natick, MA 01760, USA
stephanie.a.brown88.civ@mail.mil

Abstract. Marksmanship, a key cornerstone of military training, is one area of military assessment that includes standardized quantifiable measures. However, assessment of marksmanship in a traditional live-fire setting can be costly, time consuming, and dangerous, while frequently only providing rudimentary objective measures of performance. This research created an enhanced combined marksmanship assessment methodology, which builds on earlier static and dynamic methodologies. The successful portions of previous methods, to include static and dynamic shooting with acquisition assessments, were integrated and additional pertinent assessment areas were added (i.e., targets of varying height and increased distance to force gross movements in transitions across engagements), while minimizing execution time and still using a mobile, low-cost weapon simulator. This methodology is executable in any setting, is easy to assemble, provides streamlined metrics on the entire marksmanship process across two critical shooting styles, and can track changes in marksmanship across various performance periods throughout a training cycle.

Keywords: Human-systems integration · Human factors · Military ·
Marksmanship · Simulation and training · Test and evaluation · Lethality ·
Soldier performance · Soldier readiness

1 Introduction

Military training replicates operational tasks and missions at various levels in order to focus on skills required to maximize performance, minimize risk and increase probability of mission accomplishment. Marksmanship is one area of military assessment that includes standardized quantifiable measures. Shooting proficiency has been utilized for both training qualifications and assessment of military equipment [1–11].

This is a U.S. government work and not under copyright protection in the U.S.;
foreign copyright protection may apply 2020
D. N. Cassenti (Ed.): AHFE 2019, AISC 958, pp. 3–13, 2020.
https://doi.org/10.1007/978-3-030-20148-7_1

However, assessment of marksmanship in a traditional live-fire setting can be costly, time consuming, and dangerous, while frequently only providing rudimentary objective measures of performance such as count of shot hits, misses, and shot group dispersion [12]. Methodologies to assess marksmanship performance have recently been established using the Fabrique National (FN) American simulator systems (formerly Noptel Oy), but most of these methods focus on the effects of clothing and individual equipment (CIE) [13–18]. These methodologies provide a cost effective and efficient alternative to evaluate marksmanship performance, while still being operationally relevant, and allow for objective measures without the safety concerns and risks that are associated with live-fire. Other research has utilized similar simulator systems to assess Soldier performance in a marksmanship scenario while evaluating a variety of physical [19–22], physiological [23, 24], or psychological [25, 26] attributes and conditions in a lab-based setting. All of these systems are limited by their source of recoil (i.e., tethered to a compressed gas tank or power source), thus restricting the methodologies from being transferrable to the field setting. There is a need for a streamlined methodology that is easy to set up, evaluate, and can be utilized in the field where the Soldiers are training and fighting.

This current study was conducted to evaluate a refined marksmanship methodology that combines multiple shooting styles and target distances, and incorporates streamlined metrics that assess the entire marksmanship process in a scenario that can be executed in under five minutes. In addition, this methodology utilizes the FN Expert weapon simulator and a demilitarized M4 (although other weapons can be used), allowing it to be flexible and able to be set up in a field setting. This study assessed the methodology to see if it can detect performance changes over the course of a 72-hour field training exercise. This quick, combined methodology provides the benefit of mobile and dynamic marksmanship assessment in the middle of other active operational field performance tests without any delay or break in testing. Additionally, this enhanced methodology integrates a static self-paced shooting task with a dynamic fast-paced shooting task in order to capture both suppressive and combat shooting requirements in one scenario. This paper reports the initial findings, focusing on the overarching metrics of lethality, mobility, and stability across the all target engagements.

2   Methods

2.1   Study Participants

Study volunteers consisted of forty-six active duty Soldiers (primarily infantry). These participants were predominantly males (4 females) between the ages of 18 and 37 years ($M = 24.5$, $SD = 4.2$). There were five participants who were left-handed (11%), and five who wore glasses (11%). All were qualified "marksman" through the Army Basic Marksmanship qualification process using the M4 carbine. Three (6.5%) were classified as Marksmen (score of 23–29 out of 40 on the standard marksmanship test), three (6.5%) were Sharpshooters (score of 30–35), and the additional forty (87%) were Experts (score of 36+).

2.2   Test Apparatus

All testing utilized the FN Expert simulator and associated NOS pro software. The system's optical unit was mounted to the picatinny rail on the right side of the barrel of a de-militarized M4 carbine with an integrated carbon dioxide recoil simulation system manufactured by LaserShot, Inc. A M68 close combat optic (CCO) sighting system was also utilized in this testing scenario. This system was mounted on the picatinny rail section on top of the weapon receiver. The FN Expert optical unit was mechanically zeroed to the CCO, utilizing the standard procedures as laid out in the product manual [27].

Five paper ring targets with diamond graded reflector rings were used in this methodology, four scaled to 75-meters and one 150-meters when placed at an actual distance of 5-meters. These special targets reflect the infrared beam from the FN Expert optical unit, providing x, y coordinates for aiming points and shot locations to the NOS pro software via wireless Bluetooth. The targets were set up as depicted in Fig. 1 below. Pre- and post-mission testing occurred in an enclosed hanger bay facility. Mid-mission testing occurred outside in a level grassy field at the training exercise location.

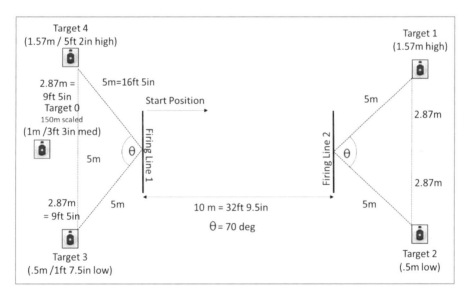

Fig. 1. Diagram of the combined static and dynamic marksmanship methodology layout, with distance and heights labeled for each target. Target 0 is 150-meter scaled, while targets 1–4 are 75-meter scaled.

2.3   Test Procedures

Marksmanship performance was assessed at three time points across a 2-week training period containing a 72-hour mission exercise (with multiple assessments at pre-, mid-, and post-mission exercise) utilizing traditional measures of lethality (e.g., accuracy).

Supplemental measures of mobility (e.g. acquisition and engagement time) and weapon handling (e.g., stability) were also collected in order to assess the entire marksmanship process of locating, moving, positioning, and engaging the targets.

At the beginning of each testing session, the weapon was mechanically zeroed by the data collector. This acted as gross adjustment and zeroing for the weapon sights, with minor zeroing adjustments for each individual made via the software. The individual software zeroing consisted of shooting 3 shots as accurately as possible at a target in the prone position. The simulator's associated software then moves the shot grouping to the center of the target, adjusting the subsequent shots during the scenario. This process simulates the zeroing process that would occur when using a live weapon and sighting system.

Next, the scenario begins with a static task of firing three series of five shots (15 shots total) at the 150-meter simulated distance target, with a priority on accuracy over speed. Upon completion of the third series of static shooting, the test participant is given a ten-second countdown to initiate part two. The second part of the scenario is a fast-paced dynamic task, requiring the participant to sprint 10-meters to a second firing line, engage two targets spread across a 70° arc with two sets of controlled pair shots (4 shots per target total) at a simulated distance of 75-meter at various heights. Upon completion of engagement, the participant must sprint back to the original firing line, and engage two more targets with the same shot requirements. Order of target engagement was based on the participant's dominant shooting side in order to ensure assessment of a weak-side transition (right handed individuals engage the left target first and move to the right, and left handed individuals engage the right target first and move to the left). Upon completion of the first set, the participant is given a 60 s recovery period and the entire combined scenario is then repeated in a different firing position as randomly assigned (i.e., unsupported standing or unsupported prone). The required table of fires is shown in Table 1 below. Figure 1 shows a diagram of the target placement and heights. This combination of tasks not only quantifies marksmanship performance across the entire marksmanship process from acquisition to engagement and transition, it can also focus in on additional skills such as low to high and high to low transitions.

Table 1. Table of fires for combined marksmanship methodology

Trial no.		Scenario section	Firing position[a]	No. of trials	No. shots/trial	Total no. shots
I	A.	One target	Standing unsupported	3	5	15
	B.	Four targets	Standing unsupported	1	16	16
II	A.	One target	Prone unsupported	3	5	15
	B.	Four targets	Prone unsupported	1	16	16

[a]Order of firing position randomized across trial

2.4   Measures of Marksmanship Performances

Lethality. Shooting lethality was measured utilizing the shot accuracy, or the distance of a single shot to the target center. This is calculated using the Euclidian distance in millimeters of the shot to the bull's eye, or center of mass of the target [28] as seen in Fig. 2.

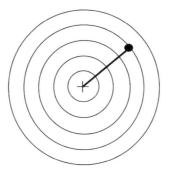

Fig. 2. Visual image of shooting accuracy, or distance from shot to the center of the target.

Mobility. The marksmanship process includes threat detection, movement, positioning, sighting, aiming, and engagement. The mobility metric utilized here includes the target acquisition (i.e., move, detect, position) and target engagement (i.e., aim, shoot, adjust, shoot, etc.) as depicted in Fig. 3.

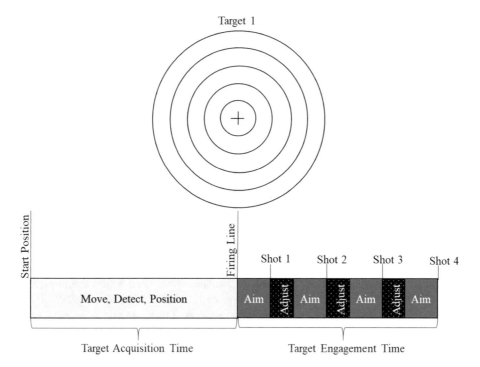

Fig. 3. Visual representation of target acquisition (i.e., move, detect, position) and target engagement (i.e., aim, shoot, adjust, aim, shoot, etc.) measurements per target.

Stability. Weapon handling and barrel stability describes the aiming and degree of movement prior to shot execution. The FN Expert software records aim trace data points every .15 s prior to shot (up to 2.99 s). Stability is measured here as the area of aiming during the critical aiming window, or the last .60 s to .20 s prior to shot [17], as depicted in Fig. 4.

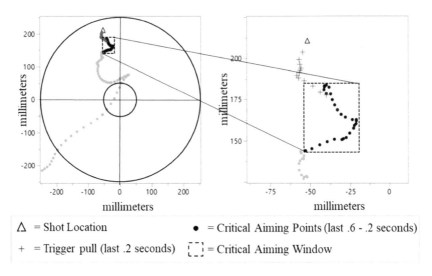

△ = Shot Location   ● = Critical Aiming Points (last .6 - .2 seconds)

+ = Trigger pull (last .2 seconds)   ⌐ ¬ = Critical Aiming Window

Fig. 4. Visual image of overall stability, as measured by the area of movement within the critical aiming window (last .6 to .2 s of aiming prior to engagement)

2.5   Data Analysis

The study used a repeated measures experimental design across a 72-h training mission (pre-, mid-, and post-) with counterbalancing to control for order of firing position exposure (i.e., order effects). The statistical analyses were conducted to investigate mean shooting performance across the entire marksmanship process in the areas of lethality, mobility, and stability utilizing within-subjects repeated measures analysis of variance (ANOVAs). Tests of multiple comparisons were conducted using the Tukey Honestly Significant Differences (HSD). Confidence intervals were set at 95% (alpha = .05).

3   Results

Significant differences in marksmanship performance were seen across the three mission time points for the lethality, mobility, and stability measures. Lethality and mobility performance declined at mid-mission compared to pre-mission, but rebounded at post-mission; whereas, stability measures degraded in the post-mission session. These differences indicate that the methodology is able to identify changes in marksmanship performance over time due to mission activities and recovery.

3.1   Lethality

Analysis of marksmanship lethality as measured by mean accuracy across all targets revealed a main effect of session, $F(2, 84.1) = 6.365, p = .003$. As seen in Fig. 5, the Soldiers had significantly greater shot accuracy during the pre- ($M = 916$ mm, $SD = 202$ mm) and post- ($M = 884$ mm, $SD = 210$ mm) mission test sessions as compared to the mid-mission session ($M = 1021$ mm, $SD = 328$ mm).

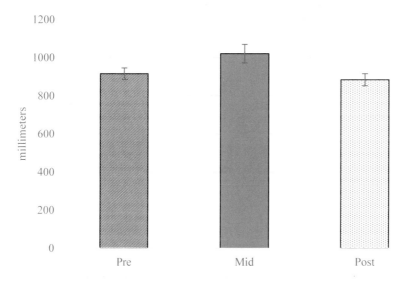

Fig. 5. Lethality as measured by shot accuracy across the pre-, mid-, and post-mission time points (error bars represent Standard Error).

3.2   Mobility

Analysis of mobility in the marksmanship process as measured by combined mean target acquisition and target engagement across all the targets revealed a main effect of session, $F(2, 85.11) = 3.59, p = .032$. As seen in Fig. 6, Soldiers took significantly less time to acquire and engage the targets during the pre- (M = 6.06 s, SD = 1.27 s) and post- (M = 6.01 s, SD = 1.97 s) mission test sessions as compared to the mid-mission session (M = 6.48 s, SD = 1.32 s).

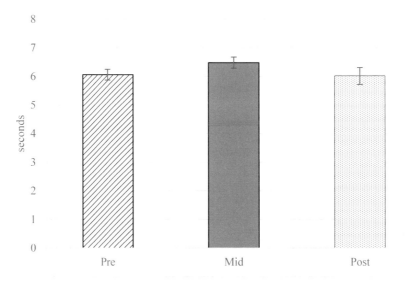

Fig. 6. Mobility, as measured by combined target acquisition time and target engagement time, across the pre-, mid-, and post-mission time points (error bars represent Standard Error).

3.3   Stability

Analysis of weapon handling stability in the marksmanship process as measured by mean area of aiming movement during the last .60 s prior to shot across all the targets revealed a main effect of session, $F(2, 85.29) = 4.15$, $p = .019$. As seen in Fig. 7, the Soldiers moved more during aiming, covering significantly larger areas and were less stable handling their weapon during the post-session (M = 116890 mm^2, SD = 90471 mm^2) as compared to the pre- (M = 86092 mm^2, SD = 46029 mm^2) and mid-mission sessions (M = 83360 mm^2, SD = 40660 mm^2).

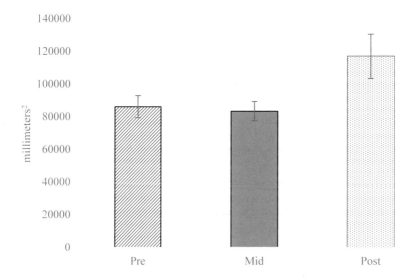

Fig. 7. Stability, as measured by the overall area of aiming points, within the last .60 to .20 s of aiming prior to engagement across the pre-, mid-, and post-mission time points (error bars represent Standard Error).

4 Discussion

The differences seen across the marksmanship measures of lethality, mobility and stability indicate that the methodology is able to identify changes in marksmanship performance over time due to mission activities and recovery. The ability to streamline three high level outputs of marksmanship performance that cover the entire process from acquisition through engagement is also very important for researchers in order to understand areas of degradation over time and to provide training feedback. However, the differences in performance could have been influenced by a variety of environmental factors which we had limited control over (e.g., time of day during test execution, test location variances, temperature and humidity variances). For example, although the outdoor testing area was flat, level, and hard-packed, it was not as hard and smooth as the cement floors indoors. This could account for the slower target acquisition and engagement times during the mid-mission test session rather than fatigue-related degradations in performance. Future testing of this methodology should utilize identical locations in order to limit the external noise when modeling performance over time.

Additionally, although a simulator system provides many benefits (e.g., flexibility, reduced cost, increased safety), there are limitations to consider as well (e.g., reduced realism and limited recoil effects resulting in potential changes in user behavior and performance). Future applications of this methodology could incorporate live-fire training with minor modifications while still integrating the various key methodological elements that provide streamlined output metrics in the areas of lethality, mobility, and stability.

This enhanced methodology is fast to execute, yet still provides sufficient information for accurate assessment of mission-related marksmanship performance and fatigue. This methodology is unique as it combines multiple shooting skills as required by the Army, in order to provide a single assessment of marksmanship performance in the areas of lethality, mobility, and stability. Additionally, the design of this methodology allows for additional in-depth analysis to pinpoint the areas of deficiency as necessary (i.e., target acquisition, target transition, transition type, approach shot versus transitional shot, firing position, etc.).

Acknowledgments. This research was funded by Combat Capabilities Development Command, Soldier Center research program 18-101. The authors would also like to thank the data collection team and the program coordinators for ensuring smooth execution of these events. Most importantly, the authors would like to acknowledge the Soldiers who participated in this study.

References

1. Adams, A.A.: Effects of extremity armor on metabolic cost and gait biomechanics. Master's thesis, Worcester Polytechnic Institute, MA (2010)
2. Bensel, C.K.: Soldier performance and functionality: impact of chemical protective clothing. Mil. Psychol. **9**(4), 287–302 (1997)
3. Bewley, W.L., Chung, G.K., Girlie, C.D.: Research on USMC Marksmanship Training Assessment Tools, Instructional Simulations, and Qualitative Field-Based Research. (UCLA CSE/CRESST N00014-02-1-0179). Los Angeles, CA (2003)
4. Carbone, P.D., Carlton, S.D., Stierli, M., Orr, R.M.: The impact of load carriage on the marksmanship of the tactical police officer: a pilot study. J. Aust. Strength Cond. **22**(2), 50–57 (2014)
5. Garrett, L., Jarboe, N., Patton, D.J., Mullins, L.L.: The effects of encapsulation on dismounted warrior performance. Technical report No. ARL-TR-3789. Aberdeen Proving Ground, M.D: U.S. Army Research Laboratory (2006)
6. Headquarters DOA: Rifle and Carbine. (TC 3-22.9 C1). Washington, DC (2017)
7. Johnson, R.F., Kobrick, J.L.: Effects of wearing chemical protective clothing on rifle marksmanship and on sensory and psychomotor tasks. Mil. Psychol. **9**(4), 301–314 (1997)
8. Johnson, R.F., McMenemy, D.J., Dauphinee, D.T.: Rifle marksmanship with three types of combat clothing. In: Proceedings of the Human Factors Society 34th Annual Meeting, pp. 1529–1532 (1990)
9. Krueger, G.P., Banderet, L.E.: Effects of chemical protective clothing on military performance: a review of the issues. Mil. Psychol. **9**(4), 255–286 (1997)
10. Son, S., Xia, Y., Tochihara, Y.: Evaluation of the effects of various clothing conditions on firefighter mobility and the validity of those measurements made. J Hum. Environ. Syst. **13**(1), 15–24 (2010)
11. Taylor, H.L., Orlansky, J.: The effects of wearing protective chemical warfare combat clothing on human performance. (No. IDA Paper P-2433). Alexandria, VA: Institute for Defense Analysis (1991)

12. Department of the Army: Rifle marksmanship M16-/M4-series weapons (FM 3-22.9). Washington, D.C. (2008)
13. Brown, S.A.T., McNamara, J.A., Choi, H.J., Mitchell, K.B.: Assessment of a marksmanship simulator as a tool for clothing and individual equipment evaluation. In: Proceedings of the Human Factors Society 60th Annual Meeting, pp. 1424–1428 (2016)
14. Choi, H.J., Mitchell, K.B., Garlie, T., McNamara, J., Hennessy, E., Carson, J.: Effects of body armor fit on marksmanship performance. In: Advances in Physical Ergonomics and Human Factors, pp. 341–354 (2016)
15. McNamara, J., Choi, H.J., Brown, S.A.T., Mitchell, K.: Evaluating the effects of clothing and individual equipment on marksmanship performance using a novel five target methodology. In: Proceedings of the Human Factors Society 60th Annual Meeting, pp. 2043–2047 (2016)
16. Brown, S.A.T, McNamara, J., Mitchell, K.B.: Dynamic marksmanship: a novel methodology to evaluate the effects of clothing and individual equipment on mission performance. In: Proceedings of the Human Factors Society 61th Annual Meeting, pp. 2020–2024 (2017)
17. Brown, S.A.T., Mitchell, K.: Shooting stability: a critical component of marksmanship performance as measured through aim path and trigger control. In: Proceedings of the Human Factors Society 61st Annual Meeting, vol. 61, no. 1, pp. 1476–1480 (2017)
18. Brown, S.A.T., Villa, J., Hancock, C.L., Hasselquist, L.: Stance and transition movement effects on marksmanship performance of Shooters during a simulated multiple target engagement task. In: Proceedings of the Human Factors Society 62nd Annual Meeting, pp. 1444–1448 (2018)
19. Baca, A., Kornfeind, P.: Stability analysis of motion patterns in biathlon shooting. Hum. Mov. Sci. 31(2), 295–302 (2012)
20. Hawkins, R.N., Sefton, J.M.: Effects of stance width on performance and postural stability in national-standard pistol shooters. J. Sports Sci. 29(13), 1381–1387 (2011)
21. Lawson, B.D., Ranes, B. M., Thompson, L.B.I.: Smooth moves: Shooting performance is related to efficiency of rifle movement. In: Proceedings of the Human Factors and Ergonomics Society Annual Meeting, pp. 1524–1528 (2016)
22. Shorter, P.L., Morelli, F., Ortega, S.: Shooting Performance as a Function of Shooters' Anthropometrics, Weapon Design Attributes, Firing Position, Range, and Sex (No. ARL-TR-7135). Army Research Lab Aberdeen Proving Ground MD Human Research and Engineering Directorate (2014)
23. Tharion, W.J., Shukitt-Hale, B., Lieberman, H.R.: Caffeine effects on marksmanship during high-stress military training with 72 hours sleep deprivation. Aviat. Space Env. Med. 74(4), 309–314 (2003)
24. Warber, J.P., Tharion, W.J., Patton, J.E.: The effect of creatine monohydrate supplementation on obstacle course and multiple bench press performance. J. Strength Cond. Res. 16(4), 500–508 (2002)
25. Chung, G.K.W.K., O'Neil, H.F., Delacruz, G.C., Bewley, W.L.: The role of anxiety on novices' rifle marksmanship performance. Educ. Assess. 10(3), 257–275 (2005)
26. Smith, C.D., et al.: Sleep restriction and cognitive load affect performance on a simulated marksmanship task. J. Sleep Res. (2017). https://doi.org/10.1111/jsr.12637
27. FN Expert: FN Expert Application User's Guide, pp. 9–10. Noptel Oy, Oulu, Finland (2017)
28. Johnson, R.F.: Statistical measures of Marksmanship. (USARIEM Technical Note TN-0 1/2/). U. S. Army Research Institute of Environmental Medicine, Natick, MA 01760-5007 (2001)

Common Ground and Autonomy: Two Critical Dimensions of a Machine Teammate

Corey K. Fallon[1(✉)], Leslie M. Blaha[2], Kris Cook[1], and Todd Billow[1]

[1] Pacific Northwest National Laboratory, Richland, WA, USA
{corey.fallon, kris.cook, todd.billow}@pnnl.gov
[2] Air Force Research Laboratory, Pittsburgh, PA, USA
Leslie.Blaha@us.af.mil

Abstract. We propose common ground and autonomy are the two critical dimensions necessary for intelligent machine agents to make the transition from tool to teammate. Existing models delineate a number of teammate characteristics. We explore how these teammate characteristics can be distilled into common ground and autonomy and suggest research steps to test our proposal.

Keywords: Human-machine teaming · Autonomy ·
Human-autonomy teaming · Trust · Sensemaking · Common ground

1   Introduction

Technological advances are improving the tools designed to assist humans in complex task domains. As the capabilities of these systems improve, machines can begin to transition from tools to teammates. To guide this transition, the human factors community has begun to identify the characteristics needed for a machine to function as a teammate. For example, Klein, Woods, Bradshaw, Hoffman and Feltovich [1] identified ten characteristics necessary for making automation a team player. More recently, Lyons, Mahoney, Wynne and Roebke [2] identified seven factors that cause a machine to be viewed as a teammate. We propose these characteristics, along with others identified as important for the design of human-machine teaming (HMT), are more than what is necessary to design an effective system. They introduce too many dimensions to track. Instead, we propose there are two high-level dimensions necessary for the design of effective teams: (Establishing and Maintaining) Common Ground, and Autonomy.

2   Common Ground

2.1   What Is Common Ground?

One of the most important factors to consider when designing machine teammates is how to establish and maintain common ground. According to Baber et al. [3], common ground is the knowledge and assumptions that agents share and know that they share. Common ground includes shared knowledge of the roles and functions of the team

D. N. Cassenti (Ed.): AHFE 2019, AISC 958, pp. 14–26, 2020.
https://doi.org/10.1007/978-3-030-20148-7_2

members, skills and competencies both individually and for the team as a whole, goals of the participants (individually and as a team), and awareness of each team member's concerns [4]. It is important to note that common ground does not imply team members have a complete shared awareness. Understanding does not need to be total; it needs to be sufficient for the specific task [3]. The type of information and level of detail necessary for common ground depends on the goals of the task.

2.2   Benefits of Common Ground

The ability to establish and maintain common ground with a teammate provides several advantages. First, common ground lowers coordination costs between teammates [5]. Coordination cost is the cost associated with keeping the team functioning as a coordinated unit. Low coordination costs save resources, so the team can focus more energy on the mission. Coordination cost can increase if a teammate's communication is unnecessary. For example, a teammate may provide detailed directions or explanations not realizing that the other party either does not need or already has the information. Coordination cost can also increase if a teammate uses vague or abbreviated language that does not sufficiently communicate his or her message in the initial attempt. This initial unsuccessful attempt at communication will likely lead to additional exchanges between team members for clarification [4]. Each party's ability to maintain shared awareness affects the likelihood that they will over- or under-estimate what information needs to be shared.

Common ground allows teammates to predict each other's behavior [1]. Machine teammate predictability is important because it helps the human calibrate trust and appropriately rely on the technology [6, 7]. Additionally, teammates can predict each other's future needs, proactively provide appropriate resources, minimizing coordination costs [8].

Common ground facilitates task interdependence. Task interdependence exists when the human and machine work on different aspects of the same task, and task components are allocated according to each's unique strengths [2]. In HMT, task interdependence can exist without common ground, but achieving this interdependence places greater burden on the human to allocate, direct, and monitor task execution. Common ground provides the human and machine with a shared understanding of the task that needs to be accomplished and the relative strengths and weaknesses of each. Armed with this shared awareness, the team can demonstrate task interdependence with less direction from the human.

Common ground makes teams more robust. Robustness is the team's ability to adapt to changes in the environment; it is one of several factors impacting HMT resiliency [9]. When new insights shift the focus of an analysis or environmental factors require an adjustment to the plan, common ground provides the ability to adapt. Adaptation may involve reprioritizing tasks, one teammate providing backup to another, or simply redirecting each other's attention [1]. A team on common ground can more effectively and efficiently adjust to changes that may otherwise lead to a break down in team performance.

2.3   Common Ground Breakdown

Common ground poses a major challenge for HMT; not only is it critical for effective team work, it is in a regular state of decay [4]. Common ground breakdowns occur when there are differences in understanding and the team members are unaware of these differences or unmotivated to resolve them [1]. Lack of shared understanding can be based on differences in reasoning, data synthesis and analysis, and/or access to information. Changes in the task environment can spur differences. Perhaps one team member notices the change and adjusts although another teammate does not. Or both teammates may interpret a change differently, resulting in different mental models of the situation. Differences in understanding can also stem from different goals. This type of breakdown is common if the team leader fails to clearly state the team's or individual members' goals.

A breakdown in common ground occurs when the differences in understanding are coupled with ignorance of these differences [4]. Unaware of the differences in goals or understanding, team members will continue to behave under the assumption that common ground exists. Consequently, behaviors will be out of sync and may hurt team performance. The breakdown can only be restored when asynchronous behavior is identified and remedied. Unfortunately, it is typically salient events such as errors that alert team members to the breakdown. Generally, teams lacking experience working together are more susceptible to breakdowns in common ground [4].

2.4   Maintaining Common Ground

Communication. Teams should practice common ground maintenance as a default activity in an attempt to identify breakdowns before they hurt performance [4]. One way to maintain common ground is through communication. Designing technology that is capable of a rich dialog with the human teammate is important for the perception of machines to transition from tools to teammates [2]. Machine teammates must be able to communicate their status, including task progress and future intentions. Additionally, the machine teammate should direct the human's attention to important information. For example, if the machine detects a change in an investigation, it should make the analyst aware before common ground begins to decay. A rich communication dialog also includes the machine's ability to receive and interpret verbal and nonverbal signals from the user [1]. When receiving a message, the machine should both acknowledge receipt and communicate its understanding of the message to help prevent common ground breakdowns [4].

Design for common ground and communication must account for coordination cost. Although the capability for a rich dialog is important, designing for a smooth interaction is also important. Communication with the machine teammate imposes a cost on the team. If establishing and maintaining common ground carries too great a coordination cost, the human may stop attempting to establish common ground. HMT is especially vulnerable to this threat when under time pressure, or experiencing high task load, duress, or fatigue. Designers must consider the amount, form, and granularity of information that is exchanged between the human and machine. A machine requiring a large amount of detailed information in a format unfamiliar to the human places a high burden on the human. Similarly, a machine communicating at inappropriate times using unnecessary details or unfamiliar jargon increases the difficulty of maintaining common ground.

One way designers might be able to lower coordination costs while still maintaining common ground is to leverage cognitive artifacts. Cognitive artifacts are physical or digital representations such as lists, schedules, and analytic visualizations that reflect the thoughts and goals of the creator [10]. These representations are compact representations enabling efficient coordination [11]. These artifacts may be at least partially comprised of symbols and visualizations that depart from traditional linguistic communication. It is important to note that common ground maintenance is not always verbal [3]. In some cases, it may be more natural and efficient for a human to rely on cognitive artifacts when coordinating with the machine teammate. For example, one can imagine an interactive visualization that allows both the human and machine to manipulate the artifact as they coordinate and build a shared understanding. Designers should strive to develop machines that can detect the human's artifacts and correctly translate them into meaning for the machine. Placing some coordination burden on the machine will reduce coordination costs for the human and maximize the likelihood that the user will work to maintain common ground with the technology. Perhaps interactive machine learning could be employed to fine tune the machine's understanding of the human's cognitive artifacts over time [12].

Another technique for keeping coordination costs low is to use language and concepts that are familiar to the user [13]. Unfamiliar information creates confusion and requires additional explanation. Miller's [14] human-computer etiquette concept highlights the importance of familiarity during human-machine communication. According to [14], etiquette in complex task environments is behavior consistent with the norms, expectations, and terminology used by the team. An AI that behaves in a way familiar to the user helps reduce coordination costs and improves HMT performance.

Monitoring. In addition to engaging in regular communication, the human-machine team must also continually monitor itself for signs of common ground decay [4]. The goal of vigilant monitoring is to detect decay *before* it leads to performance errors or a complete breakdown. Monitoring involves attending to the teammate's behavior, the operational environment, and one's own performance. For a machine to function as a teammate, it must monitor the human's behavior and cognitive state for signs of a common ground breakdown. If the human is behaving in a manner inconsistent with the machine's expectations, this inconsistency could reflect a change in the human's goals, analysis of the situation, or access to information unavailable to the machine [3]. Inconsistent behavior may be a sign of early common ground decay. Additionally, the operational environment should be monitored for changes. Changes in mission or investigation make the team vulnerable to common ground decay. Finally, a machine teammate must be aware of decline in its own performance. If the machine's performance is slipping, the human may not be aware of it, and this lack of awareness could lead to a breakdown in common ground.

2.5 Trust

As we build more adaptable, intelligent tools, they become less predictable [1]. This lack of predictability can result in a loss of trust [6, 15]. Common ground may be one way to protect against teammate unpredictability. A machine attempting to maintain common ground is more transparent about its actions. The human is able to build a shared understanding of the task while also gaining insight into the system's functioning. Common ground increases machine transparency, which in turn allows the human to calibrate his/her trust in the technology [16].

While building and maintaining common ground can eliminate under-trust due to unpredictability, common ground breakdowns can lead to over-trust. When a breakdown occurs, the human may be unaware that goals and knowledge are no longer shared with the machine. The human incorrectly trusts the machine to continue behaving consistently with his/her own situation understanding. Restoring common ground can repair trust. However, trust in a teammate is multifaceted. In the case of restoring common ground after a breakdown, the human will likely consider both the machine's ability to perform the task as well as its ability to maintain common ground. These are two separate abilities, and trust in one does not automatically generalize to the other. The human's loss of trust in the machine's ability to maintain common ground may lead to underutilization of the tool.

3 Autonomy

The second critical dimension for HMT is autonomy. Autonomy is a machine's ability to function independently of the human. Machines with little autonomy require explicit direction from the human; highly autonomous machines require little direction. This continuum is captured in Parasuraman, Sheridan, and Wickens, levels of automation concept [17]. There are ten levels of automation, from a machine that offers no assistance (level 1) to a machine that behaves completely independently disregarding any human direction or intervention (level 10).

Machines that establish and maintain common ground without autonomy have a clear understanding of shared goals and tasks but are completely dependent on the human's direction. These machines neither take initiative nor adapt performance to changes in demands unless specifically directed by the human. The burden is on the human to provide clear instructions to the machines at all times. On the other end of the continuum, level 10 automation [17] would also not be suitable for a machine teammate. Good team performance requires the machine be adaptable to the preferences and needs of its human teammate.

3.1 Dynamic Leadership

Sufficient autonomy renders the machine and the human equally responsible for task performance. However, leadership and responsibility may shift depending on the unique strengths of each teammate and the changing demands of their tasks. This interaction is referred to as dynamic leadership [18]. In light of the need for shifting roles and responsibilities, perhaps the most ideal configuration is a machine that is adaptable and/or adaptive to the level of autonomy appropriate for the situation [18]. For example, in certain environments the machine may have access to better quality information than the human and may need to assert a high level of autonomy to ensure optimal task performance. In environments where the machine's capabilities are more limited, it would be ideal for the machine to recognize its own limitations and reduce its own level of autonomy.

3.2   Agency

Autonomy allows machines to exhibit agency by observing and acting on their environment [18]. The ability to take action independently of the human's specific direction allows the machine the freedom to engage in good teammate behavior. For example, the machine can help redirect the human's attention to important changes in the mission and provide backup if the human's performance is degrading. A machine with a high autonomy may be afforded the freedom to correct the human's errors or suggest changes to an existing plan that it deems problematic. According to Lyons et al. [2], humans' perception of a machine's agency influences the extent to which it is viewed as a teammate. Haslam's [19] model of dehumanization identifies agency as one of several characteristic separating humans from machines.

3.3   Trust

Trust is the willingness to be vulnerable to another party when that party cannot be controlled or monitored [20]. Low-autonomy machines can and must be controlled, rendering trust less relevant in the HMT relationship. Instead of trust, humans establish and maintain confidence in the technology through control. Organizational controls (e.g., strict policies and procedures) instill confidence that team members will behave in predictable ways [21]. Similarly, humans may rely on their control over low-autonomy machines to establish confidence in the technology. The ability to direct the specific actions of a machine enables predictable behavior patterns that instill confidence.

As a machine's autonomy increases, the human's level of control over the machine decreases. The absence of control increases the human's vulnerability, placing a greater demand on the importance of trust in the machine. For highly autonomous machines, the human's trust in the technology may play a more significant role in predicting the human's confidence and reliance on the technology.

4   Relationship Between Common Ground and Autonomy

We propose that for a machine to be a teammate it must behave both *autonomously* while striving to *establish and maintain common ground* with its human partner. Machines lacking in either dimension will place additional burdens on the human to manage the HMT. Machines demonstrating a high degree of autonomy without attempting to maintain common ground perform tasks independently; however, they lack awareness of the human's goals and actions. Additionally, these machines are not attempting to communicate their goals, progress, or reasoning with the human. The burden is therefore on the humans to closely monitor these machines and correct their actions if necessary. Machines that establish and maintain common ground but lack autonomy have shared knowledge of the goals, responsibilities, and task demands. However, these machines do not take initiative or adapt their performance and must be directed by the human. The burden is on the human to provide clear instructions at all times. Both cases are examples of poor teamwork. Both examples over-burden the human with team management, diverting resources that could be devoted to task goals.

4.1 Trust Calibration and Risk

Both common ground and autonomy impact the likelihood and severity of machine error, two major components of risk [22]. We propose that as a machine's autonomy increases, the level of risk increases. The machine has more freedom to initiate activities, and it assumes greater responsibility for the team's performance. We also assert that common ground can impact risk via trust calibration. A lack of common ground can lead to a miscalibration of trust in a machine. This miscalibration can, in turn, lead to inappropriate reliance on the technology and degrade HMT performance [6].

For designers interested in developing machines with high levels of autonomy, the impact of autonomy on risk can be mitigated by developing a machine that strives to establish and maintain common ground. Machines that are ineffective at maintaining common ground and are highly autonomous pose the biggest risk to performance. The human has less control over the machine's actions and his/her inability to maintain common ground with the machine makes trust calibration very challenging and (see Fig. 1).

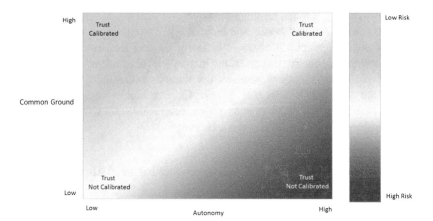

Fig. 1. Conceptual model of risk levels in HMT. The degree of risk (color coded) varies as a function of the two critical dimensions, Autonomy and Common Ground.

5   Models and Characteristics of Human-Machine Teams

We are not the first group to attempt to identify the dimensions critical for transforming computational agents into teammates. A number of models and frameworks have been proposed for HMT. These can be found across the literatures addressing "human-machine teaming", "human-autonomy teaming", "human-robot teaming", and related topics. Regardless of the physical form factor for the computational agents, each model suggests important characteristics[1] to consider in the design of a good teammate. We summarize three of these models in Table 1, with particular attention to the definitions for each of the characteristics or dimensions. For additional models or characteristics, we refer the reader to [16, 17, 23–25].

Reviewing the definitions in Table 1, we can start to see some key themes in the characteristics considered desirable in any team member and which could be particularly important for machines to be perceived as teammates, not just tools. But in this table alone, only accounting for three published models, there are 20 characteristics that should inform machine teammate design. The design space could get intractably large very quickly, for both design and evaluation considerations.

However, the themes of these characteristics are consistent with the critical dimensions of common ground and autonomy. We hypothesize that many of the constructs discussed in the human-machine team literature are highly associated with one or both of these dimensions. If supported, it would mean that the individual characteristics could be "collapsed" onto the critical dimensions common ground and autonomy. And if someone is designing a teammate for the critical dimensions, the resulting system should support many of the characteristics in Table 1 because it was designed to support common ground and autonomy. Table 1 therefore includes a proposed mapping of these individual constructs onto common ground and/or autonomy.

Table 1. Characteristics of machines as teammates

Citation	Characteristic	Definition	Common Ground	Autonomy
Klein, Woods, Bradshaw, Hoffman and Feltovich [1]	Engage in common-grounding activities	Ability to comprehend messages and signals to coordinate joint activities, pertinent knowledge, beliefs, and shared assumptions	X	
	Model others' intentions/actions	Ability to model the intentions-actions vis-à-vis joint activity's state and evolution	X	

(continued)

---

[1] We note that between papers or domains, "characteristics" are also referred to as dimensions, factors, features, constructs, traits, challenges. We use characteristics herein as a blanket term covering all these.

Table 1. (*continued*)

Citation	Characteristic	Definition	Common Ground	Autonomy
Klein, Woods, Bradshaw, Hoffman and Feltovich [1]	Mutual Predictability	Being predictable and able to predict other's actions	X	
	Directability	Capacity for deliberately assessing and modifying other parties' actions in a joint activity as conditions and priorities change		X
	Revealing status & intentions	Agents make their own targets, states, capacities, intentions, changes, and upcoming actions available/obvious to other agents who are supervising or coordinating with them	X	X
	Interpreting signals	Ability to send, receive, and interpret signals to form models of teammates, including non-verbal cues	X	X
	Goal negotiation	Ability to enter into negotiation when a situation changes and team must adapt; convey current and potential goals		X
	Collaboration	Given and take; processes of understanding, and task execution are necessarily incremental, subject to negotiation, and forever tentative		X
	Attention management	Ability to direct each other's attention to important signals, activities, changes in an intelligent and context-sensitive manner	X	X
	Cost control	Achieving an economy of effort, including time and energy		X

(continued)

Table 1. (*continued*)

Citation	Characteristic	Definition	Common Ground	Autonomy
Ososky, Schuster, Jentsch, Fiore, Shumaker, Lebiere, Kurup, Oh and Stentz [26]	Shared Mental Models	Knowledge structures held by a team about relevant capabilities, knowledge, and interactions; enable predicting future system states given a set of inputs	X	
	Shared Situation Awareness	Ability to share task-relevant information after gathering information independently	X	X
	Dynamic Task Allocation	Human and Machine are assigned roles or functions according to abilities; Machine dynamically adjusts its behavior according to Human's action		X
	Coordination	Ability to interact dynamically, interdependently and adaptively; tasks are sequenced, synchronized, integrated and completed within established constraints		X
Lyons, Mahoney, Wynne and Roebke [2]	Perceived agency	Effective agents should be able to observe the environment, process relevant goal-oriented information, and act on the environment	X	X
	Perceived benevolence	Team mates have your best interests in mind, support each other, provide back up when needed	X	

(*continued*)

Table 1. (*continued*)

Citation	Characteristic	Definition	Common Ground	Autonomy
Lyons, Mahoney, Wynne and Roebke [2]	Perceived task interdependency	Tasks are divided into components to be worked on separately by the human and machine to maximize effective use of capabilities		X
	Relationship building	Interactive affordances move from one-sided information-centric transmissions to more naturalistic dialogue-based interactions	X	
	Communication richness	Team mates are capable of rich dialog to convey task & team-based information, including social cues	X	
	Synchrony	Shared, synchronized mental models, including common perception of team/members' capabilities, task, context; facilitates joint adaptation and anticipation of each other's actions	X	

Distilling the characteristics of a machine teammate into two dimensions offers a simplicity that can aid technology design to support HMT. Certainly, the numerous factors that comprise each high-level dimension are important to consider. However, reducing many characteristics to two high-level dimensions may help eliminate possible redundancies among the low-level characteristics, enabling researchers and practitioners to more easily explore and visualize the relationships between these two critical dimensions.

6   Future Research

Our hypothesis that the many design characteristics for HMT reduce down to the two critical dimensions proposed herein, establish and maintain common ground and autonomy, remains to be tested. Future research will pursue both conceptual mappings between characteristics and critical dimensions and empirical usability testing with HMT systems. The former will entail directly asking people to consider the importance of each potential machine teammate characteristic in multiple potential contexts (e.g., everyday smart home appliances, intelligent driving assistants, job-related technologies). The latter, longer effort, will entail designing variations in actual systems for human user experiments. Over this series of evaluations, we will assess the degree to which common ground and autonomy are both necessary and sufficient for a machine to be perceived as a true teammate. As outlined in Table 1, designers interested in developing machine teammates have a variety of characteristics to consider. As we examine the potential to collapse those onto common ground and autonomy, we may identify additional characteristics that do not fit either dimension, or we may identify application domains where more nuance is required. Empirical support that common ground and autonomy act as two critical dimensions would provide much needed clarity and organization for the design community. Such work would also pave the way for research to more deeply explore the relationship between these two critical dimensions of a machine teammate.

References

1. Klein, G., Woods, D.D., Bradshaw, J.M., Hoffman, R.R., Feltovich, P.J.: Ten challenges for making automation a "Team Player" in joint human-agent activity. IEEE Intell. Syst. **19**, 91–95 (2004)
2. Lyons, J.B., Mahoney, S., Wynne, K.T., Roebke, M.A.: Viewing machines as teammates: a qualitative study. In: AAAI Spring. AAAI, Menlo Park (2018)
3. Baber, C., Cook, K., Attfield, S., Blaha, L., Endert, A., Franklin, L.: A conceptual model for mixed-initiative sensemaking. In: CHI Sensemaking Workshop, pp. 1–8 (2018)
4. Klein, G., Feltovich, P.J., Bradshaw, J.M., Woods, D.D.: Common ground and coordination in joint activity In: Rouse, W.R., Boff, K.B. (eds.) Organizational Simulation, pp. 139–178. Wiley, New York (2005)
5. Klein, G.: Streetlights and Shadows: Searching for the keys to Adaptive Decision Making. MIT Press, Cambridge (2009)
6. Lee, J.D., See, K.A.: Trust in automation: designing for appropriate reliance. Hum. Factors **46**, 50–80 (2004)
7. Madsen, M., Gregor, S.: Measuring human-computer trust. In: Proceedings of the 11th Australasian Conference on Information Systems, vol. 53, pp. 6–8 (2000)
8. Jonker, C.M., Riemsdijk, M.B., Vermeulen, B.: Shared mental models: a conceptual analysis. In: Proceedings of the 9th International Conference on Autonomous Agents and Multiagent Systems (2014)
9. Woods, D.D.: Essential characteristics of resilience. In: Hollnagel, E., Woods, D.D., Leveson, N. (eds.) Resilience Engineering, pp. 21–34. Ashgate Publishing, Aldershot (2012)
10. Nemeth, C., O'Connor, M., Klock, P.A., Cook, R.: Discovering healthcare cognition: the use of cognitive artifacts to reveal cognitive work. Organ. Stud. **27**(7), 1011–1035 (2006)
11. Attfield, S., Fields, B., Baber, C.: A resources model for distributed sensemaking. Cogn. Technol. Work **20**, 651–664 (2018)
12. Jasper, R.J., Blaha, L.M.: Interface metaphors for interactive machine learning. In: Proceedings of Human-Computer Interaction International: Augmented Cognition, Vancouver, Canada (2017)
13. Kass, R., Finin, T.: A general user modelling facility. In: Proceedings of CHI, pp. 145–150 (1988)

14. Miller, C.A.: Human-computer etiquette: managing expectations with intentional agents. Commun. ACM **47**(4), 31–34 (2004)
15. Muir, B.M.: Operators' trust in and use of automatic controller in a supervisory process control task. Doctoral Dissertation. University of Toronto, Canada (1989)
16. Lyons, J.B.: Being transparent about transparency: a model for human-robot interaction. In: Sofge, D., Kruiff, G.G., Lawless, W.F. (eds.) Trust and Autonomous Systems: Paper from the AAAI Spring (Technical Report SS-13-07). AAAI, Menlo Park (2013)
17. Parasuraman, R., Sheridan, T.B., Wickens, C.D.: A model for types and levels of human interaction with automation. IEEE Trans. Syst. Man Cybern. Part A Syst. Hum. **30**(3), 286–297 (2000)
18. Bruemmer, D.J., Walton, M.C.: Collaborative Tools for Mixed Teams of Humans and Robots. Idaho National Engineering and Environmental Lab, Idaho Falls (2003)
19. Haslam, N.: Dehumanization: an integrative review. Pers. Soc. Psychol. Rev. **10**(3), 252–264 (2006)
20. Mayer, R.C., Davis, J.H., Schoorman, F.D.: An integrative model of organizational trust. Acad. Manag. Rev. **20**(3), 709–734 (1995)
21. Das, T.K., Teng, B.: Between trust and control: developing confidence in partner cooperation in alliances. Acad. Manag. Rev. **23**(3), 491–512 (1998)
22. Young, S.L., Wogalter, M.S., Brelsford Jr., J.W.: Relative contribution of likelihood and severity of injury to risk perceptions. Proc. Hum. Factors Ergon. Soc. Annu. Meet. **36**(13), 1014–1018 (1992)
23. Christoffersen, K., Woods, D.D.: How to make automated systems team players. In: Advances in Human Performance and Cognitive Engineering Research, pp. 1–12. Emerald Group Publishing Limited (2002)
24. Parasuraman, R., Miller, C.A.: Trust and etiquette in high-criticality automated systems. Commun. ACM **47**(4), 51–55 (2004)
25. Parasuraman, R., Riley, V.: Humans and automation: use, misuse, disuse, abuse. Hum. Factors **39**(2), 230–253 (1997)
26. Ososky, S., Schuster, D., Jentsch, F., Fiore, S., Shumaker, R., Lebiere, C., Kurup, U., Oh, J., Stentz, A.: The importance of shared mental models and shared situation awareness for transforming robots from tools to teammates. In: Unmanned Systems Technology XIV, vol. 8387, pp. 838710-1–838710-12. International Society for Optics and Photonics (2012)

Exploring the Effect of Communication Patterns and Transparency on the Attitudes Towards Robots

Shan G. Lakhmani[1]([✉]), Julia L. Wright[1], Michael Schwartz[2], and Daniel Barber[2]

[1] Army Research Laboratory, 12423 Research Pkwy, Orlando, FL 32826, USA
{shan.g.lakhmani.civ,julia.l.wright8.civ}@mail.mil
[2] UCF Institute for Simulation and Training, 3100 Technology Pkwy, Orlando, FL 32826, USA
{mschwart,dbarber}@ist.ucf.edu

Abstract. Robots' increased autonomous capabilities necessitate human-robot communication. Exploration of this communication, including the pattern of and the content of such, is relevant to the development of these robots and the understanding of how they can interact with human teammates. This study compares two different patterns of communication and two approaches to transparent interaction, looking at their effects on human team members' attitudes towards robots with which they are communicating. Participants found robots using a bidirectional communication pattern to be more animate, likeable, and intelligent than robots using a unidirectional communication pattern.

Keywords: Human factors · Transparency · Attitudes towards robots · Unmanned vehicles · Human-robot teaming · Bidirectional communication

1 Introduction

Future battlefields will be complex, multidomain environments, where soldiers will require new strategies to effectively face emerging challenges [1]. As such, the U.S. military is developing robots with greater autonomous capabilities and teaming them with soldiers [1, 2]. In order to effectively perform as teammates, these future robots will be required to have their own decision making processes [3]. This increased autonomy, however, can lead to out-of-the-loop situations, where the human may be baffled by the robot's behavior, which may have disastrous results [4]. To prevent these negative outcomes, these robots must explicitly facilitate their human teammates' understanding of their decision making process, keeping their human counterparts in-the-loop [4, 5].

An understanding of how human-robot communication affects the human's understanding of the robot must be established. Since robots are designed entities, it stands to reason that the manner in which they communicate with both their operators and their teammates must also designed [6]. Currently, robots either push information to their user or they are controlled directly by an operator, negating the need for further

D. N. Cassenti (Ed.): AHFE 2019, AISC 958, pp. 27–36, 2020.
https://doi.org/10.1007/978-3-030-20148-7_3

communication [7, 8]. However, future robots will be expected to communicate in a manner similar to the way human team members communicate–i.e. bidirectionally, where information is pushed or pulled between teammates as needed [5, 9]. Consequently, explorations of human-robot communication must include not only the kind of information that the robot needs to share with its human counterparts, but also how the pattern of communication between them should be designed [10–12]. How the communication patterns in human-robot interaction will influence human teammates is still unknown, and as such, is an area of research that must be explored to further the current understanding of human-robot communications.

2   Background

The development of autonomous robots changes the existing human-robot relationship from that of operator-tool to one more akin to a human-human relationship [13]. Robots that once were predictable in their actions will now be capable of gathering, processing, and acting upon information independently of, and much more quickly than, their human counterparts [13, 14]. Unfortunately, these internal processes will be hidden from the human team mate unless explicitly shared by the robot. Hence, the need for the robot's internal reasoning processes to be transparent to its team mates, thus facilitating their understanding of the robot [15].

2.1   Transparency

Transparency can be described as an emergent property of the interaction between a human and a robot that facilitates the human's comprehension of the robot's actions, reasoning, future plans, etc. [16, 17]. When the human does not comprehend their robotic teammates actions, goals, etc., there is an increased likelihood of an out-of-the-loop situation, which can result in error, poor performance, reduced situation awareness, and reduced trust in the robot [18, 19]. To create the necessary understanding to avoid these consequences, the robot needs to provide information to the human about its internal processes, i.e. become more transparent. While simple, discrete information can be useful to support understanding of simpler robots, more complex robots require a model-based approach to transparency [20, 21]. The Situation awareness-based Agent Transparency (SAT) model uses a psychological framework, based on Endsley's [22] model of SA, to determine the kind of information a human might require from a robot to maintain awareness of a robot's internal processes [11]. The SAT model suggests that a robot communicate three levels of information: Level 1 information describes the robot's current actions, or what it is doing; Level 2 information describes the underlying rationale for those actions, or why it's doing what it's doing; Level 3 information describes the robot's planned actions and uncertainties, or what are its expected outcomes [11]. Autonomous robots engaged in shared tasks with humans will also be expected to understand the human decision making processes to effectively contribute [10, 23]. To maintain this mutual understanding, robots can be facilitate a mutual transparency between itself and its human counterparts [24]. This study seeks to explore an approach to supporting this mutual transparency by exposing a human teammate to relevant decision making information from the robot. The information that robots share with humans can be used to further the human-robot team's goals and build the humans' relationship with their robotic teammates [25, 26].

2.2 Communication Patterns

Robots convey relevant information to their human operators so that those operators can stay in the loop and effectively accomplish their goals [5]. However, with a teamwork-based approach to human-robot interaction, humans and robots are pursuing shared goals; to effectively accomplish these shared goals, humans and robots must communicate [5, 26, 27]. One of the major barriers to this communication is that the communicators have to do so in a manner that the recipient can understand [26]. Robotic teammates' capabilities are determined by human design, so they can be tailored to meet specific needs [6, 28]. The difficulty, however, is determining the communication needs of the human-robot team.

Robots can be designed to push information to their human operators without the expectation of receiving information back [8, 9]. Robots can also be designed to be manipulated by commands pushed by humans [7, 29]. These unidirectional patterns of communication can be a practical choice for human-robot interaction, and can consist of the robot pushing information to the human, the human pushing information to the robot, the robot pulling information from the human, or the human pulling information from the robot [7, 8]. However, in situations where robots have greater autonomous capabilities and the work they do has informational components, then a bidirectional flow of information may better support the goals of the team [5, 7]. Bidirectional communication—where information can be sent back and forth between a human and a robot—allows for a greater shared understanding to be established and maintained [7, 30]. In this study, we seek to explore how people respond to a bidirectional communications loop using two of afore-mentioned communication patterns (i.e. Robot push to human and robot pull from human).

2.3 Attitudes Toward Robots

The ways in which humans and robots interact—e.g. asking questions, drawing attention to objects in the environment—can influence the attitudes that humans have about the robot [25, 26]. Humans' attitudes towards a robot, in turn, can affect the way that they respond to that robot [31]. A robot that seems more animate, for example, can be presumed to have an intentionality that a more static-seeming robot may lack [32]. People are more likely to respond to that supposed intentionally, acquiescing to the supposed desires of that animate robot [32, 33]. Given the influence that a human's attitude towards a robot can have on their expectations of that robot, the factors that influence that attitude must be explored [34, 35]. This study will also examine whether communication patterns of a robot in a human-robot team influence the human's attitudes regarding the robot.

2.4   Current Study

While factors, such as the modality of human-robot interactions or the robot's physical form, has been shown to affect the human-robot relationship, fewer studies have examined how the sharing information and the patterns of interaction between humans and robots affect their relationship [14, 36, 37]. We hypothesize that the directionality of human-robot communication and the type of information provided by a robot will influence a human's relationship with their robotic teammate.

3   Methods

3.1   Experimental Design

The study used the context of a simulated cordon-and-search-like task, where both a human and robot had individual tasks as well as a shared task, in a 2×2 within-subjects design, where communication pattern and transparency were manipulated. For the communication pattern variable, participants were exposed to a robot that only pushed information to them (unidirectional) or pushed information to them and pulled information about their decision making process (bidirectional). For the transparency variable, participants either received the three levels of SAT information about the robot's decision-making process (Agent Transparency) or the prior information as well as the robot's understanding of the human's decision making process (Team Transparency).

3.2   Participants

The participants included 40 individuals (13 men, 27 women, M_{age}= 21.13, SD = 3.95), recruited locally from the central Florida area. No participants had prior experience with cordon-and-search tasks or had any prior military experience.

3.3   Task/Apparatus

Participants completed the experiment on a standard desktop computer and two 22" monitors, each presenting different stimuli. On the left-hand screen, participants were shown a simulated environment, developed in Unreal Engine 4. This environment showed the backside of a building, in which the robotic teammate was searching. Between the human and this building was an area with cross-traffic, both human and vehicular. Participants had to monitor the traffic and alert their robotic teammate when a person approached the entrance to the building. The right-hand screen displayed software adapted from a multimodal interface prototype developed for the Robotics Collaborative Technology Alliance [38]. This screen displayed the robot's point of view, a map of the surrounding area, the modules that present the SAT information, and the queries that the robot used to pull information from the participant. Figure 1 shows the contents of the screens on which participants completed the experiment.

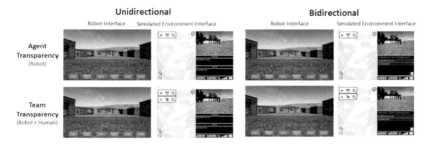

Fig. 1. Left and right screens presented to participants, for each condition.

After completing a scenario in each condition, participants were asked to rate their attitude toward the robot they interacted with using the Godspeed questionnaire [39]. This questionnaire asks participants to describe their feelings of anthropomorphism, animacy, likeability, perceived intelligence, and perceived safety towards the robot.

4 Results

Participants did find the robot to be more animate if they had a bidirectional communication pattern ($M = 2.87$, $SD = 0.64$) with the robot than if they had a unidirectional one ($M = 2.67$, $SD = 0.75$), $F(1, 39) = 5.90$, $p = .02$. No interaction effect was found, $F(1, 39) = 1.38$, $p = .25$, and transparency did not seem to affect participants, $F(1, 39) = 0.76$, $p = .39$ (Fig. 2).

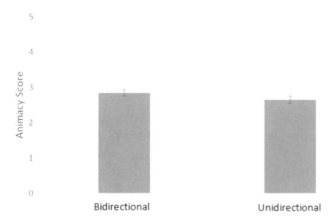

Fig. 2. Mean animacy score of bidirectional and unidirectional communication pattern. Error bars represent standard error.

Participants did find the robot to be more likeable when they had a bidirectional communication pattern ($M = 3.33$, $SD = 0.08$) with the robot than when they had a unidirectional one (M = 3.14, SD = 0.10), $F(1, 39)$ = 4.17, p = .048. No interaction effect was found, $F(1, 39)$ = 0.08, p = .78, and transparency did not seem to affect participants, $F(1, 39)$ = 3.15, p = .08 (Fig. 3).

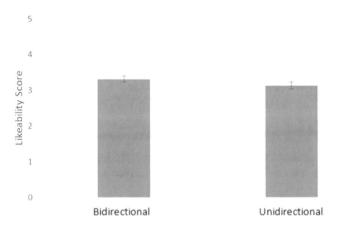

Fig. 3. Mean likeability score of bidirectional and unidirectional communication pattern. Error bars represent standard error.

Participants did rate the robot to be more intelligent when they had a bidirectional communication pattern (M = 3.91, SD = 0.08) with the robot than if they had a unidirectional one (M = 3.67, SD = 0.11), $F(1, 39)$ = 5.49, p = .02. No interaction effect was found, $F(1, 39)$ = 0.80, p = .38 and transparency did not seem to affect participants, $F(1, 39)$ = 2.30, p = .14 (Fig. 4).

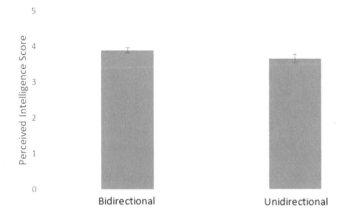

Fig. 4. Mean perceived intelligence score of bidirectional and unidirectional communication pattern. Error bars represent standard error.

Participants did not perceive the robot to be more or less safe in any of the four conditions. No interaction effect was found, $F(1, 39) = 0.26$, $p = .61$, nor any effects for either transparency, $F(1, 39) = 0.40$, $p = .53$, or communication pattern $F(1, 39) = 0.40$, $p = .53$.

5 Discussion

While participants did not respond differently to robots providing information about more than itself, we found that when the robot queried the participant to confirm its understanding of the human's rationale (a bidirectional flow of information) participants considered the robot to be more animate, intelligent, and likeable than a similar robot that did not request confirmation (a unidirectional flow of information). These findings suggest that the more explicitly interactive element of answering queries affected participants' attitudes towards the robot, but the robot's understanding of the human's decision making process may have been unnecessary, either due to the nature of the participants' tasks or their complexity. Wright and associates [40] found something similar when comparing transparency modules that described the robot's decision making process at either a surface level or at a more in-depth level.

The findings here have implications for the design of communication protocols for robots meant to team with humans on shared tasks. A robot that exhibits a bidirectional communication pattern, specifically a robot-push, robot-pull pattern, is considered smarter, more likeable, and more animate than a robot that communicates unidirectionally. A robot that is considered more animate and more intelligent commands more respect for its agency, and so in instances where a robot needs to be taken seriously or its advice must be taken under consideration, the robot may benefit from a bidirectional communication pattern implemented in this way [32, 41]. A more likeable robot may be used more and be less likely to be abandoned, which can be useful if the robot is intended for long term use.

Not only are the attitudes humans hold towards robots immediately relevant, these attitudes have implications for future human-robot teaming. In human teamwork, the attitude that people have towards their teammates is an important factor for effectively completing tasks with those teammates [42]. As robots gain the ability to act more like humans, the way we interact with human team members becomes more relevant to human-robot interaction. Consequently, the attitudes that people hold towards their robotic teammates becomes a greater interest as well. With that in mind, future research can explore other factors that influence human attitudes towards robots with which they share tasks, which can be used to more finely tune the treatment of robots in military situations.

Acknowledgments. This research was funded by the U.S. Army Research Laboratory's Human-Robot Interaction program. The authors would like to thank Jason English, Christopher Miller, Thomas Pring, and Dr. Jessie Chen for their contributions to this project.

References

1. U.S. Army: The U.S. Army Robotic and Autonomous Systems Strategy. In: Maneuver, A., Soldier Division Army Capabilities Integration Center, (ed.) TRADOC. Fort Eustis, VA (2017)
2. David, R.A., Nielsen, P.: Defense science board summer study on autonomy. Defense Science Board Washington United States (2016)
3. Fan, X., Yen, J.: Modeling and simulating human teamwork behaviors using intelligent agents. Phys. Life Rev. **1**, 173–201 (2004)
4. Stubbs, K., Wettergreen, D., Hinds, P.H.: Autonomy and common ground in human-robot interaction: a field study. IEEE Intell. Syst. **22**, 42–50 (2007)
5. Chen, J.Y.C., Barnes, M.J.: Human–agent teaming for multirobot control: a review of human factors issues. IEEE Trans. Hum. Mach. Syst. **44**, 13–29 (2014)
6. Fiore, S.M., Wiltshire, T.J.: Technology as teammate: examining the role of external cognition in support of team cognitive processes. Front. Psychol. **7**, 1531 (2016)
7. Kaupp, T., Makarenko, A., Durrant-Whyte, H.: Human–robot communication for collaborative decision making—a probabilistic approach. Robot. Auton. Syst. **58**, 444–456 (2010)
8. Sweet, N.: Semantic Likelihood Models for Bayesian Inference in Human-Robot Interaction (2016)
9. Héder, M.: The machine's role in human's service automation and knowledge sharing. AI & Soc. **29**, 185–192 (2014)
10. Chen, J.Y., Lakhmani, S.G., Stowers, K., Selkowitz, A.R., Wright, J.L., Barnes, M.: Situation awareness-based agent transparency and human-autonomy teaming effectiveness. Theor. Issues Ergon. Sci. **19**, 259–282 (2018)
11. Chen, J.Y., Procci, K., Boyce, M., Wright, J.L., Garcia, A., Barnes, M.J.: Situation Awareness-Based Agent Transparency. U.S. Army Research Laboratory, Aberdeen Proving Ground, MD (2014)
12. Lyons, J.B., Havig, P.R.: Transparency in a human-machine context: interface approaches for fostering shared awareness/intent. In: 6th International Conference on Virtual, Augmented, and Mixed Reality: Designing and Developing Virtual and Augmented Environments, pp. 181–190. Springer International Publishing, Las Vegas, NV (2014)
13. Phillips, E., Ososky, S., Grove, J., Jentsch, F.: From tools to teammates toward the development of appropriate mental models for intelligent robots. In: Proceedings of the Human Factors and Ergonomics Society Annual Meeting, pp. 1491–1495. SAGE Publications (2011)
14. Ososky, S., Schuster, D., Phillips, E., Jentsch, F.G.: Building appropriate trust in human-robot teams. In: 2013 AAAI Spring Symposium Series (2013)
15. Cramer, H., Evers, V., Ramlal, S., Someren, M., Rutledge, L., Stash, N., Aroyo, L., Wielinga, B.: The effects of transparency on trust in and acceptance of a content-based art recommender. User Model. User-Adapt. Inter. **18**, 455–496 (2008)
16. Maass, S.: Why systems transparency? In: Green, T.R.G., Payne, S.J., van der Veer, G.C. (eds.) The Psychology of Computer Use, pp. 19–28. Academic Press Inc, Orlando (1983)
17. Ososky, S., Sanders, T., Jentsch, F., Hancock, P., Chen, J.Y.C.: Determinants of system transparency and its influence on trust in and reliance on unmanned robotic systems. In: SPIE Defense+ Security, pp. 90840E-90841–90840E-90812. International Society for Optics and Photonics (2014)
18. Grote, G., Weyer, J., Stanton, N.A.: Beyond Human-Centred Automation–Concepts for Human–Machine Interaction in Multi-layered Networks. Taylor & Francis, London (2014)

19. Kilgore, R., Voshell, M.: Increasing the transparency of unmanned systems: applications of ecological interface design. In: International Conference on Virtual, Augmented and Mixed Reality, pp. 378–389. Springer (2014)
20. Dzindolet, M.T., Peterson, S.A., Pomranky, R.A., Pierce, L.G., Beck, H.P.: The role of trust in automation reliance. Int. J. Hum.-Comput. Stud. **58**, 697–718 (2003)
21. Helldin, T., Falkman, G., Riveiro, M., Dahlbom, A., Lebram, M.: Transparency of military threat evaluation through visualizing uncertainty and system rationale. In: International Conference on Engineering Psychology and Cognitive Ergonomics, pp. 263–272. Springer (2013)
22. Endsley, M.R.: Toward a theory of situation awareness in dynamic systems. Hum. Factors: J. Hum. Factors Ergon. Soc. **37**, 32–64 (1995)
23. Allen, J.E., Guinn, C.I., Horvtz, E.: Mixed-initiative interaction. IEEE Intell. Syst. Their Appl. **14**, 14–23 (1999)
24. Lakhmani, S., Abich IV, J., Barber, D., Chen, J.: A proposed approach for determining the influence of multimodal robot-of-human transparency information on human-agent teams. In: International Conference on Augmented Cognition, pp. 296–307. Springer (2016)
25. Bütepage, J., Kragic, D.: Human-robot collaboration: from psychology to social robotics. arXiv preprint arXiv:1705.10146 (2017)
26. Sycara, K., Sukthankar, G.: Literature review of teamwork models. In: Institute, R. (ed.) Carnegie Mellon University, Pittsburgh, PA (2006)
27. Yen, J., Fan, X., Sun, S., Hanratty, T., Dumer, J.: Agents with shared mental models for enhancing team decision makings. Decis. Support Syst. **41**, 634–653 (2006)
28. Cooke, N.J., Demir, M., McNeese, N.: Synthetic Teammates as Team Players: Coordination of Human and Synthetic Teammates. Cognitive Engineering Research Institute Mesa United States (2016)
29. Sheridan, T.B.: Teleoperation, telerobotics and telepresence: a progress report. Control. Eng. Pract. **3**, 205–214 (1995)
30. Marko, H.: The bidirectional communication theory–a generalization of information theory. IEEE Trans. Commun. **21**, 1345–1351 (1973)
31. Hancock, P.A., Billings, D.R., Schaefer, K.E., Chen, J.Y.C., De Visser, E.J., Parasuraman, R.: A meta-analysis of factors affecting trust in human-robot interaction. Hum. Factors: J. Hum. Factors Ergon. Soc. **53**, 517–527 (2011)
32. Jones, K.S., Schmidlin, E.A.: Human-robot interaction toward usable personal service robots. Rev. Hum. Factors Ergon. **7**, 100–148 (2011)
33. Morrow, P.B., Fiore, S.M.: Supporting human-robot teams in social dynamicism: an overview of the metaphoric inference framework. In: Proceedings of the Human Factors and Ergonomics Society Annual Meeting, pp. 1718–1722. SAGE (2012)
34. Williams, T., Briggs, P., Scheutz, M.: Covert robot-robot communication: Human perceptions and implications for human-robot interaction. J. Hum.- Robot Interact. **4**, 24–49 (2015)
35. Norman, D.A.: How might people interact with agents. Commun. ACM **37**, 68–71 (1994)
36. Schillaci, G., Bodiroža, S., Hafner, V.V.: Evaluating the effect of saliency detection and attention manipulation in human-robot interaction. Int. J. Soc. Robot. **5**, 139–152 (2013)
37. Fink, J.: Anthropomorphism and human likeness in the design of robots and human-robot interaction. In: International Conference on Social Robotics, pp. 199–208. Springer (2012)
38. Barber, D.J., Abich IV, J., Phillips, E., Talone, A.B., Jentsch, F., Hill, S.G.: Field assessment of multimodal communication for dismounted human-robot teams. In: Proceedings of the Human Factors and Ergonomics Society Annual Meeting, pp. 921–925. SAGE Publications, Los Angeles, CA (2015)

39. Bartneck, C., Kulić, D., Croft, E., Zoghbi, S.: Measurement instruments for the anthropomorphism, animacy, likeability, perceived intelligence, and perceived safety of robots. Int. J. Soc. Robot. **1**, 71–81 (2009)
40. Wright, J.L., Chen, J.Y.C., Lakhmani, S.G., Selkowitz, A.R.: Agent transparency for an autonomous squad member: depth of reasoning and reliability. U.S. Army Research Laboratory, Aberdeen Proving Ground, MD (in press)
41. Sandoval, E.B.: Reciprocity in human robot interaction. Human Interface Technology. University of Canterbury (2016)
42. Mathieu, J.E., Heffner, T.S., Goodwin, G.F., Salas, E., Cannon-Bowers, J.A.: The influence of shared mental models on team process and performance. J. Appl. Psychol. **85**, 273–283 (2000)

The Influence of Signal Presentation Factors on Performance of an Immersive, Continuous Signal Detection Task

Rhyse Bendell[✉], Gabrielle Vasquez, and Florian Jentsch

University of Central Florida, Orlando, FL 32826, USA
{rbendell, gvasquez, fjentsch}@ist.ucf.edu

Abstract. Conventional independent trial signal detection tasks are representative of many real-world tasks, such as identifying flavors or spotting abnormalities on an x-ray film; however, there is also a wide range of task types that are not accurately modeled by the standard yes/no decision making paradigm. Active monitoring and driving hazard detection tasks, for example, do not occur as a set of independent events but are executed in response to a continuous flow of incoming stimuli. Identification of a potential threat is therefore dependent on a multitude of factors that may be influenced by not only the characteristics of the threat (the relevant signal) but also by ongoing events (dynamic noise) and one's environment. Although signal detection theory is typically applied to trials of static signal-noise constructs, it may be extended to many more task types by relaxing the temporal independence of trials and allowing both noise and signal stimuli to be presented dynamically. The primary drawback of these modifications is the reduction of control over the noise that accompanies any given signal. Fortunately, a reduction in that control does not necessitate a loss of control; instead, it changes the tactics that must be employed to ensure the production of usable data. Rather than attend to signal-noise characteristics between trials, designers of a continuous signal detection task must focus on those characteristics as a function of time and also consider the potential lifespan and evolution of any given signal. This paper considers the presentation of signals in a continuous monitoring task which was administered in two experiments conducted in immersive virtual reality. The factors that were found to most strongly influence participants' detection tendencies are discussed with regards to their impacts and the means by which their effects may be mitigated or controlled. Both signal and noise characteristics were manipulated across the two experiments; the influence of each is evaluated in the context of the resulting signal-noise construct which was presented to participants. Additionally, standard signal detection task parameters such as event rate, signal salience, signal origin type, signal-to-noise ratio, etc. are discussed with regards to their extension into the continuous domain.

Keywords: Signal detection theory · Perception ·
Fuzzy signal detection theory · Virtual environments ·
Dynamic signal detection · Continuous signal detection

© Springer Nature Switzerland AG 2020
D. N. Cassenti (Ed.): AHFE 2019, AISC 958, pp. 37–48, 2020.
https://doi.org/10.1007/978-3-030-20148-7_4

1 Introduction

Signal detection theory (SDT) is one of the most useful tools for investigating stimulus detection and decision making and is accordingly one of the most widely applied theories in the study of human behavior [1–3]. Conventional SDT assumes that task stimuli are clearly separated into one of two categories, signal or noise, and that a given stimuli set may receive one of two responses: signal present or absent. This simple model allows for the extraction of impressively complex behavioral metrics such as perceptual sensitivity (d': one's ability to detect a signal stimulus in the presence of noise stimuli) and response criterion (c: one's tendency to classify a signal stimulus correctly given task demands). Elegant experimental design, however, is at least partly responsible for the success of the model, as its metrics are only meaningful when the noise and signal + noise events presented to participants are well designed [4]. Some practitioners have noted that while the conventional SDT model is an excellent tool, its dependence on the simple two-level decision framework is a weakness that does not allow for the consideration of nuanced or evolving signal stimuli nor for any temporal fluidity [1, 2, 5, 6]. Fuzzy signal detection theory (FSDT) was developed as a response to many of the shortcomings of conventional SDT; it considers the possibility that signal versus noise characteristics may exist along a spectrum rather than as crisp, non-overlapping definitions as well as the reality that an individual's response behavior may not be constant under identical conditions. Relaxing the binary categorizations of conventional SDT allows detection theory to be applied to more realistic tasks, such as those involved in air traffic control or hazard perception during driving scenarios [2, 7]. Hazard detection research in particular benefits greatly from FSDT given that many tasks involving the detection of threats rely on the presentation of continuously changing search displays that necessarily include stimuli that exist on a spectrum between noise and signal events [7, 8].

   The overarching goal of the research efforts discussed in this manuscript is to study human-robot interaction (HRI) topics (e.g., situation awareness, usability, and trust) relevant to human-robot teaming involving interactions with highly autonomous unmanned ground vehicles (UGVs). As described in our previous work [9], the first step we took towards investigating complex HRI scenarios was to develop an immersive virtual reality simulation tool which would provide a meaningful context for a dual-task testing paradigm. The studies discussed here are concerned with the primary task that is used in that tool to induce workload and support ecological validity for future studies of human-robot teaming in dismounted military scenarios: that task is a continuous threat monitoring task meant to emulate a portion of a cordon-and-search operation [10]. The secondary task of interacting with an autonomous teammate (receiving reports, transmitting commands/decisions, maintaining situational awareness, etc.) may be modified inside the simulation tool to investigate a multitude of HRI research questions under similar workload conditions induced by a standard task. These tasks were chosen to be relevant to teaming of dismounted soldiers with an autonomous teammate under high workload conditions. However, the ecologically valid threat-monitoring task does not neatly fit under the umbrella of conventional SDT (temporally independent 2 alternative forced choice tasks). Rather, the task more closely aligns with models of FSDT because, although the distinction between noise and signal events is both clearly defined and constant, the temporal evolution of the search display (and therefore the presence and strength of any given noise or signal stimulus) is dynamic. A result of the time dependence of stimuli presentation is that the factors which influence detection and decision making are somewhat different from the major parameters which affect conventional signal detection task performance (e.g., event rate, signal to noise ratio, signal discriminability). Additionally, performing the threat detection task in virtual 3D introduces factors, such as presentation distance and visibility considerations into the evolution of the strength of a signal stimulus over its lifetime. Therefore the outcomes of our studies will be analyzed with conventional signal detection metrics though the interpretation of those metrics may differ slightly from conventional SDT or FSDT. The two primary factors that we investigated in these studies are the distance at which signals are presented (an operational stand-in for visual angle due to the lack of constancy in the more rigorous metric) and the salience of signal stimulus with respect to its accompanying noise; these will each be discussed separately in brief followed by a more general evaluation of the factors that should be considered in designing continuous signal detection tasks.

2 Experiment 1: The Effect of Signal Distance in a 3D Environment

2.1 Purpose

The task paradigm most similar to ours was developed initially by Szalma et al. [11]: their 3D virtual simulation investigated signal detection performance by moving the participant along a fixed path through an environment and asking them to identify static targets. The search display experienced in their study was necessarily dynamic, as participants were required to make real-time decisions to a flow of changing stimuli. Our study reversed mobility from Szalma et al.'s approach, such that the participant inhabited a fixed space (i.e., stood in one spot) within which they could adjust their gaze, while viewing signal and noise targets that moved through the environment. Targets were presented as humanoid figures (actors) walking from a start point to an end point along a line while carrying an object-of-interest (signal) or a miscellaneous object (noise) (see Figs. 1). The scenario was modeled after a cordon-and-search operation during which some team members would be tasked with monitoring an area (outer or inner section of the area that is cordoned off) for potential threats among the civilian population, in order to ensure the safety of the team members conducting a search. Participants in this study were informed during a training period that they would be tasked with executing the monitoring task through the eyes of a robotic team member (i.e., they would be in control of their team member's gaze and targeting decisions). The goal of this experiment was primarily to investigate the effect of target distance on participant's performance tendencies during a continuous signal detection task in a 3D virtual environment.

2.2 Sample Demographics

Participants in this study were undergraduate students recruited at the University of Central Florida through the university's SONA system (an online portal for facilitating research participation). A total of 20 individuals were in the sample that is described here: 7 males (35%) and 13 females (65%). Many participants chose not to indicate their age upon request, and this information was not pursued post-hoc in the interest of maintaining participants' anonymity; however, those who did indicate their age reported a range between 18 and 22 years old, with an average of 19 years. The study was approved by the UCF IRB and received secondary review by the Army Research Laboratory IRB, and participants were rewarded with class credit as compensation for their participation.

2.3 The Continuous Signal Detection Task

Participants were tasked with controlling a camera system that was equipped with a laser targeting system for marking actors within a virtual environment. Participants searched the display until they identified a signal actor, at which time they were instructed to point the beam of a laser system at that actor and click the left mouse button to indicate their detection decision (see Fig. 2). Signal actors were defined by the object that they carried in their hands: actors carrying guns (an M4 assault rifle) were defined as signals and those carrying other objects were defined as noise (see Fig. 1). The task was continuous in that actors did not cease motion upon being marked; however, correctly identified signal actors highlighted for 500 ms to indicate a hit (shown in Fig. 2, top). Three trials were administered to each participant - each trial lasted for 5 min and employed one of three Center Points (20 m, 30 m, 40 m; see Fig. 2, bottom). Each center point was associated with 3 spawn points per side (all actors appeared in front of the participant, sides are considered relative to their forward vector) such that for the 20-meter Center Point actors may spawn at 10, 20, or 30 m on either side and then proceed at a given movement speed to the opposite side to despawn. Actors could be targeted at any point along their path except when they were occluded by the environment or other actors. 515 actors were spawned for each trial: 55 signals and 460 noise actors yielding a signal-to-noise ratio of 0.12. Correct hits, misses, false positives, and correct rejections were tracked throughout the task to render performance metrics for each participant.

To ensure that enough signals were presented during a trial, six origin points (and corresponding end points) were positioned around a given Central Point placed at a specified distance from the participant as indicated in Fig. 2. The use of multiple start points for actors allowed for the search field to remain uncluttered while maintaining an appropriate signal-to-noise ratio.

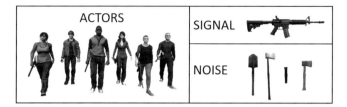

Fig. 1. The continuous signal detection task featured 6 actors who spawned repeatedly (a total of 515 times per trial, roughly 33 times each) carrying one of the five objects shown on the right. Any one of the actors could spawn with any one of the objects, i.e. though the female actor on the far right is shown above as carrying a rifle (making her a "signal actor") she does not always spawn with a rifle.

This experiment made use of two primary task administration tools: an online Qualtrics survey, and a custom immersive virtual reality simulation. All participants began the experiment in an online survey which presented them with an informed consent as well as questions regarding demographics and potentially relevant experience. The survey was also used to administer training materials for the continuous signal detection task which consisted of familiarization with the object-of-interest (the rifle indicated in Fig. 1) as well as thorough explanations of the targeting system and task requirements.

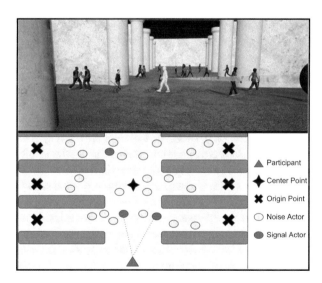

Fig. 2. *Top*: Participant's view of noise and signal actors walking through a virtual environment. The laser targeting system used to indicate signal detection decisions can be seen on the far right-its motion matched participants' gaze adjustments to create the illusion of being strapped to a helmet. *Bottom*: A top-down diagrammatic view of the layout of an individual trial - three origin points per side (doubling as end points for each origin on the opposite side) generate actors which walk along a line until they pass out of the participant's field of view.

Actors spawn on one side of the visual field (occluded at their origin so as not to distract participants) at origin (or spawn) points whose distance from the participant is determined by the Center Point for that trial. Center Points of 20 m, 30 m, and 40 m were used across three trials such that signal and noise actors ranged between 10–50 m from the participant across the experiment (see Fig. 3). In this experiment actors in the simulation traveled along vectors roughly orthogonal to the participant's forward vector (i.e. if they spawned at 10 m forward on the left, their end point would be 10 m forward on the right by walking from left-to-right).

Completion of training prompted the transition of participants from the online survey into the custom simulation. All participants used an Alienware Aurora desktop computer to interact with the Qualtrics survey and simulation; however, completion of the simulation was accomplished with one of two control sets - typical desktop monitor (Dell 24") with mouse control and input (using an Alienware mouse), or with an HTC Vive head mounted display (HMD) with mouse input. This manipulation was part of a separate investigation; only the participants in the group which utilized the HMD will be discussed in this manuscript.

The custom simulation administered three experimental trials to each participant in a randomized order. Trials differed only in the center point of actor origins and otherwise spawned signal and noise actors according to a scheme which maintained a stable signal-to-noise ratio and quantity in the search field at any given point in time. The duration of each trial was 5 min resulting in a spawn rate of 103 actors per minute.

2.4   Results

A repeated-measures ANOVA was conducted to investigate the effect of signal + noise center point distance on perceptual sensitivity (d'). The results evidenced a significant effect of relative distance, $F(2,38) = 25.72$, p < .001, $eta^2 = 0.575$. Additionally, post-hoc analysis using the Bonferroni correction indicated significant differences in perceptual sensitivity between the 20-meter ($M = 1.932$, $SD = 0.370$) and 30-meter ($M = 1.526$, $SD = 0.388$) center points, $t(19) = 3.351$, Cohen's $d = 0.749$, $p = 0.01$, $MDifference = 0.407$, $SE = 0.121$, as well as between the 20 m and 40 m ($M = 0.993$, $SD = 0.565$) distance centers, $t(19) = 6.70$, Cohen's $d = 1.498$, $p < .001$, $MDifference = 0.940$, $SE = 0.140$. Similarly, post-hoc analysis using the Bonferroni correction indicated significant differences in perceptual sensitivity between the 30-meter and 40-meter center points, $t(19) = 4.041$, Cohen's $d = 0.904$, $p = .002$, $MDifference = 0.533$, $SE = 0.132$.

A repeated-measures ANOVA was conducted to investigate the effect of signal + noise center point distance on normalized response criterion (c). The results did not evidence a main effect of relative signal distance, $F(2,38) = 0.188$, $p = .892$, $eta^2 = 0.010$. On the other hand, post-hoc analysis using the Bonferroni correction indicated significant differences in response criterion between the 20 m ($M = 0.591$, $SD = 0.144$) and 30-meter ($M = 0.903$, $SD = 0.320$) center point distances, $t(19) = -4.341$, Cohen's $d = -0.971$, $p = 0.001$, $MDifference = -0.312$, $SE = 0.072$, though interestingly not between the 20-meter and 40-meter ($M = 0.914$, $SD = 3.316$) center points, $t(19) = -0.440$, Cohen's $d = -0.098$, $p = 1.00$, $MDifference = -0.332$, $SE = 0.733$. Additionally, post-hoc analysis using the Bonferroni correction did not indicate significant differences in response criterion between the 30 m and 40 m center points, $t(19) = -0.015$, Cohen's $d = -0.003$, $p = 1.00$, $MD = -0.011$, $SE = 0.726$.

Figure 3 below illustrates the targeting behavior displayed by the participants in this study: proximal signal actors were preferentially targeted during each trial regardless of spawn center point while overall more signals were correctly identified the closer the center point was to the participant's location.

Correct Hit Density: Effect of Center Point and Row

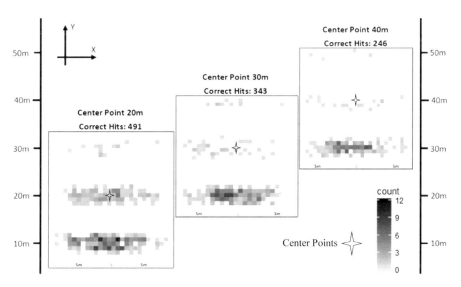

Fig. 3. Top down view of the density of correct hits made within the 3D virtual environment as a function of the center point of a given trial. While the most proximal row receives the greatest density of correct hits for each of the center points, the overall number of hits as well as number of medial and distal row hits decreases with increasing distance. Note that the participant is positioned at (0, 0) for each of the trials.

2.5   Discussion

Increasing the average distance of signal-noise actor presentation revealed two interesting tendencies in participants' targeting behavior: first that there are obvious detrimental effects of displaying stimuli at distances that significantly reduce discriminability (with little effect on response criterion), and second that closer targets will more often be targeted as shown by the density of hits in the most proximal row for each Center Point. The first outcome is unsurprising as it is true in both virtual and material reality that distal objects are more difficult to discern, however this finding does highlight the importance of ensuring that signals are presented in a manner that does not unduly reduce participants' ability to perform (unless that is the intent of the task designer). The second outcome is intriguing because it is unlikely that the tendency to target the most proximal row follows from an inability to discern targets in the remaining rows. Consider the trend exhibited by the hit density of Center Point 1; it is clear from the propensity of correct hits around 20-meters and 30-meters - evidenced in the most proximal rows of Center points 2 and 3 respectively - that participants were capable of distinguishing signal targets at those distances, yet still directed their focus to the most proximal row. Although the source of that trend may be attention, visual occlusion, or strategy it is evident that arrangement of target paths should be carefully considered when designing (virtual) 3D monitoring tasks.

3   Experiment 2: Signal Salience

3.1   Purpose

This second experiment sought to investigate parameters related to the signal-noise distinction that influence task difficulty (in terms of workload as well as SDT metrics) during a continuous signal detection task. Two variables were manipulated in this study: the rate at which new signal and noise events were spawned (spawn rate), and the magnitude of the distinction between signal and noise events (signal salience). Only the effects of the salience manipulation will be considered here, due to the fact that the highest spawn rate investigated in the full study was nearly identical to that used in experiment 1 (105 versus 103 per minute). Considering the effect of salience under similar spawn rate conditions as Experiment 1 is intended to lend clarity to the display factors that should be carefully designed when building a continuous signal detection task without complicating matters by introducing an additional manipulation.

3.2   Sample Demographics

A total of 54 participants were recruited through the University of Central Florida's SONA system to participate in this experiment. The sample contained self-reported males (63%), and 20 self-reported female (37%), with an average age of 20.24 years (Range = [18–30]). Participants were rewarded with class credit as compensation for their participation.

3.3   The Continuous Signal Detection Task

Similar to Experiment 1, the signal detection task was presented to participants as a cordon and search mission being executed in the context of a military operation. The signal-to-noise ratio used was a roughly equivalent 0.13 (experiment 1 used 0.12), and the same actors, noise objects, and M4 rifle stimulus signal were employed (refer to Fig. 1). The primary difference between the two simulations/tasks was that this experiment did not require that signal or noise actors follow a path orthogonal to the participant. Instead actors could appear at any of the spawn points arranged around a square search display/field (20 m × 20 m) and proceed to any point on the opposite side of the field. Therefore, actors still walked across the search field, but as shown in Fig. 5 there was less rigidity in the potential location of actors when they were viewed by participants.

Previous investigations into signal salience have suggested that reducing the salience of a signal can have a deleterious effect on performance [12, 13] particularly as the rate of signal events and search display clutter increase [14]. The salience of a signal is primarily a commentary on the degree to which any given signal may be readily perceived as well as the difficulty of distinguishing a signal stimulus from accompanying noise. Experiment 1 utilized an M4 rifle as its signal object in order to present a relevant threat in a cordon-and-search context; however, that signal is readily distinguishable from the noise objects used in the experiment (refer to Fig. 1) and is - by that metric - quite salient. This experiment manipulated the salience of the signal object by presenting one group with an M4 assault rifle as their signal stimuli, and another with a Beretta handgun as their signal (see Fig. 4).

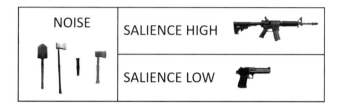

Fig. 4. By switching out the M4 rifle for a Beretta the signal stimuli is made more difficult to discern as well as less distinguishable from the noise objects, particularly the pipe and hatchet. Note that the noise objects are not to scale – the Beretta is closer in size to the hatchet, whereas the M4 rifle is closest in size to the large ax.

3.4   Procedure

Participants were instructed to complete a demographics questionnaire followed by a series of training scenarios that were completed on a desktop computer and within the virtual reality environment. During training, participants were explicitly told about the cordon and search mission, their role in the mission, and the characteristics of the enemy to be identified. The training sequence was designed as such to have stopping points where participants would be given the chance to become familiar with operating the HTC Vive VR equipment and performing the SDT; a practice trial was given to all participants to ensure they understood the point-and-click nature of the SDT. Participants completed three experimental trials with each trial employing a different signal-noise spawn rate (while holding the ratio constant), and signal salience was manipulated as a between subjects variable (i.e. one group searched for the M4 rifle, the other for the Beretta). Presentation of the experimental trials was counterbalanced to ensure that order effects did not occur. Each trial was 10 min in length. In between trials, participants completed the NASA-TLX and Multiple Resource Questionnaires regarding the event rate condition they had just experienced in order to capture perceived workload as well as to reduce the probably of simulation sickness. Participants were dismissed after the questionnaires pertaining to the last experimental trial were completed.

3.5   Results

An independent samples t-test was conducted to investigate the effect of signal salience on participant's perceptual sensitivity (d'). Results evidenced a large effect of signal salience between the Low ($M = 2.361$, $SD = 0.424$) and High ($M = 2.790$, $SD = 0.471$) salience groups, $t(52) = -3.520$, $p < .001$, Cohen's $d = -0.958$.

Additionally, an independent samples t-test was conducted to determine the effect of signal salience on participant's response criterion (c). A similarly strong effect was evidenced between the Low ($M = 0.535$, $SD = 0.143$) and High ($M = 0.409$, $SD = 0.136$) salience groups, $t(52) = 3.323$, $p = .002$, Cohen's $d = 0.904$.

Correct Hit Density: Effect of Signal Salience

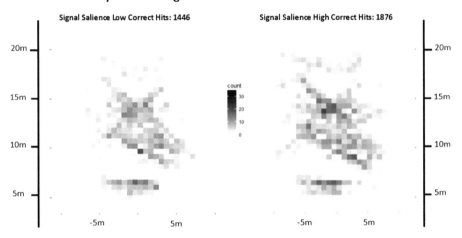

Fig. 5. Correct hit density over the search field in experiment 2 for both Low and High signal salience conditions. Note that the participant would be located at (0, 0).

3.6   Discussion

The presentation of signals of differing salience levels had a strong impact on participants' performance as evidenced by a significant decline in perceptual sensitivity; however, the trend of targeting proximal actors that was evidenced in Experiment 1 was less clear if not wholly absent from the hit densities displayed in Fig. 5. Rather than focus attention on proximal targets, participants seem to have directed their efforts towards the center of the search field. It is notable that the probability of a signal actor passing near the center of the field was higher given the reduced path rigidity that was employed in this experiment. Coupled with the fact that participants successfully accomplished correct hits at the furthest distances employed in this study, and considering also that the furthest distance an actor could appear in this experiment was 20 m (the same distance as the nearest Center Point in experiment 1) it seems most credible that this trend followed from an attentional strategy which prioritized areas most likely to contain a signal actor at any given time. It is unclear from these studies whether or how rapidly participants would be capable of adapting strategies to changing probability fields (say, if the task were to only present signals within a certain distance range and then at some point in the task shift that band) because we failed to capture the temporal adaptation of correct hit locations; however, these findings further emphasize the importance of attending to the presentation of signals not only with regards to their salience but also to the cumulative probability of presentation location.

4   General Discussion

Conventional signal detection tasks are ideal for studying human behavior in many domains, and even when they aren't practitioners regularly find ways of adapting their studies to accommodate SDT. The task used in the experiments described above deviates somewhat from both conventional and fuzzy signal detection theory in its incorporation of signal and noise stimuli which develop over time before exiting the search field. Many of the parameters which are often used to manipulate the difficulty (a stand-in for the expected value of participant performance metrics) in SDT and FSDT are maintained in this altered paradigm: conventional "display clutter" for example is in many ways equivalent to the spawn rate used in this continuous task as both determine the number of signal/noise stimuli that may be visible to participants at any given time. That said, the introduction of temporal evolution of signal and noise stimuli does increase the number of parameters that may be adjusted as well as the interplay between parameters because factors such as salience (display homogeneity) and spawn rate (display clutter) cannot be disambiguated as easily as they may be with conventional static frame trials.

The simulation and VR setting developed for these projects may provide a starting point for the extension of signal detection theory into time-dependent domains. Incorporating the dimension of time and temporal evolution into behavioral research tasks may complicate design and interpretation, but the benefit to real-world relevance is more important - particularly for applied domains such as HRI, Product Design, or Occupational Ergonomics. For example, the studies reported here laid the groundwork for future work in the domain of HRI by developing and testing an ecologically valid and psychometrically sound task that is sensitive to changes with respect to task load, stimulus characteristics, and noise employed. Results showed sensitivity of the task to changes in target rate, signal-to-noise ratio, stimulus salience, and secondary task loads. Additionally, participant responses after task completion showed that they considered the task engaging, challenging, and not predictable, thereby being less likely to be affected by vigilance decrement or practice effects that would change results in repeated-measures designs due to habituation, strategy acquisition, or experience. Real-world replication of the threat detection task would be required to confirm that participants' behaviors in virtual reality map directly onto real-world performance; however, this task is decidedly a meaningful step away from artificial laboratory tasking and towards naturalistic tasks that elicit real-world behaviors.

Acknowledgments. The research reported in this document was performed in connection with Contract Number W911NF-10-2-0016 with the U.S. Army Research Laboratory. The views and conclusions contained in this document are those of the authors and should not be interpreted as presenting the official policies or position, either expressed or implied, of the U.S. Army Research Laboratory, or the U.S. Government unless so designated by other authorized documents. Citation of manufacturers or trade names does not constitute an official endorsement or approval of the use thereof. The U.S. Government is authorized to reproduce and distribute reprints for Government purposes notwithstanding any copyright notation hereon. The authors would like to thank Dr. Valerie Sims and Dr. Corey Bohil for their assistance in analyzing and interpreting the results of these studies.

References

1. Balakrishnan, J.D., MacDonald, J.: Performance measures for dynamic signal detection. J. Math. Psychol. **55**, 290–301 (2011)
2. Masalonis, A., Parasuraman, R.: Fuzzy signal detection theory: analysis of human and machine performance in air traffic control, and analytic considerations. Ergonomics **46**(11), 1045–1074 (2003)
3. Szalma, J., Hancock, P.: A signal improvement to signal detection analysis: fuzzy SDT on the ROCs. J. Exp. Psychol. **39**(6), 1741–1762 (2013)
4. Kim, J.H., Rothrock, L., Laberge, J.: Using Signal Detection Theory and Time Window-based Human-In-The-Loop simulation as a tool for assessing the effectiveness of different qualitative shapes in continuous monitoring tasks. Appl. Ergon. **45**(3), 693–705 (2013)
5. Hancock, P., Masalonis, A., Parasuraman, R.: On the theory of fuzzy signal detection: theoretical and practical considerations. Theor. Issues Ergon. Sci. **1**(3), 207–230 (2000)
6. Murphy, L., Szalma, J., Hancock, P.: Comparison of fuzzy signal detection and traditional signal detection theory: analysis of duration discrimination of brief light flashes. Proc. Hum. Factors Ergon. Soc. **48**(21), 2494–2498 (2004)
7. Burge, R., Chaparro, A.: An investigation of the effect of texting on hazard perception using fuzzy signal detection theory (fSDT). Transp. Res. Part F: Traffic Psychol. Behav. **58**, 123–132 (2018)
8. Malone, S., Brunken, R.: The role of ecological validity in hazard perception assessment. Transp. Res. Part F: Traffic Psychol. Behav. **40**, 91–103 (2016)
9. Vasquez, G., Bendell, R., Talone, A., Nguyen, B., Jentsch, F.: The use of immersive virtual reality for the test and evaluation of interactions with simulated agents. In: Cassenti, Daniel N. (ed.) AHFE 2018. AISC, vol. 780, pp. 15–25. Springer, Cham (2019). https://doi.org/10.1007/978-3-319-94223-0_2
10. United States. Air Land Sea Application Center. Cordon and search: multi-service tactics, techniques, and procedures for cordon and search operations, 2006. (Distributed by Air Force Publishing Wed Site)
11. Szalma, J., Schmidt, T., Teo, G., Hancock, P.: Vigilance on the move: video game-based measurement of sustained attention. Ergonomics **57**(9), 1315–1336 (2014)
12. Abich, J., Reinerman-Jones, L., Taylor, G.: Establishing workload manipulations utilizing a simulated environment. In: Shumaker, R. (ed.) VAMR 2013. LNCS, vol. 8022, pp. 211–220. Springer, Heidelberg (2013). https://doi.org/10.1007/978-3-642-39420-1_23
13. Helton, W.S., Warm, J.S.: Signal salience and the mindlessness theory of vigilance. Acta Psychol. **129**(1), 18–25 (2008)
14. Sawyer, B.D., Finomore, V.S., Funke, G.J., Mancuso, V.F., Funke, M.E., Matthews, G., Warm, J.S.: Cyber vigilance: effects of signal probability and event rate. In: Proceedings of the Human Factors and Ergonomics Society Annual Meeting, 58(1), 1771–1775. SAGE Publications, Los Angeles (2014)

Naturalistic Decision-Making Analogs for the Combat Environment

Thom Hawkins[✉]

United States Army, Arlington, USA
jeffrey.t.hawkins10.civ@mail.mil

Abstract. The combat environment is a challenging space in which to conduct observations relevant to analysis of organizational and cognitive processes, with a low frequency of task repetition, a lack of natural or imposed constraints, and features hostile to external study (e.g., mission secrecy, inaccessible terrain, explosions). An environmental analog assists researchers by providing parallel models and data from a more readily accessible and observable space that can be used for developing concepts and models relevant to the more hostile environment (e.g., space, exoplanets, extreme mountaineering, overwintering in Antarctica). Co-analogs can also be used to understand features and relationships better even in the more accessible space. While there is no commonly accepted process or framework for establishing an environmental analog for naturalistic decision-making, a cognitive task analysis combined with functional abstraction may be used to derive a taxonomic description to demonstrate the appropriateness of establishing an analog relationship.

Keywords: Naturalistic Decision-Making (NDM) · Environmental analog · Cognitive task analysis · Combat environment

1 Introduction

In contrast with traditional decision-making research, which is focused on optimizing structures and models to aid decision-makers, naturalistic decision-making (NDM) describes how practitioners make decisions in the context of their environment. For example, NDM might study how a police officer decides whether or not to pull over a vehicle, or how doctors and nurses make decisions related to patient care. One key feature of NDM is that it eliminates the assumption that the decision-making environment is well-structured, a necessary factor for implementing quantification-based decision support tools such as multi-criteria decision-making or the analytic hierarchy process. In many environments, key information is missing or the quality of the data is not assured. The experience level of the decision-maker is also not consistent —while a decision aid may provide structure that improves the decision-making capability or reduces the cognitive load of an inexperienced practitioner, that same structure may constrain a more experienced practitioner.

In some environments, our ability to study NDM could be constrained by the conditions inherent in the environment. For example, in situ observations may be subject to the Hawthorne effect, where participants modify their behavior because they

D. N. Cassenti (Ed.): AHFE 2019, AISC 958, pp. 49–56, 2020.
https://doi.org/10.1007/978-3-030-20148-7_5

know that are being observed, requiring either obscuration of the study, use of proxy measures, or post–decision follow-up studies. However, there are other study-limiting factors in the combat environment. The dynamic nature of the combat environment means that there is a low frequency of task repetition that inhibits the examination of the same decision made by different individuals or the same individual making multiple decisions. There are also few natural or imposed constraints to serve as study controls. As opposed to a training environment, where one might adhere strictly to doctrine to justify one's actions in an after-action review, the commander knows that he or she has more leeway on judgment in a combat situation where there is no "schoolbook" answer. Finally, the combat environment also includes features that make it hostile to real-time study—mission secrecy, inaccessible terrain, explosions. Embedding academics in this environment to study decision–making is not feasible.

Analog environments have been used to simulate the physical conditions of extreme environments for the purpose of testing and training—for example, using rocky desert terrain to simulate the surface of the Moon or Mars. Analog environments can also be identified for NDM studies by identifying environments with similar NDM factors. However, no standard protocol has been developed for establishing an analog environment.

2 Literature Study

2.1 Naturalistic Decision–Making

NDM is "the way people use their experience to make decisions in field settings" [1]. This is often contrasted with "traditional decision-making," a more deterministic approach that focused on optimization of rational choice methods to improve decision–making against an assumed standard. This alternative to quantitative decision-making approaches predates introduction of the NDM terminology. Rasmussen, as far back as 1983, identified that "what we need is not a global quantitative model of human performance but a set of models which is reliable for defined categories of work conditions together with a qualitative framework describing and defining their coverage and relationships" [2]. Further, Rasmussen tied the concept to behaviorism and Weiner's work on teleology (in terms of interaction between state and goal) in the 1940s.[1]

Quantitative decision-making applications have proven hard to resist, especially in a military context, due to the temptation to reduce decision-making opportunities into algorithmically recognizable situations and to standardize the output of those decisions. Because NDM's aims have been descriptive rather than prescriptive, automated

---

[1] One can find evidence acknowledging NDM even earlier; for example, in Tolstoy's novel *War and Peace* [3]: "A commander in chief always finds himself in the middle of a shifting series of events, and in such a way that he is never able at any moment to ponder on the meaning of the ongoing event. Imperceptibly, moment by moment, an event is carved into its meaning, and at every moment of this consistent, ceaseless carving of the event, a commander in chief finds himself in the center of a most complex play of intrigues, cares, dependency, power, projects, advice, threats, deceptions, finds himself constantly in the necessity of responding to the countless number of questions put to him, which always contradict each other."

decision aids have been focused on improving situational awareness of decision–makers, as well as reducing potential for judgment bias or automation to reduce decision timelines [4].

2.2 Cognitive Task Analysis

Cognitive task analysis (CTA) is a thinking-focused method for documenting a process in terms of how cognitive resources are managed and processed—for example, what information a decision-maker needs and how they consider and communicate that information. Zachary et al. [5] identifies two approaches to establishing an introspection-based CTA model—interviewing subjects as they are executing the activity [6], or interviewing them after the activity is complete [7]. They note the challenges with both methods—mainly the intrusion of the in situ interviews and challenges with recall in the post-interview methods.[2]

The level of detail in the CTA documentation depends on the methods, participants, and how the CTA output will be used [8]. Stanton et al. [9] provides a protocol for using Rasmussen's levels of abstraction to aid in the affinization of cognitive tasks. However, no methods have been described for comparing and contrasting CTAs between different functional areas.

2.3 Military Decision-Making Models

From the late 1980s to the late 1990s, the US Department of Defense spent $25 to $35 million on NDM research [1].[3] Despite this, the U.S. Army still relies on the highly structured and step-intensive military decision–making process (MDMP) as its doctrinal approach to decision–making [10]. MDMP is a rational choice methodology focused on the elucidation and comprehensive quantification of options.

Drillings and Serfaty [11] note that the U.S. military has not completely embraced this rational choice approach, preserving the prerogative of an officer's particular style of command, setting up a dual standard for what is taught and what is practiced. Some of this dichotomy is attributable to the assumed level of experience of the commander [1]; the structure of the MDMP is a substitute for the lack of operational experience in the novice Warfighter. Van Trees [12] found that decision support tools and training were focused on rational choice methods. Thunholm [13] provides an overview of the efforts by western militaries to provide a prescriptive model for decision-making.

The Army's current focus is on improving the availability of situational awareness to the commander in the field, and increasing the speed at which that information can be ingested [14], an approach that may tacitly acknowledge the influence of NDM.

---

[2] The military uses the latter method in training, with observer-controllers conducting after-action reviews, but the training conditions are more controlled than in an actual combat environment.

[3] No updated estimates are available to show the spending since the late 1990s.

2.4   Environmental Analogs

Extreme environments are "an external context that exposes individuals to demanding psychological and/or physical conditions, and which may have profound effects on cognitive and behavioral performance" [15]. Orasanu and Lieberman [16] identify environmental analogs for space as an extreme environment—such as submarines and overwintering in Antarctica. Orasanu and Connolly [17] characterize environments conducive to NDM, including a low level of problem structure, an uncertain or dynamic environment, time stress, and high stakes.

According to Hammond, establishing formal analogs is necessary to validate testable theories, "otherwise results simply become retrospective products subject to multiple ad hoc interpretations that cannot be falsified" [18]. The challenges of conducting studies in the combat environment has led researchers to draw conclusions from studies that may mirror only some of the NDM–representative factors identified by Orasanu and Connolly [17]. For example, Lieberman et al. [19] examined the effect of caffeine on cognitive decline in military training. As noted by Jones et al., "simulation is only an approximation of the real world and can be ineffective if it deviates significantly from reality" [20].

3   Analysis

Analogs for extreme environments consider exactly the physical features most relevant to the research at hand—for example, terrain, temperature, isolation. In "The Taxonomy of Man in Enclosed Space" [21], Sells provides an analogy framework for extreme environments by characterizing one such feature, isolation, according to a series of ten factors, including voluntary versus involuntary, planned versus unplanned, length of duration, space restriction, threat, etc. An analog for an NDM environment must rely on a taxonomy relevant to the subject, such as the eight features identified by Orasanu and Connolly [17] (see Table 1).

A key difference between Sells' taxonomy for isolation and Orasanu and Connolly's taxonomy for an NDM environment is that Sells' elements are more strictly defined with binary choices and quantifiable comparisons—is the isolation voluntary or involuntary? For how long is the subject in isolation? Orasanu and Connolly, in contrast, are qualitatively described—e.g., "information may be ambiguous or simply of poor quality" [17]. By this rubric, it is certainly easy to argue that both hospital emergency rooms and combat operations are "uncertain dynamic environments," but also a trip to the grocery store with a list made by one's spouse. These criteria are to determine whether or not NDM is an appropriate decision-making methodology for that situation. It is less effective as a mechanism to relate two NDM environments to each other, although it may provide some cross–domain insight for a particular feature, such as action and feedback loops. This NDM framework may provide the top-level hierarchy for a taxonomy just as Sells built his taxonomy from a single element of the extreme environment—"isolation."

Table 1. Features of an NDM environment – comparison

Feature	Combat environment	Hospital emergency room
Ill-structured problems	Based on low rate of repetition for similar problems	Based on complexity of human biology and the field of medicine
Uncertain, dynamic environments	Low ability to predict inputs; high reliance on assumptions	Low ability to predict inputs; some reliance on assumptions
Shifting, ill-defined, or competing goals	Overlapping, competing goal structures	Goals are relatively stable, but prioritization varies based on inputs and resources
Action/feedback loops	Situational awareness consistently checked against goals at tactical and strategic levels	Highly instrumented
Time stress	Self-imposed to preserve decision space	Imposed by biological necessity
High stakes	Life/death of multiple operators and potential strategic gains/losses	Life/death of individual cases
Multiple players	Includes strict multi-level authority hierarchy with multiple specialty functions at some levels	Flatter (and occasionally disputed) hierarchy with multiple specialty functions
Organizational goals and norms	Tactical level organization strictly defined with pronounced norms	Organizational norms centered on patient outcomes and hygiene practices

One technique that might aid in identifying lower levels of an NDM environment taxonomy is cognitive task analysis (CTA). CTA's primary benefits have been in identifying learning objectives for training systems and elucidating system design requirements [22], but the structural description of knowledge representation and cognitive tasks could be useful for comparative analyses. The training regime of a system, developed using CTA, is essentially itself an analog of the environment it is simulating. Because the cognitive tasks at the lowest level are closely associated with function, it may be necessary to apply a hierarchy of abstraction to the CTA to elucidate the relationships between NDM environments. Potter et al. [23] identify the need to apply Rasmussen's abstraction hierarchy to generalize a process within a discipline, noting that "the explicit relationships between those nodes provide the natural structure of the work domain." Indeed, leveraging abstraction as a tool may help to generalize the CTA sufficiently from nodes to purpose and relationships to identify a level appropriate for generalization across disciplines without abstracting to the point of not being able to draw any relevant conclusions.

Functional abstraction is typically depicted in five levels (from least to most abstract): (1) physical form, (2) physical function, (3) generalized function, (4) abstract function, and (5) functional purpose [2]. Fackler et al. found in a CTA of critical care physicians five types of cognitive activity: "pattern recognition, uncertainty management, strategic vs. tactical thinking, team coordination and maintenance of common ground, and creation and transfer of meaning through stories" [24]. The cognitive activities are conceptualized to the point of abstract function (level 4) and are recognizable as potential outputs of a CTA from a military decision–making effort. Drillings and Serfaty identify "key components of the commander's task are to assess the situation, develop courses of action, make decisions, and monitor their implementation" [11]. These are generalized functions (level 3), but one could also identify their applicability to an emergency room situation. Comparative analysis of CTA from two disciplines, taken at different levels of abstraction, may help to identify the affinities that guide the applicability of findings across the domains.

While CTA and functional abstraction[4] can be used to identify positive associations between NDM environments, the proposed analog must resolve the constraints identified in the target environment, just as a proposed new scientific paradigm must resolve the anomalies presented in the previous paradigm [25]. In the case of the combat environment, the proposed analog must resolve the inherent dangers of the combat environment. At the same time, the resolution of this feature must not affect the applicability of the feature being measured—for example, the ever-present danger of military combat, the kill-or-be-killed mentality, may provide a particular kind of time pressure that is impossible to replicate and this must be acknowledged by constraining the use and caveating the conclusions drawn from the analog environment. The use of unobtrusive measures, especially with the expanded use of sensors and network analysis inherent in modern warfare, could produce data useful for the examination of NDM in the combat environment. Post hoc interviews could be aided by data produced in situ rather than relying solely on participants' memory.

Two other limiting features of the combat environment—the lack of natural or imposed constraints and the low frequency of repetition may have less of an effect on the ability to draw conclusions from an analog environment, but must be acknowledged. Both these features are a factor in any conclusions drawn from inferential statistics and probabilistic outcomes and are common caveats when transferring models from a field setting to a laboratory setting.

4   Conclusion

The case for using a data set from an analog environment to supplement or replace a data set from a less accessible environment must be made based on the specific targets and constraints of the aspect of NDM under study. Just as there are no complete environmental analogs for extreme environments because there will always be one or more features that differ in relevant ways, analogs for NDM environments must be justified based on the similarity of the feature in focus, and closely examined to determine whether differing features affect the validity of that relationship.

---

[4] Cognitive work analysis (CWA), described by Rasmussen [2], includes cognitive task analysis as work domain analysis in an expanded framework, to include elements of functional abstraction.

The assumption that one environment can be used as an NDM analog for another environment is based on the similarity of the two as described in terms of NDM structure and CTA. Finding a useful analog for the combat environment, it is not only necessary to find an environment that is structurally similar, from an NDM or CTA perspective, but also that resolves the challenges of the combat environment without nullifying its applicability as an NDM analog.

References

1. Zsambok, C.E.: Naturalistic decision making: where are we now? In: Zsambok, C.E., Klein, G.A. (eds.) Naturalistic Decision Making, pp. 23–36. Routledge, New York (2009)
2. Rasmussen, J., Pejtersen, A.M., Goodstein, L.P.: Cognitive Systems Engineering. Wiley, New York (1994)
3. Tolstoy, L.: War and Peace. HarperCollins, New York (1869)
4. Klein, G.A.: An overview of naturalistic decision making applications. In: Zsambok, C.E., Klein, G.A. (eds.) Naturalistic Decision Making, pp. 69–80. Routledge, New York (2009)
5. Zachary, W., Ryder, J.M., Hicinbothom, J.H.: Cognitive task analysis and modeling of decision making in complex environments. In: Cannon-Bowers, J.A., Salas, E. (eds.) Making Decisions Under Stress: Implications for Individual and Team Training, pp. 315–344. American Psychological Association, Washington (2000)
6. Klein, G.A., Calderwood, R., Macgregor, D.: Critical decision method for eliciting knowledge. IEEE Transact. Syst. Man Cybern. **19**(3), 462–472 (1989)
7. Ericsson, K.A., Simon, H.A.: Protocol Analysis: Verbal Reports and Data. MIT Press, Cambridge (1984)
8. Gordon, S.E., Gill, R.T.: Cognitive task analysis. In: Zsambok, C.E., Klein, G.A. (eds.) Naturalistic Decision Making, pp. 131–140. Routledge, New York (2009)
9. Stanton, N.A., Ashleigh, M.J., Roberts, A.D., Xu, F.: Levels of abstraction in human supervisory control teams. J. Enterp. Inf. Manage. **19**(6), 679–694 (2006)
10. Department of the Army: Field Manual 6-0, Commander and Staff Organization and Operations. Headquarters, Department of the Army, Washington DC (2016)
11. Drillings, M., Serfaty, D.: Naturalistic decision making in command and control. In: Zsambok, C.E., Klein, G.A. (eds.) Naturalistic Decision Making, pp. 71–80. Routledge, New York (2009)
12. Van Trees, H.L.: C3 systems research: a decade of progress. In: Johnson, S.E., Levis, A.H. (eds.) Science of Command and Control, Part 2: Coping with Complexity, pp. 24–43. AFCEA International Press, Fairfax (1989)
13. Thunholm, P.: Planning under time pressure: an attempt toward a prescriptive model of military tactical decision making. In: Montgomery, H., Lipshitz, R., Brehmer, B., and Naturalistic Decision Making Conference (eds.) How Professionals Make Decisions, pp. 43–56. Lawrence Erlbaum Associates, Mahwah (2005)
14. Army Capabilities Integration Center: The United States Army Warfighters' Science and Technology Needs (2017)
15. Paulus, M.P., Potterat, E.G., Taylor, M.K., Van Orden, K.F., Bauman, J., Momen, N., Swain, J.L.: A neuroscience approach to optimizing brain resources for human performance in extreme environments. Neurosci. Biobehav. Rev. **33**(7), 1080–1088 (2009)
16. Orasanu, J., Lieberman, P.: NDM issues in extreme environments. In: Mosier, K.L., Fischer, U.M. (eds.) Informed by Knowledge: Expert Performance in Complex Situations, pp. 3–21. Psychology Press, New York (2011)

17. Orasanu, J., Connolly, T.: The reinvention of decision making. In: Klein, G.A., Orasanu, J., Caldenwood, R., Zsambok, C.E. (eds.) Decision Making in Action: Models and Methods, pp. 3–20. Ablex Publishing Corporation, Norwood (1993)

18. Hammond, K.R.: Naturalistic decision making from a brunswikian viewpoint: its past, present, future. In: Klein, G.A., Orasanu, J., Caldenwood, R., Zsambok, C.E. (eds.) Decision Making in Action: Models and Methods, pp. 205–227. Ablex Publishing Corporation, Norwood (1993)

19. Lieberman, H.R., Stavinoha, T., McGraw, S., White, A., Hadden, L., Marriott, B.P.: Caffeine use among active duty US Army Soldiers. J. Acad. Nutr. Diet. **112**(6), 902–912 (2012)

20. Jones, R.M., Laird, J.E., Nielsen, P.E., Coulter, K.J., Kenny, P., Koss, F.V.: Automated intelligent pilots for combat flight simulation. AI Mag. **20**(1), 27 (1999)

21. Sells, S.B.: The taxonomy of man in enclosed space. In: Rasmussen, J. (ed.) Man in Isolation and Confinement, pp. 281–303. Transaction Publishers, Mahwah (1973)

22. Chipman, S.F., Schraagen, J.M., Shalin, V.L.: Introduction to cognitive task analysis. In: Schraagen, J.M., Chipman, S.F., Shalin, V.L. (eds.) Cognitive Task Analysis, pp. 3–23. London Psychology Press, New York (2000)

23. Potter, S.S., Elm, W.C., Roth, E.M., Gualtieri, J., Easter, J.: Bridging the gap between cognitive analysis and effective decision aiding. In: McNeese, M.D., Vidulich, M.A. (eds.) State of the Art Report (SOAR): Cognitive Systems Engineering in Military Aviation Environments: Avoiding Cogminutia Fragmentosa, pp. 137–168. Human Systems Information Analysis Center, Wright-Patterson AFB (2002)

24. Fackler, J.C., Watts, C., Grome, A., Miller, T., Crandall, B., Pronovost, P.: Critical care physician cognitive task analysis: an exploratory study. Crit. Care **13**(2), R33 (2009)

25. Kuhn, T.S.: The Structure of Scientific Revolutions. University of Chicago Press, Chicao (1970)

Nefarious Actors: An Agent-Based Analysis of Threats to the Democratic Process

Nicholas Stowell$^{(\boxtimes)}$ and Norvell Thomas$^{(\boxtimes)}$

Claremont Graduate University, Claremont, CA, USA
nicholas.stowell@cgu.edu, norvellthomas@gmail.com

Abstract. Recent election cycles in the United States and in other democracies have often been accompanied by accusations of hacking, meddling, disinformation, and voter suppression by non-state actors as well as elected government officials. Such threats pose a grave danger to the perceived legitimacy of democratic elections and public trust in the institutions upon which representative democracy relies. This paper utilizes an agent-based modeling approach to simulate the impact of actions by nefarious state and non-state actors on voter turnout, trust in electoral democracy, misrepresentation, and the potential for political conflict. Through repeated simulation and the development of a landscape of plausible futures, this paper uncovers means by which the electoral process may be most sensitive to attack. Given that the potential erosion of trust in elections poses an existential threat to democracy itself, this paper should provide valuable insight for those who seek to preserve and protect the most fundamental elements of liberal democracy.

Keywords: Democracy · Elections · Agent-based model · Suppression · Hacking · Conflict · Trust · Representation · Voting

1 Introduction

Around the globe, democracy is in decline. According to the most recent Polity Index, liberal democracies have declined. The 2016 Freedom House report downgraded the United States from a full democracy to a "flawed democracy", specifically noting the United States' willingness to suppress votes via gerrymandering, voter ID laws and other techniques to limit voter participation [1]. At the same time, we are witnessing social media being used as a tactical weapon of warfare by various governments to spread propaganda and misinformation, and to divide the population. To add more elements to this toxic mix, trolls and hackers are attempting to sow chaos by producing false news stories and other content in an attempt to confuse citizens and reduce the overall trust level between people and their governments.

This paper uses an agent-based model to explore the effect of nefarious state and non-state actors on the electoral process. We focus on several key processes through which liberal democracy is under attack: election hacking, disinformation campaigns, and voter suppression. An election is not a monolithic event. Rather, there are hundreds or thousands of voting precincts with varying resources and levels of security. A hacker need not even flip an election to have a significant, negative impact on a democracy.

© Springer Nature Switzerland AG 2020
D. N. Cassenti (Ed.): AHFE 2019, AISC 958, pp. 57–68, 2020.
https://doi.org/10.1007/978-3-030-20148-7_6

Merely hacking one precinct and advertising that they did so, nefarious non-state actors can erode public trust in the entire process. Disinformation campaigns on social media and by propaganda outfits act to increase overall political polarization, which has been shown to lead to social unrest and weaken democracy [2]. Voter suppression, which includes voter ID laws, restricted early voting, and other means to disallow segments of the population from engaging in the electoral process is a means through democracy is weakened by democratically elected politicians themselves. These attacks on democracy, by way of helping to increase voter polarization and decrease voter turnout, have the unfortunate effect of undermining the overall levels of trust within a democratic system, both between citizens as well as between citizens and their government.

A divergence of ideological distribution between the elected government and the population may cause a decrease in trust in the electoral system for those who are least represented as well as for those whose votes were suppressed. Distrust may also cause a decrease in voter turnout, leading to a feedback loop of distrust, low turnout, decreased representation, and further distrust. Prior research indicates that a lack of trust in the electoral process may lead to political apathy or, in a highly polarized context, potential for political conflict and social unrest. Both key potential outcomes constitute a grave threat to democracy [2]. This paper models these dynamic processes and reveals emergent outcomes of these interactions over time. One of our goals is to show how a computational framework can serve as a theoretical tool for determining how reducing trust in the electoral process works to undermine democracy through relatively simple means by both state and non-state actors.

2   Literature Review

The Pew Research Center has tracked the levels of a variety of threats to democracy as they relate to overall trust. The Director of the Institute, Jared Diamond, lists four elements of American democracy and argues for preservation of healthy debate amongst citizens and between citizens and government:

1. A healthy, respectful debate helps ensure comity as well as decreased polarization.
2. A full democracy allows for citizens to feel their voices are being heard. The absence of this may lead to general frustration and eventually resort to social unrest and political violence.
3. Compromise as an essential component of democracy with the long-held notion of majority rule, with minority rights. Without compromise, democracy begins to falter.
4. Lastly, in modern democracies, all citizens are allowed to vote. Active voter suppression, gerrymandering, and other ways in which politicians and state officials work to circumvent this crucial element of democracy serve only to weaken the very institutions of the system, while causing more distrust overall. In this paper, we see the effects of nefarious agents.

Sokhey and McClurg (2012) find that social networks help inform voters' preferences when deciding on issues and candidates in an election cycle. The researchers argue that, because most Americans have supportive social networks, they help determine the "correct" way of thinking and deciding on how to vote. However, they note that these social networks do not provide increased learning in a traditional sense [3]. We model voters' social networks as a fundamental attribute to determine how susceptible voters are to hackers and trolls when shaping political opinions. With the rise of new technology and social media, voters have increased ability to find their own social network that conforms to their own political ideology.

Political disagreement amongst citizens is a fundamental occurrence within a democracy and, as such, we should expect a wide variety of discussion on topics that are important to each individual voter. However, when political engagement becomes so severe as to create extreme polarization, this may lead to social unrest and even political violence, particularly when there is a lack of personal ties to those with different political views [4].

In terms of political polarization and social networks, current research shows that individuals who don't regularly engage with others holding opposing viewpoints are less likely to see these arguments as legitimate and less able to provide rationales for their own political decisions [5, 6].

Warren (2018) provides research that suggests voter trust in government and institutions, as well as social trust, is a fundamental underpinning of a healthy democracy. He defines trust along several dimensions including social trust, which is the individual's judgment that another person is both motivated and competent to act in their best interest and will do so without resorting to political or social harm. Institutional trust is the individual's judgement that government institutions will act in the best interests of the individual without consequence. Both types of relationships require a level of trust to ensure cooperation in a democratic society. The paradox, as Warren sees it, is that institutions are borne out of general distrust by citizens to provide certainty and security of policies from their representative governments [7]. That there are elements within the political infrastructure that will use voter suppression, gerrymandering and other actions that undermine this trust helps to either reinforce voter apathy or create conditions for social unrest and political violence. Understanding these fundamental underpinnings of democracy, one country could, if given the right tools and resources, wreak considerable havoc upon another country without having the need to engage in direct conflict.

3   A Nefarious Actor Model

The choice to utilize an agent-based model makes sense given that real-world data on the effects of nefarious actors on the electoral system is neither readily available nor reliable, and that traditional experiments in this realm are inconceivable. This section therefore outlines an agent-based model of the electoral process that is sufficiently abstract to be implemented computationally while retaining the granularity necessary for substantive value. Agent-based models have a distinct advantage as a data-generating process under such restraints. To do so, it was necessary to make decisions regarding the specific type of election to be modeled. We settled on a two-party congressional general election cycle with 36 distinct voting districts.

3.1   Agents

Each voting district has two candidates per election, one from each party. Initially, candidates from each party are distributed random-normally across an ideological continuum, with the two parties having considerable ideological overlap, which we believe to be true of most two-party democratic systems. In addition to their position on the ideological continuum, some candidates from one party are attributed with a willingness to suppress votes to gain or retain power. We view the suppression of votes to be anti-democratic and therefore conceptualize this subset of candidates to be nefarious state actors.

The second agentset in our model is the population of voting-aged adults. We control the size of this population, but their distribution is randomly assigned, which allows for significant variance in the ideological distribution of each voting district. Each voting-aged adult is also assigned an initial position on the ideological continuum which is distributed random-normally. Each voting-aged adult either trusts or does not trust that their vote will be counted. This is crucial to our model given that trust in the electoral process is fundamental to representation and therefore to a healthy liberal democracy. Adults are also assigned an initial level of political engagement, the distribution of which we control. They are also assigned a level of mobility to represent the fact that adults tend to move several times in their life and population movements affect the ideological makeup of voting districts over time. Voting-aged adults are also assigned a conflict score, as a function of their level of political engagement and position on the ideological continuum, to represent their willingness to take to the streets in protest as well as the intensity of political conflict that could take place. Finally, voting-aged adults are assigned a level of connectivity, which we control. Connectivity represents the size of social networks and determines the number of links that each adult has with other adults.

The third set of agents is nefarious non-state actors. We conceptualize this agentset as hackers, online political trolls that spread disinformation through social media, and propagandist media. While the motives and methods of individual agents within this group vary, the outcomes they seek to produce–discord, polarization, and decreasing voter turnout of those with whom they disagree–are similar enough to justify conflating them into one group. Nefarious non-state actors are assigned a level of influence, which we control, and which determines how connected they are with the population, as well as how able they are to sway those with whom they are connected.

3.2   Quantifying Agent Attributes

This section explains how agent attributes are quantified. As previously stated, each political candidate and voting-aged adult are situated along an ideological continuum. This ideology score continuum runs from −10 to +10. Candidates from "party A" begin with a mean ideology score of −5, a standard deviation of 2.5, and range from −10 to +3. Candidates from "party B" begin with a mean ideology score of 5, a standard deviation of 2.5, and range from −3 to +10. Voting-aged adults begin with a mean ideology score of zero and a standard deviation of 4.5.

The percentage of political candidates willing to suppress votes to gain or retain a seat in congress ranges from zero to 100% of only one party, set by us and assigned randomly. The effectiveness of their suppression ranges from zero to five percent. The number of nefarious non-state actors ranges from zero to 100. Their influence ranges from zero to ten. The number of links generated from nefarious non-state actors to constituents is equal to the product of the number of non-state actors and the level of influence. These links are assigned randomly, leading to variation in the number of links from each nefarious non-state actor.

The number of constituents ranges from 500 to 5000. While this leads to an unrealistically high ratio congresspeople to constituents, a realistic number of constituents is beyond the computational capacity of NetLogo. Additionally, we do not believe that this ratio has significant effect on outputs. The initial percentage of potential voters that trusts the electoral process is allowed to range from 70 to 100%. Popular mobility is allowed to range from zero to 15%. Between each election, each constituent has a probability of moving equal to this percentage. If a constituent moves, they move in a random direction and a random number of patches up to the popular mobility percentage. Given that each voting district is ten patches by ten patches in size, this allows for considerable change in district composition over time. Constituent connectivity is allowed to range from zero to ten. The total number of initial links generated is equal to one half the number of constituents times connectivity. Links are generated randomly, with 70% of links based on geographical proximity and remaining links generated randomly. The number of links varies significantly from constituent to constituent, which we believe reflects the real world in which some people are far more connected than others. Engagement ranges from zero to 100%, and each constituent's initial engagement level is assigned in a random-normal distribution around a mean set by us. Each constituent's conflict score is the product of their level of engagement and their ideology score divided by ten.

3.3   Implementing the Model

The Nefarious Actor model is written in NetLogo 6.0.4 with a two-dimensional space representing 36 distinct voting districts. The space is populated by agents as described above. Time is modeled discreetly, with each tick representing an election cycle. Nefarious actors first act on the constituents with which they are linked. Each constituent linked with a nefarious non-state actor moves, with a probability equal to the influence level of the non-state actor, increases or decreases their ideology score. If the constituent has an ideology score greater than zero, her score increases. Otherwise, it decreases. This models the polarizing effect of propagandists and those who spread disinformation. Constituents linked with a nefarious non-state actor also lose trust in the electoral process with a probability equal to the difference between the nefarious actor's influence and the engagement level of the constituent. Finally, the engagement level of constituents linked with a nefarious non-state actor increases with a probability and magnitude equal to the nefarious actor's influence. This models the sensational effects of propaganda and online trolling on a population.

After the nefarious non-state actors act upon the population, elections occur in each district. Constituents that do not trust their votes will be counted will not vote. Of those that do have trust in the electoral process, a percentage will vote that corresponds with a global input for initial voter turnout. Then, in each district, the candidate with the ideology score closest to the median voter in each district becomes a congressperson. In this way, elections are strictly based on the median voter theorem [8].

After the election, those congresspeople who are assigned the attribute of willingness to suppress votes seek to do so. First, they assess the difference between their ideology score and that of their opponent–the candidate that lost the previous election. If the difference is greater than zero, then a percentage of constituents with an ideology score that is less than that of their opponent will have their subsequent vote suppressed. The proportion of voters that have their votes suppressed in that district is a random percentage between zero and the effectiveness of suppression attribute assigned to the nefarious congressperson. Of those constituents whose votes are suppressed, a random percentage up to the effectiveness of suppression attribute will also lose trust in the electoral process. We feel that this accurately models the effects of real-world efforts by some members of congress to reduce voter turnout in a manner that disproportionately affects likely voters for the opposite party.

Next, those candidates that lost the previous election seek to realign themselves with their constituency so as to increase the probability with which they will win the next election. First, they assess the median ideology score of all constituents within their district that voted in the previous election. Second, they calculate the divergence between their ideology score and that of the median voter within their district. Third, they change their ideology score by a random number up to five percent of the divergence between their ideology score and that of the median voter within their district. Fourth, they assess the median ideology of voters within their party within their district. The party of constituents is operationalized as party A for those constituents with an ideology score less than zero, and party B for those constituents with an ideology score greater than zero. This corresponds roughly with the initial distribution of party A and party B with regard to political candidates. Fifth, candidates change their ideology score by a random number up to ten percent of the divergence between their ideology score and that of their in-party constituent voters. This movement, first toward the median voter and, second, toward the median in-party voter, models the political tug-of-war that political candidates face between representing all voters and appealing to their base constituents.

The next step in the model is population movement. Each constituent has a random probability of moving in each cycle up to their mobility score, which is generated by a random-normal distribution around the global popular mobility input. Constituents' mobility scores also affect the distance that they move, whereas the direction is random. This models the real world in which, in the case of the United States, people move roughly 11.4 times in their lives on average.

After the constituent movement step, constituents interact with their social networks. If a member of a constituent's social network has an ideology score that is two or more points different from oneself, the constituent cuts ties with that person with a random probability up to 25%. This represents the notion that people tend to associate more with likeminded people than with those of opposing views. After cutting ties with those of divergent ideologies, each constituent assesses the mean ideology score of all members of her remaining social network and changes her own ideology score, closing as much as 25% of the divergence between herself and her network. Each constituent then tabulates the ratio of members of their network that trusts in the electoral process. If more than 70% of their network trusts the process, they will also trust. If fewer than 30% of their network has trust, they will not have trust. This process is repeated for each constituent regarding whether or not they will vote in the next election. Finally, each constituent's conflict score moves up to 25% toward the average conflict score of their social network. This process of social network interaction reflects the nature of human interactions in the modern, technological age.

Finally, constituents calculate the extent to which they are misrepresented by congress. Misrepresentation for each constituent is operationalized as the difference between the mean ideology score of congress and the ideology score of oneself. If a constituent's misrepresentation score exceeds 6 or −6, their conflict score increases by a random amount up to 20% of their engagement level. They also lose trust in the system with a probability equal to the absolute value of their level of misrepresentation. On the other hand, if a constituent feels as though they are well represented by congress, operationalized as a misrepresentation score between negative one and positive one, their conflict score will decrease by a random amount up to 20% of their engagement level and, if they did not trust the electoral process prior to this step, they will gain trust with a probability of 50%. This completes one iteration of the model, conceptualized as one voting cycle.

The entire process is outlined in Figs. 1 and 2. The outputs of interest are the percent of constituents that trust the electoral process, voter turnout, mean conflict score, and the mean level of misrepresentation. The simulation ends if voter turnout dips below 25%, if the percent of constituents that trust their vote would be counted dips below 50%, or if the average level of misrepresentation exceeds four.

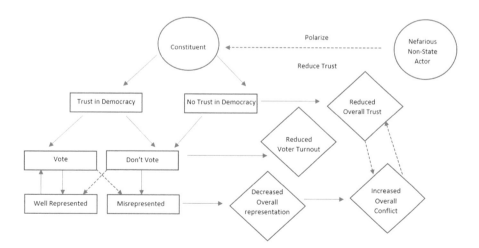

Fig. 1. Flow chart outlining major actions by constituents in the nefarious actor model

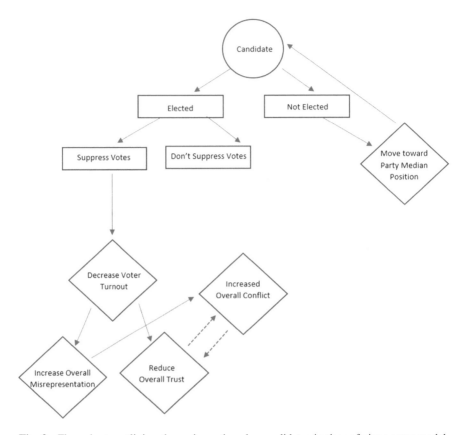

Fig. 2. Flow charts outlining the major actions by candidates in the nefarious actor model.

4   Findings

The purpose of developing the nefarious actor model was to examine the extent to which actions by nefarious politicians and non-state actors may harm the fundamental democratic institution of free and fair elections. In order to generate data on the relative impact of the various threats to democracy modeled herein, we ran 1000 simulations with randomized parameter settings. The range of parameters in our sensitivity test, such as number of people and parameters affecting the number of links, were limited by computing power. Initial trust in elections swept from 30% upward, which captures a minimum which is slightly less than the minimum voter turnout for congressional elections in the US. The number of candidates willing to subvert swept a range of values from zero to 100% (from one party only), and their effectiveness of suppression ranged from zero to five percent. Our effort was to model a range of values somewhat beyond that which we may see in the real world, while acknowledging that data on suppression and subversion is not readily available.

When run at a conservative baseline including only five nefarious non-state actors per 3000 constituents, five percent of candidates willing to suppress votes at only two percent effectiveness, and with initial voter turnout at fifty percent, the results vary. The model tends to run beyond 40 election cycles consistently, at usually beyond 100. When it does stop, it is virtually always because of voter turnout rather than trust in elections.

We then ran four OLS regressions on the results, corresponding with the four key outputs of interest: voter turnout, trust in elections, misrepresentation, and political conflict. The results are shown in Table 1.

Regarding the two types of nefarious actors that we modeled, the sensitivity analysis demonstrates that nefarious political candidates pose the largest threat to the electoral process. While both types of nefarious actors reduce voter turnout and trust in the electoral process, nefarious politicians directly contribute to misrepresentation,

Table 1. Sensitivity analysis: OLS regressions on 1000 randomized simulations.

VARIABLES	(1) Voter Turnout	(2) Percent Trust	(3) Mean Misrepresentation	(4) Mean Conflict Score
Number of People	0.000218	0.000663**	6.61e-05***	-0.000827***
	(0.000242)	(0.000269)	(1.58e-05)	(0.000123)
Nefarious Actors	-0.0329***	-0.0523***	1.67e-05	0.0586***
	(0.0109)	(0.0121)	(0.000710)	(0.00554)
Nefarious Actor Influence	-0.336***	-0.607***	-0.00500	0.685***
	(0.101)	(0.112)	(0.00657)	(0.0513)
Initial Trust in Elections	0.298***	0.0806**	0.00895***	0.0950***
	(0.0363)	(0.0403)	(0.00236)	(0.0184)
Network Connectivity	0.684***	2.115***	-0.0690***	-0.461***
	(0.101)	(0.112)	(0.00654)	(0.0510)
Nefarious Candidates	-0.114***	-0.0859***	0.00749***	-0.0498***
	(0.0112)	(0.0124)	(0.000725)	(0.00566)
Engagement	0.0154	0.0376***	0.000130	0.270***
	(0.0111)	(0.0124)	(0.000723)	(0.00564)
Initial Voter Turnout	0.582***	-0.513***	0.0114***	0.0919***
	(0.0153)	(0.0170)	(0.000997)	(0.00778)
Popular Mobility	0.0570	0.0769	-0.0100**	-0.0577*
	(0.0683)	(0.0758)	(0.00444)	(0.0346)
Effectiveness of Suppression	-3.479***	-2.895***	0.181***	-0.705***
	(0.189)	(0.210)	(0.0123)	(0.0958)
Constant	-15.26***	98.92***	-1.085***	-3.740**
	(3.583)	(3.980)	(0.233)	(1.817)
Observations	993	993	993	993
R-squared	0.677	0.618	0.382	0.759

Standard errors in parentheses
*** $p<0.01$, ** $p<0.05$, * $p<0.1$

which we view as damaging to democracy. The model is particularly sensitive to the effectiveness with which nefarious politicians are able to suppress votes. While hackers, trolls, and propagandists are a threat, their influence appears to pale in comparison to that of nefarious politicians.

One finding of particular interest is the effect of network connectivity on the key electoral outputs. Contrary to what we assumed prior to simulation, larger social networks seem to have a moderating effect on political behavior and ideology. The amount of person-to-person connectivity is directly associated with voter turnout, and trust, and is inversely related to misrepresentation and conflict. These findings are statistically and substantively significant. If the nefarious actor model is valid, social networks and the increased connectivity afforded by modern technology and social media are not responsible for polarization and political conflict in and of themselves. Rather, it is nefarious actors working through social networks that may lead to polarization and political conflict.

5   Conclusion

Our results show that nefarious actors have a potentially devastating effect on democracy by undermining trust in elections as well as increasing polarization and the potential for political conflict. State actors have a more direct effect in limiting voter turnout, through gerrymandering, voter ID laws and other voter suppression techniques. Non-state actors work to divide citizens and increase political polarization, which in turn may lead to further erosion of democracy, albeit indirectly. Taken together, these actions have a compounding effect of decreasing voter enthusiasm for participating in elections, increasing polarization, increasing voter apathy, and increasing social and political conflict. Our model shows that with just the right amount efforts by non-state actors polluting the system with disinformation and hacking efforts, along with efforts by state actors administering voter suppression protocols, the overall trust in democracy could reach critical levels below 25% within a generation. Democracy has always relied upon its citizens being informed of their choices, and actively participate in their own governance. If enough citizens are not able to vote, or dissuaded to participate in voting, then the foundations of democracy itself are in grave danger.

5.1   Policy Implications

Voters across the political divide are uniting in demands to end gerrymandering by politicians [10]. In the past two years, several states have stripped the ability of politicians to draw political districts for partisan advantage. Many states, like California, have either created non-partisan redistricting committees or allowed independent judges to draw political maps to help ensure fairness. This is a step in the right direction and helps support the doctrine of free and fair elections. More states will need to place limits on partisan gerrymandering activities or simply abolish them altogether to increase voter turnout and decrease apathy. Most recently, President Obama launched an initiative to combat gerrymandering, calling it an "issue of singular importance," [11] and will work with state legislatures and judicial groups to fight for basic electoral fairness. More work needs to be done to ensure free and fair elections, and fighting gerrymandering is a step in the right direction.

It is clear that, when politicians try to maximize their own utility of winning elections by using subversive tactics like voter suppression, they do so at the risk of undermining democracy itself. Voter suppression, as our model shows, has a devastating effect on voter turnout and electoral trust, leading to and increased misrepresentation and potential conflict, all of which poses a severe threat to democracy. Quite simply, any efforts to limit citizens from fully voting is a direct violation of free and fair elections, and as such, proper consideration by legislative authorities to outlaw these tactics needs to be seriously considered.

By now it is abundantly clear that the rise is social media, has both empowered communities to find common ground among social and political issues and has even helped facilitate mass protests against authoritarian regimes. Unfortunately, the same technology has been used by governments and shadowy forces to spread disinformation and "fake news" to pervert free political discourse and polarize the population. Any efforts to combat online manipulation would have to involve cooperation between governments, civil society, and big-tech companies to ensure fair and accurate information is dispersed and not be used for nefarious reasons. Within the United States, lawmakers should pass the Global Online Freedom Act, which would impose penalties on countries that restrict internet freedom. Policymakers should introduce further legislation to combat the intentional disbursement of propaganda and fake news.

5.2   Areas for Future Research

There are several ways in which this model could be extended or refined. First, an additional layer could be added to model a presidential election cycle in addition to the congressional election cycle. Doing so would more accurately reflect the American political system. Second, the model could be altered to allow for multi-party, parliamentary systems. This would elucidate different sensitivities between various electoral systems. Third, as opposed to the current setup, which has initial candidate and constituent distributions hard-coded, the model could be more easily validated by allowing the user to set the distributions to reflect specific, real world scenarios. Finally, the model would benefit from the introduction of political and economic shocks. We believe that it would be a useful addition to examine the resiliency of a political system to stochastic events such as terrorist attacks or a run on the banks. While it does demonstrate the emergence of some startling phenomena, these extensions would add considerable utility to the nefarious actor model.

References

1. Freedom House. https://freedomhouse.org/
2. Hetherington, M.: Putting polarization in perspective. Br. J. Polit. Sci. **39**(2), 413–448 (2009)

3. Sokhey, A., McClurg, S.: Social networks and correct voting. J. Polit. **74**(3), 751–764 (2012). https://doi.org/10.1017/s0022381612000461

4. Esteban, J., Schneider, G.: Polarization and conflict: theoretical and empirical issues: introduction. J. Peace Res. **45**(2), 131–141 (2008)

5. Huckfeldt, R., Mendez, J., Osborn, T.: Disagreement, ambivalence, and engagement: the political consequences of heterogeneous networks. Polit. Psychol. **25**(1), 65–95 (2004)

6. Price, V., Cappella, J., Nir, L.: Does disagreement contribute to more deliberative opinion? Polit. Commun. **19**(1), 95–112 (2002)

7. Warren, M.: Trust and Democracy. The Oxford Handbook of Social and Political Trust. Oxford University Press, Oxford (2018)

8. Downs, A.: An Economic Theory of Democracy. Harper, New York (1957)

9. FiveThirtyEight.   https://fivethirtyeight.com/features/how-many-times-the-average-person-moves/

10. New York Times. https://www.nytimes.com/2018/07/23/us/gerrymandering-states.html

11. Huffington   Post.   https://www.huffingtonpost.com/entry/obama-gerrymandering_us_5c1de4b08aaf7a8826f5

Steady-State Analysis of Multi-agent Collective Behavior

Hui Wei, Mingxin Shen, and Xuebo Chen[✉]

School of Electronics and Information Engineering,
University of Science and Technology Liaoning, Anshan 114051,
Liaoning, People's Republic of China
xuebochen@126.com

Abstract. Currently, many scholars research the interactions of collective behaviors of human beings according to the rules of collective behaviors of animals and make great progress in theory and practical application. However, in the case of multi-agent system consisting of large number of individuals, the existing control strategies cannot perfectly meet the actual demand of collective motions, those issues would cause the failure of large-scale collective motion. Thus, the research on the steady-state analysis of collective motion is more crucial. Based on the Lennard-Jones potential function and a self-organization process, this paper proposes a topological communication model to simulate the collective behaviors of real society. The structure of the collective motion at stable state are analyzed systematically by changing the number of agents in the group, the number of interconnections and the initial position. This work could provide a theoretical basis for the establishment of mass events and the prevention of mass incidents caused by social collective behavior.

Keywords: Multi-agent system · Collective behavior ·
Lennard-Jones potential function · Steady-state structure

1 Introduction

Collective behaviors are a universal phenomenon in various biological species which live together in groups and swarms such as the flocking of birds, swimming of schools of fish, motion of herds of quadrupeds migrating bacteria, molds, ants or pedestrians and so on [1–4]. These fascinating natural phenomena have attracted growing attentions among researchers from different fields of science to discover the principles of interactions among individuals in system and the relationship between local interactions and global emergences.

Research on collective behavior tends to adopt a self-organization perspective, with investigations of how global-level collective behaviors emerge from local interactions among individuals [5]. The majority of this work is informed by laboratory experiments which tend to combine mathematical simulations, or agent-based models [6]. For instance, Aoki [7] proposed an individual behavior model to simulate fish schooling toward fisheries science, where the very important interactions among biological individuals are emphasized, that is, alignment, repulsion and attraction. These points

© Springer Nature Switzerland AG 2020
D. N. Cassenti (Ed.): AHFE 2019, AISC 958, pp. 69–77, 2020.
https://doi.org/10.1007/978-3-030-20148-7_7

play an essential role in the study of biological groups. Vicsek et al. [8] established a minimal model to study how cohesive clustering motion is organized under external disturbance, and enlightened for statistical mechanical researches on collective behaviors. Couzin et al. [9] studied the spatial information of patterns and reported an interesting idea of "collective memory". Measurements and observation data are also stimulating the studies of group dynamics. Katz et al. [10] tried to discover the interrelations among individuals by using observation data of fish schooling, and Ballerini et al. [11] directly measured three-dimensional position of individual birds within flocks and found that birds in flocks interact with the closest 6–7 neighbors, rather than with all neighbors within a fixed metric distance. By taking into account both the interrelations and the psychological factors affecting the interactions among individuals, Gao et al. [12] founded a network-based model for simulating human collective behaviors. In [13], Chen et al. simulated the aggregation, spiral and bifurcation phenomena in the evolution of social collective behavior, and provided a theoretical basis for the establishment of mass events and the prevention of mass incidents caused by social collective behavior. These numerical models show us rich information about collective behaviour, however, no study has been done on such stable state for collective behavior. In the case of multi-agent system consisting of large number of individuals, the existing control strategies cannot perfectly meet the actual demand of collective motions, those issues would cause the failure of large-scale collective motion. To address this problem, we focus primarily on the structure of collective motion at stable.

This paper is organized as follows. In Sect. 1, we describe the recent advances on collective behavior. In addition, Sect. 2 introduces the related concepts and proposes a topological communication model to simulate the collective behaviors of real society. Moreover, in Sect. 3 the experimental results of our proposed model, show that the structure of collective behavior at stable satisfies certain geometric regularity. Finally, Sect. 4 concludes our work in this paper.

2 Methods

In the modeling of fish group, two basic types of interaction are usually considered: the interactive relationships of metric distance and topological distance [14]. Under the rule of the interactive relationship of metric distance, every individual has a perception radius and can only communicate with other individuals within the range of the perception radius. Each individual is able to obtain information from other individuals in the process of interactions and will change its behavior according to the information received regarding the opinions and behavior of other individuals. However, under the rule of the interactive relationship of topological distance, each individual does not need a perception radius and pays more attention to its nearest neighbors. Compared with the interactive relationship of metric distance, each individual has a larger perception range, the constraint on its perception ability is the number of nearest neighbors that they can perceive. The crucial difference between metric and topological interaction really comes in how the perception ability of individuals in the model is restricted. In order to construct a more intelligent model, we choose the method of topological interaction, which has been successfully adapted to describe collective behavior in the world, to explain phenomena in real society.

Under the background of multi-agent systems, the interactive relationship is only considered to the effects of collective motion by removing the attractive effects to individuals from the target, taking the interactive relationship between two nearest neighbors as interconnected objects, three-dimensional interconnection rule is modeled.

$$f_i = \sum_{j \in N_i} f(||v_i - v_j||) \cdot e_{ij} \tag{1}$$

In Eq. (1), $e_{ij} = a_{ij} \cdot \frac{v_j - v_i}{||v_j - v_i||}$ describes the unit vector of interconnection (v_i, v_j) with the right value a_{ij}, where (v_i, v_j) is considered as the edge set with a direction from v_i to v_j, $i, j \in N = \{1, 2, \ldots, N\}$, $i \neq j$. and the interaction unit vector in the direction of force can be described by Lennard-Jones potential function.

Based on the Lennard-Jones potential function [15] and a self-organization process [9], our paper proposes a topological communication model to simulate the collective behaviors of real society. Different from traditional agent-based model [16], this paper ignores velocity-based alignment, discovers the collective behavior of multi-agent under a rule that only considers repulsion and attraction (long rang attraction, short rang repulsion). The interaction function among individuals is defined as follows.

$$f(x) = \begin{cases} k_1(\frac{1}{x} - \frac{1}{r^*}), & \text{if } r_l < x < r^* \\ k_2 \sin[\frac{(x - r^*)\pi}{r_h - r^*}], & \text{if } r^* < x < r_h \\ 0, & \text{others} \end{cases} \tag{2}$$

In Eq. (2), k_1 and k_2 are coefficients which define interconnection force constant respectively, r^* represents the equilibrium distance. The parameter r_l defines the lower bound on the distance between two individuals that can generate repulsive interaction (r_l is ignored in this article), and the parameter r_h represents the higher bound on the distance between two individuals that can generate attractive interaction.

3   Results

In this section, we use MATLAB to simulate the model and analyze these results. In the simulations, a group of representative initialization parameters are listed in Table 1. The number of agents N is from 1 to 55 and the number of topology K is from 0 to 54. Many simulations are performed to discover different kinds of collective motions, which are based on interactions among individuals in the group under different initial conditions.

Firstly, in order to better analyze the collective behaviors of multi-agent system, in three-dimensional space, the multi-agent based on the topological interaction relationship of the collective motion, the number of agents N randomly distributed at the initial position (here take $N = 10$), the number of topology $K = 1$–9 changes in turn. As shown in Fig. 1, the red line in the figure represents the average position of ten agents (or the group trajectory) when the system is stable. Then, in order to explore the steady-state of multi-agent collective behavior more closely, we further analyze the distance and force between neighboring agents with topological interaction in the system, as shown in Figs. 2, 3, 4, 5, 6 and 7. In the distance graph, the following part of the red line represents the maximum distance of seven neighboring agents with interconnections, the black line describes the maximum distance between all the agents, while the middle portion of the red and black lines represents a distance relationship that have no interconnection but actually exists. In the force diagram, the red line represents the resultant force of all the agents with topological interconnections, while the Blue line represents the maximum forces of the agent with topological interconnection. Compared with the above two graphs, it is shown that the distance of all the agents with topological interaction is consistent with the time point that the resultant force is stable, and the resultant force of the system is stable when the topological structure of the cluster system is globally stable. More importantly, the consistency and stability of the cluster system are synchronized by comparing the graph of the collective motion and the distance and force, that is, the stable interaction forms a stable structure, and then the stable consistency is formed.

Table 1. Parameter values in the simulations.

Parameter	The mean of parameter	Parameter value
r_l	Lower bound on the distance that can generate repulsion	0
r^*	Equilibrium distance	5.2
r_h	Higher bound on the distance that can generate attraction	12
k_1	Coefficient	1
k_2	Coefficient	0.3
t	Simulation steps	2×10^4
N	Number of agents	1–55
K	Number of topology	0–54

Secondly, comparing graph (i) with other graphs in Fig. 1, those ten agents obviously have an aggregate trend, while the ten agents in the other graphs are more or less distributed in three-dimensional space in a decentralized form. What is the cause of this phenomenon? It is noted that the number of topological interaction K in the graph (i) is 9, just satisfies the formula $K = N - 1$, so we carry out a further simulation research. In order to further study the structure model of multi-agent clusters when the number of topological interaction $K = N - 1$, we have traversed all the collective motion instances of the number of agents from 10 to 55, eventually choosing some the most representative models, as shown in Fig. 8. According to the simulation results above, when the number of agents is certain, and all the neighboring agents are interconnected, the aggregation trend will spring up. In general, as the number of agents gradually increased, the aggregation trend of the system became more and more obvious, and finally formed a kind of structure model which is similar to clusters. This model reflects that everyone in real life is able to obtain enough information from others, and the

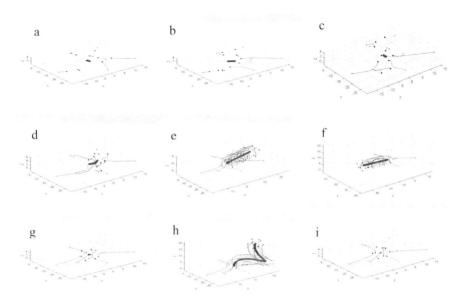

Fig. 1. Various collective motions under different combinations of N and K (N = 10, (a) K = 1; (b) K = 2; (c) K = 3; (d) K = 4; (e) K = 5; (f) K = 6; (g) K = 7; (h) K = 8; (i) K = 9).

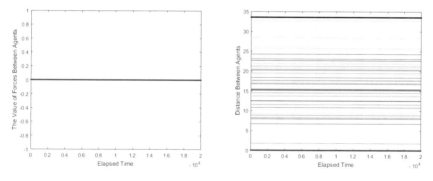

Fig. 2. Distance and force of cluster system based on topological interconnection relationship (K = 1, N = 10).

whole group quickly reaches a perfectly stable state because the information in this circumstance is open and highly transparent. Everyone can achieve the maximum satisfaction and will not want to change their behavior.

Finally, according to [17], in the rustic analysis, when the number of atoms N is 13, 19, 25, 55, 71, 87 and 147, the intensity of the atomic clusters shows a peak value, indicating that the stability of these clusters is particularly strong, and the number of contained atoms is called magic number. The stable structure of its clusters is also related to the way atoms are stacked in shells (atomic position), respectively corresponding to some regular geometry graphs. Such as N = 13 is a positive 20-body with

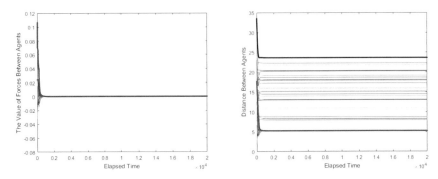

Fig. 3. Distance and force of cluster system based on topological interconnection relationship ($K = 3$, $N = 10$).

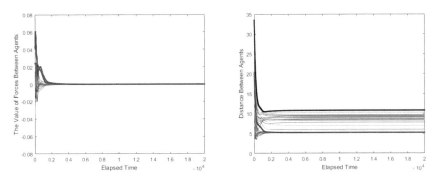

Fig. 4. Distance and force of cluster system based on topological interconnection relationship ($K = 4$, $N = 10$).

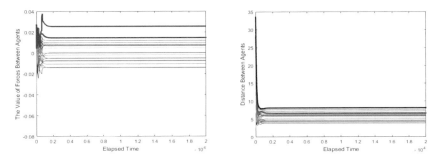

Fig. 5. Distance and force of cluster system based on topological interconnection relationship ($K = 6$, $N = 10$).

5 symmetry axis symmetry, that is, the center point to each vertex distance is all equal. Therefore, we make a further research about the structure and state of the cluster $N = 13$, $K = N - 1$. As shown in Fig. 9, the distance of the center point (-1.2308,

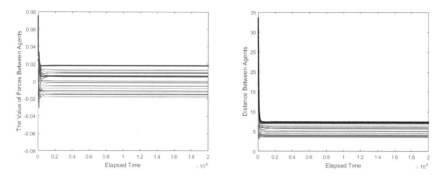

Fig. 6. Distance and force of cluster system based on topological interconnection relationship ($K = 7$, $N = 10$).

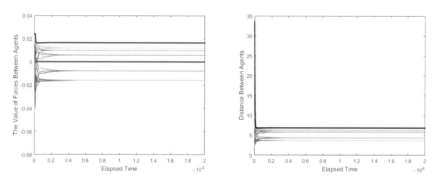

Fig. 7. Distance and force of cluster system based on topological interconnection relationship ($K = 9$, $N = 10$).

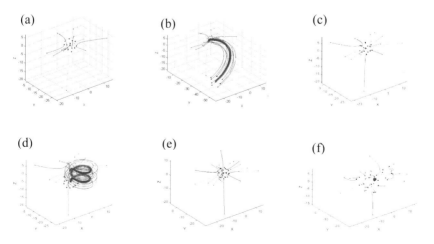

Fig. 8. Special collective motion model under different combinations of N and K ((a) $N = 13$, $K = 12$; (b) $N = 13$; $K = 8$; (c) $N = 15$, $K = 14$; (d) $N = 15$, $K = 10$; (e) $N = 25$, $K = 24$; (f) $N = 25$, $K = 4$).

−17.5077, 1.6615) to the remaining vertices is all equal to 3.5538, fully conform to the geometric properties of the positive 20-body, it can be said that the cluster structure is stable at this time.

⊞ 13x13 double												
1	2	3	4	5	6	7	8	9	10	11	12	13
1 0	7.1076	6.0461	3.7367	6.0461	3.7367	6.0461	3.7367	3.5538	3.7367	6.0461	6.0461	3.7367
2 7.1076	0	3.7367	6.0461	3.7367	6.0461	3.7367	6.0461	3.5538	6.0461	3.7367	3.7367	6.0461
3 6.0461	3.7367	0	3.7367	3.7367	3.7367	6.0461	6.0461	3.5538	7.1076	3.7367	6.0461	6.0461
4 3.7367	6.0461	3.7367	0	6.0461	3.7367	7.1076	3.7367	3.5538	6.0461	3.7367	6.0461	6.0461
5 6.0461	3.7367	3.7367	6.0461	0	3.7367	3.7367	7.1076	3.5538	6.0461	6.0461	6.0461	3.7367
6 3.7367	6.0461	3.7367	3.7367	3.7367	0	6.0461	6.0461	3.5538	6.0461	6.0461	7.1076	3.7367
7 6.0461	3.7367	6.0461	7.1076	3.7367	6.0461	0	6.0461	3.5538	3.7367	6.0461	3.7367	3.7367
8 3.7367	6.0461	6.0461	3.7367	7.1076	6.0461	6.0461	0	3.5538	3.7367	3.7367	3.7367	6.0461
9 3.5538	3.5538	3.5538	3.5538	3.5538	3.5538	3.5538	3.5538	0	3.5538	3.5538	3.5538	3.5538
10 3.7367	6.0461	7.1076	6.0461	6.0461	6.0461	3.7367	3.7367	3.5538	0	6.0461	3.7367	3.7367
11 6.0461	3.7367	3.7367	3.7367	6.0461	6.0461	6.0461	3.7367	3.5538	6.0461	0	3.7367	7.1076
12 6.0461	3.7367	6.0461	6.0461	6.0461	7.1076	3.7367	3.7367	3.5538	3.7367	3.7367	0	6.0461
13 3.7367	6.0461	6.0461	6.0461	3.7367	3.7367	3.7367	6.0461	3.5538	3.7367	7.1076	6.0461	0

Fig. 9. The distance between thirteen neighboring agents.

4   Conclusion

The emergence of multi-agent collective behavior is a specific ordered stable state, and its stability and consistency are achieved at the same time. Moreover, the number of agent and interconnection, and the initial position of the agent have an effect on the emergence of multi-agent collective behavior. The cluster structure can be stabilized only when all the agents in the intelligent group have interconnected relationships and the structure of collective behavior at stable also satisfies certain geometric regularity.

Acknowledgments. This research reported herein was supported by the NSFC of China under Grants No. 71571091 and 71771112.

References

1. Allison, C., Hughes, C.: Bacterial swarming: an example of procaryotic differentiation and multicellular behaviour. Sci. Prog. **75**, 403–422 (1991)
2. Rappel, W.J., Nicol, A., Sarkissian, A., Levine, H., Loomis, W.F.: Self-organized vortex state in two-dimensional dictyostelium dynamics. Phys. Rev. Lett. **83**, 1247–1250 (1999)
3. Rauch, E.M., Millonas, M.M., Dante, C.: Pattern formation and functionality in swarm models. Phys. Lett. A **207**, 185–190 (1995)
4. Helbing, D., Keltsch, J., Molnar, P.: Modelling the evolution of human trail systems. Nature **388**, 47–50 (1997)
5. Vicsek, T., Zafeiris, A.: Collective motion. Phys. Rep. **517**, 71–140 (2012)
6. Dell, A.I., et al.: Automated image-based tracking and its application in ecology. Trends Ecol. Evol. **29**, 417–428 (2014)

7. Aoki, I.: A simulation study on the schooling mechanism in fish. Bull. Jpn. Soc. Sci. Fish. **48**, 1081–1088 (1982)
8. Vicsek, T., Czirok, A., Ben-Jacob, E., Cohen, I., Shochet, O.: Novel type of phase transition in a system of selfdriven particles. Phys. Rev. Lett. **75**, 1226–1229 (1995)
9. Couzin, I., Krause, J., James, R., Ruxton, G., et al.: Collective memory and spatial sorting in animal groups. J. Theor. Biol. **218**, 1–11 (2002)
10. Katz, Y., Tunstrøm, K., Ioannou, C., Huepe, C., Couzin, I.: Inferring the structure and dynamics of interactions in schooling fish. Proc. Natl. Acad. Sci. U.S.A. **108**, 18720–18725 (2011)
11. Ballerini, M., Cabibbo, N., Candelier, R., et al.: Empirical investigation of starling flocks: a benchmark study in collective animal behaviour. Anim. Behav. **76**, 201–215 (2008)
12. Gao, C., Liu, J.: Network-based modeling for characterizing human collective behaviors during extreme events. J IEEE Trans. Syst. Man Cybern. Syst. **47**, 171–183 (2017)
13. Chen, X.B., Sun, Q.B., Huang, T.Y.: Prospect for social aggregation based on simulation of communication topology. J. Univ. Sci. Technol. Liaoning **40**(4), 312–320 (2017). (in Chinese)
14. Yuan, Y., Chen, X.B., Sun, Q., et al.: Analysis of topological relationships and network properties in the interactions of human beings. J. Plos One. **12**, 1–22 (2017)
15. Jones, J.E. (ed.): On the determination of molecular fields. II. From the equation of state of a gas. Proc. R. Soc. Lond. A: Math., Phys. Eng. Sciences. R. Soc. **25**, 125–130 (1924)
16. Reynolds, C.W.: Flocks, herds and schools: a distributed behavioral model. ACM SIGGRAPH Comput. Graph. **21**, 25–34 (1987)
17. Wang, G.H.: Stable structures and magic numbers of atomic clusters. J. Prog. Phys. **20**(1), 53–92 (2000). (in Chinese)

The Urban Renew Case Study on a Feasibility Profit Sharing in Different Development Intensity by Game Theory

Yuan-Yuan Lee[1(✉)], Zining Yang[1], and Oliver F. Shyr[2]

[1] School of Social Science, Policy & Evaluation,
Claremont Graduate University, Claremont, CA, USA
{yuan-yuan.lee,zining.yang}@cgu.edu
[2] National Cheng Kung University, Tainan, Taiwan
ofshyr@mail.ncku.edu.tw

Abstract. There have been many urban renewal projects in Taiwan since the Urban Renewal Act issued in November 1998. Taipei is an early developed city with a large population and higher density living population in Taiwan. This will cause public safety issues as large-scale landslide disasters are caused by earthquake or fire. Hence, Taipei city government needs to promote urban renewal projects and encourage citizen participation. This research study will discuss which combination of an urban renewal project is optimal. Using game theory, this study is looking to offer a solution in a different way for the long-term problem of uneven profit distribution in an urban renewal project. In this case study, with three areas of different land development intensity and four players in this urban renewal project owing different size of the land. Among the seven different participation combinations in this project, we identify the optimal outcome for this urban renewal project using backward induction.

The result shows the best outcome exists when all players cooperate, but Player D's profit is the smallest. However, this result still shows co-partnership is more effective than the independent operation.

Keywords: Sequential game · Game theory · Urban planning · Urban renew · Building coverage ratio · Floor area ratio

1 Introduction

This issue, urban renew in Taiwan, has been promoted more than 20 years since Urban renew Act had been issued in November 1998. Especially in Taipei and New Taipei city since these are the early development also are most politic, business, and economically active cities.

There are higher business and political activities concentrated in Taipei Metropolitan area so that there are many job opportunities which will be one of the key reasons to attract people and cause higher population density in those cities. The causes there are many, 4, and sufficient, 0, years. We can see Fig. 1 [1] shows the housing area in Taipei. There are 61% over 30 years. In those old buildings which do not have sufficient public facilities to provide better living surroundings also the quality of

© Springer Nature Switzerland AG 2020
D. N. Cassenti (Ed.): AHFE 2019, AISC 958, pp. 78–86, 2020.
https://doi.org/10.1007/978-3-030-20148-7_8

shockproof is not as good as the newest building since early technology, not an advantage as nowadays.

Moreover, also there is disaster prevention problem as fire. Another reason urban renew project is popular in Taipei Metropolitan Area since the housing price is increasing every year which we could see in Fig. 2 also available land used for construction is getting less. We suppose to find the way to offer more livings area. Also resolving zoning controlling problems and restrictions legally without affecting the rights of landlords. Therefore, that is why urban renew is important in Taiwan.

Fig. 1. Taipei city housing age

Fig. 2. Taipei city housing price index in 2009 to 2015

Since urban planning implements in Taiwan, we have a restriction on land use and zoning controlling. Hence, in different zoning, we have different regulations for building coverage ratio and floor area ratio that affect how many areas you could have a building. Moreover, it is possible in the same urban renew project which may be able to exist different zoning regulation. Since different zoning regulation which we can get different floor area for our building that also means everyone should get different benefit sharing form, an urban renew project. However, what is the suitable sharing rate in an urban renew project when it exists different zoning?

There are many study about urban renewal this topic, Zheng et al. [2] argues that although, the growing body of research covers the areas mentioned above, the mechanism for achieving sustainable urban renewal has yet to be clarified (278). In urban renewal there are more complex problems and construction in the earlier stage. We could do our best if we know which position and responsibilities we are needed to take [3]. Since the renewal of cities is an inherently more complex activity than their initial construction, it demands a clear understanding of the proper roles and responsibilities of the public and private sectors so that each can concentrate on what it can best contribute [3].

Therefore, most studies related to urban renew that have been done in the past have more focus on past experience of the positions of each role in an urban renew project. There is lack of information about research on the distribution of benefits in an urban renew project, so this study will target the distribution of reasonable profits between landlords and implementers.

2   Research Design

First, this study will introduce the current and overall situation in urban renew, the policy to understand the problems of urban renew. Second, through the exploration of problems, it will summarize the issues that need to be solved and constructed through "Game Theory," also to find reasonable compensation acceptable to the landlords and implementer. Finally, calculating how much building bulk ratio bonus and finance projects in each of them, solving the game, and giving the policy suggestions.

Game theory was invented by John von Neumann and Oscar Morgenstern "The theory and economic behavior of the game" in 1944. Game theory is about how rational players act in a game and is there possible exist reasonable results for each player. In other words, game theory is the predictive behavior and actual behavior of individuals in the game and study's the optimization strategies. Different interactions on the surface may exhibit similar incentive structures [4].

In a game, participants hold different goals or interests and how to achieve them, also how could you get the most beneficial solution for yourself.

Therefore, the game theory which studies to predict whether the game behavior of each players have the most reasonable behavior and optimization strategy.

Moreover, there are 5 basic elements in game theory which are the number of participants (players), all the strategies that the entrants can choose (Strategies), all possible combinations of strategies (Actions) and each entrant in each strategy. The combination of payoff (Payoff) and the preference generated by remuneration are represented by mathematical symbols as follows [5]:

1. Player
   $N = \{Player1, Player2, ...Player n\}\, i = 1, 2, ..., n$
2. Strategies of each players
   $S = \{S1, S2, ..., Sn\}\, i = 1, 2, ..., n$
3. Interactions
   $s = \{s1, s2, ..., sn\}$
4. Payoff
   $\pi = (s1, s2...sn)\, for\, all\, s1 \in S1,\, s2 \in S2, ..., sn \in Sn$

3   Case Studying

The case study is located in the fifth section of Zhongshan North Road, Zhongshan District, Taipei City, and the urban renewal project of the three sections of the Fulin section. The total area of the base is 3,083 square meters. It is divided into three areas due to different development intensity: area A is 1,033 square meters, area B is 1,011 square meters, and area C is 1,039 square meters.

There is 2 different land zoning in the same block in this urban renew project. The partition closer to Zhongshan North Road which has a higher floor area ratio is 400% and building coverage ratio is 45%. Moreover, another partition in this block which has lower floor area ratio is 225% and the same building coverage ratio.

Furthermore, there are 4 players, A is player 1 (P1), B is player 2 (P2), C is player 3 (P3) and the construction company is player 4 (P4) (Figs. 3 and 4).

Fig. 3. Project location   **Fig. 4.** Urban planning

3.1 Available Redevelopment Area and Floor Area

Table 1 shows available redevelopment area and floor area base on zoning regulation, in this renew project. There are 2 different floor area ratios so that there is different available redevelopment floor area.

Table 1. Available redevelopment area and floor area

Projects	Building coverage ratio(%)	Floor area ratio(%)	Gross building area(M^2)	Gross floor area(M^2)
P1	45	225 and 400	464.85	3,588.10
P2	45	225	565.95	2,274.75
P3	45	225 and 400	467.55	3,610.88
P1 + P2	45	225 and 400	919.80	5,862.85
P1 + P3	45	225 and 400	932.40	7198.98
P2 + P3	45	225 and 400	922.50	5885.63
P1 + P2 + P3	45	225 and 400	1387.35	9473.73

3.2   Building Bulk Ratio Bonus

Base on Urban Renewal Act of the urban renew project would obtain extra floor areas in rational reason which increases participants' interests [6].

In this study, the case study located in Taipei city so that need to review the urban renew regulation by Taipei City Government. In the regulation, there is a formula to calculate [6] (Table 2 and Fig. 5):

$$F = F0 + \triangle F1 + \triangle F2 + \triangle F3 + \triangle F4 + \triangle F5 + \triangle F6$$

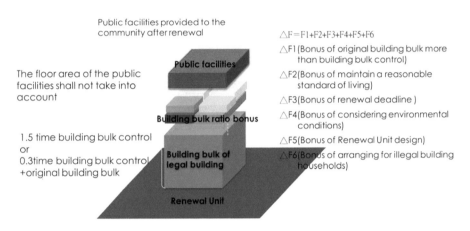

Fig. 5. Building bulk ratio bonus

F	: Total Building bulk ratio bonus
F0	: Building bulk of legal building
$\triangle F1$	: Bonus of original building bulk more than building bulk control
$\triangle F2$	: Bonus of maintain a reasonable standard of living
$\triangle F3$	: Bonus of renewal deadline
$\triangle F4$	: Bonus of considering environmental conditions
$\triangle F5$	: Bonus of Renewal Unit design
$\triangle F6$	: Bonus of arranging for illegal building households

3.3   Total Profit for Each Project

Table 3 lists all financial parameters we would use for the calculation for total profit for each project. To avoid too many different financial parameters that affect the decisions of projects since using the same amount of expenditures, but profits changing base on the area and additional conditions. For example, in Project 1, construction cost would be $1,360 USD/M2, however, project 2 construction cost would be $1,613 USD/M2.

Table 2. Building bulk ratio bonus in projects

Projects	Building bulk ratio (%)	Bonus of building bulk area (M²)	Gross floor area (M²)
P1	27.69	993.54	4581.64
P2	24.42	555.49	2830.24
P3	24.40	881.05	4491.93
P1 + P2	26.05	1,527.27	7,390.12
P1 + P3	26.04	1874.61	9073.59
P2 + P3	24.41	1436.68	7322.31
P1 + P2 + P3	28.29	2680.12	12153.85

Comparing project 1 and 2, that would affect player 1 may be able to require more benefits than player 2. There are 7 projects in this case study, based on the financial calculation, we could see in Table 4 the combination of players will require higher benefits than the individual.

4 Game Theory Application

In this paper, we use sequential-move game to solve those urban renew projects. We set up three different P4 expectation benefits: 25%, 28%t and 30% since in marking experience P4 we consider the benefits at least 25% remuneration in a project.

Table 3. Financial parameter

Parameter	Description
1. Land cost	Cooperate with landlords so land cost is 0
2. Construction cost	$4,500/per sq ft
3. Project Fee	Depends on the architect and regulations
4. Management cost	3%
5. Construction financing interest	Annual interest rate 3.5% (adjusted by case and bank)
6. Demolition compensation	$3,333/per apartment(agreement between construction company and landlords)
7. Rental compensation	$1,000/per month/per apartment(agreement between construction company and landlords)
8. Project integration commission fee	Based on marketing prices
9. Trust fee	Based on the agreement with bank
10. Operation and advertising fee	3% (including advertising, sales commission fee)
11. Management funds	0.6%
12. Urban renewal project application fee	Based on marketing prices

(continued)

Table 3. (*continued*)

Parameter	Description
13. Real estate appraiser fee	Based on marketing prices
14. Measurement fee	Based on marketing prices
15. Land administration agents fee	Registration and registration cancellation fees
16. Geological prospecting fee	Based on marketing prices

Table 4. Total profit for each projects

Projects		Total profit (Unit: USD)
1	P1(AD)	2,131,072.73
2	P2(BD)	1,313,743.37
3	P3(CD)	1,692,016.77
4	P1 + P2(ABD)	3,728,171.87
5	P1 + P3(ACD)	4,073,570.73
6	P2 + P3(BCD)	3,644,376.83
7	P1 + P2 + P3(ABCD)	5,616,745.53

However, we could not set up more than 30% remuneration in a project which would be denudation landlords' (P1, P2 and P3) benefits than affect their intentions to participate renew project. Another condition is construction company (P4) has to involve each project since most of landlords cannot afford the initial fund. From the results, we could know P4 has 1/3 opportunity to cooperate with the other player A, B and C. Multitude players in projects which benefits are greater than a single player. From those combinations, we can observe.

We can see in Fig. 6, in player A(P1), there are 4 combinations and the best payoff would be in P1 + P2 + P3 + P4. Realistically, P1 floor area ratio is 225% and 400% so that in P1 project would require more gross floor area that also may be able to affect how much bonus of building bulk area could be approved by the Taipei government. Moreover, Player 3 also has the same condition as P1. They both could obtain more gross floor area than they may have more power to decide that they want to collaborate with other players.

Moreover, we could see player B (P2) also has 4 different combinations and the best payoff would be in P1 + P2 + P3 + P4 also. However, in the real situation, when each player going on negotiation, P2 doesn't have the same in gross floor area as other players so that P2 may not have the same power to dominate the project (Figs. 7 and 8).

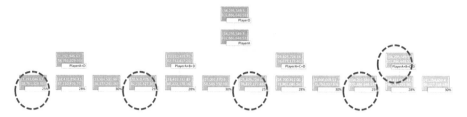

Fig. 6. Sequential-move games in player A

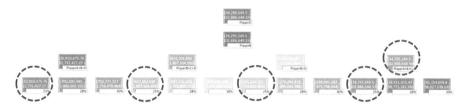

Fig. 7. Sequential-move games in player B

Fig. 8. Sequential-move games in player C

5   Discussions

5.1   Conclusion

From the results, when players collaborate that would require the best payoff than single player. Moreover, P4 has 1/3 chances to corporate with other players. When all players work together, which is the best outcome of P4. Therefore, P4 should try to collaborate with as many possible players as they can.

5.2   Limitation

Alliance Components Cannot without Construction Company. Considering most of the landlords may not afford construction cost so that most of urban renew projects landlords choose to cooperate with a construction company. Therefore, this study is based on the alliance as cooperate with a construction company.

Theory Application May Not be Able to Match the Public Expectations. The distribution of profits between which calculated by the game theory that may not be easily accepted by the landlords. Even if we offer unbiased and interest information for all players. Moreover, each player still seeks to require possible as more benefits from the urban renew project that may cause those players not to work together. However, game theory still provides some suggestions for all players to make decisions.

5.3   Extension

Urban renew project considers how people make decisions and seek the best benefits. Even we can find the best combination for those players but there are many factors that affect people making decisions. Hence, in future research we can apply to Agent based model to see how each player acts in different situations.

References

1. http://cloud.5pa.com.tw/cloud/house40y.html
2. Zheng, H.W., Shen, G.Q., Wang, H.: A review of recent studies on sustainable urban renewal. Habitat Int. **41**, 272–279 (2014). https://doi.org/10.1016/j.habitatint.2013.08.006
3. Adams, D., Hastings, E.M.: Urban renewal in Hong Kong: transition from development corporation to renewal authority. Land Use Policy **18**(3), 245–258 (2001). https://doi.org/10.1016/S0264-8377(01)00019-9
4. Aumann, R.J.: Game theory. New Palgrave Dict. Econ., 1–40 (2017). https://doi.org/10.1057/978-1-349-95121-5_942-2
5. Wu, Z.-G.: 賽局原來這麼生活 (It is very related daily life in a game theory), Taipei, Taiwan (2012)
6. Taipei City Urban Regeneration Office. https://uro.gov.taipei/Default.aspx

Diffusion of Environmental Protectionism: Single-Use Plastic Bags Ban Policy in California

Zining Yang[(✉)] and Sekwen Kim[(✉)]

Claremont Graduate Univeristy, Claremont, USA
{zining.yang, se-kwen.kim}@cgu.edu

Abstract. As California became the first state to ban the usage of single-use plastic bags due to popular vote, investigating the influence of social networks together with social economic attributes on the implementation of environmental regulatory policy becomes increasingly important. By incorporating different methods, this study provides a comprehensive analysis of how social network and socio-economic attributes contributed to the change of people's choice on environmental regulatory policy adoption and its implementation. This study uses Pooled OLS model with centrality measures from the network, demographics, and locations to generalize insights to policy diffusion across cities in California, as the core attributes influence population's the preference on policy adoption. This paper offers insights on environmental policy adoption, and the same methodological approaches can be applied to other policy studies.

Keywords: Environmental policy · Policy diffusion · Social network analysis · Leader and laggard model · Dynamic network

1 Introduction

California became the first state to ban the usage of single-use plastic bags and charge reusable plastic bags, as the majority of the population voted for the policy during the state referendum held in 2016. In 2014, California Governor Jerry Brown signed a state law which banned the usage of single-use plastic bags and required shoppers be charged at least 10 cents for paper bags and reusable plastic bags at grocers in California [1]. However, the plastic industry supported Proposition 67, which could have repealed the ban, on the ballot in 2016 [1]. The referendum vote showed that 52% of voters voted to affirm the statewide policy banning the use of single-use plastic bags and California became the first state to ban the use of single-use plastic bags in the United States in 2016 [2].

However, the 2014 statewide ban was not the first environmental regulatory policy on single-use plastic bags usage in California. The cities of Malibu and Fairfax both adopted policies banning single-use plastic bag in 2008 [3]. After 3 years of the inert activity of diffusion, the number of cities who adopted the similar regulatory measures on single-use plastic bags increased rapidly.

© Springer Nature Switzerland AG 2020
D. N. Cassenti (Ed.): AHFE 2019, AISC 958, pp. 87–94, 2020.
https://doi.org/10.1007/978-3-030-20148-7_9

The city level diffusion of the policy, which influenced the statewide regulation on plastic bags, encouraged this study to examine the diffusion of the policy by incorporating social network analysis and regression analysis. Following Kim [4, 5], this study aims to strengthen the finding from previous social network analysis by identifying how socio-economic attributes and social network attributes cities as micro-level actors, contributed to the diffusion of the environment policy in California by influencing the motivation and choice of cities. By incorporating social network attributes into regression analysis of socio-economic attributes, this study seeks to answer 'how social network and socio-economic attributes influenced the preference on degree of environmental regulatory policy adoption in California.'

2   Literature Review

By incorporating dynamic network theory and diffusion theory, this study addresses the influence of social network and socio-economic attributes on the preference on degree of environmental regulatory policy. Dynamic network theory highlights not only the progress of diffusion, but also the role of the network itself on influencing stakeholder's behavior or decisions. Diffusion theory connects the socio-economic attributes to the willingness to adopt innovative policy. By incorporating these two theories, this study aims to provide a comprehensive understanding on behavior changes or policy adoption of cities.

2.1   Dynamic Network Theory

The importance of the rate of the change within network and inter-connectedness between stakeholders and the network itself are core ideas of dynamic network theory [6, 7]. Dynamic network theory allows this study to address the importance of the pattern and network attributes of significant subgroups and to capture the dynamic changes of networks over time [6]. The theory defines the behavior change and decision as "purposive but carried out under conditions that set both opportunities for and restrictions on the achievement of these purposes," and highlights the importance of rate of change [6]. Dynamic network modeling is also closely related to diffusion modeling since the "rate of diffusion of ideas across a whole network may be significantly affected by relatively small local-level changes that have these macro-level effect[s]" [6]. The way in which social networks influence the behavior and decision of stakeholders within the network is another key feature of dynamic network theory [7]. Network influences stakeholder's motivation to regulate or modify their behavior or decision and dynamic model proves how people changes their preference, behavior, or decision accordingly [7, 9, 10]. As a result, dynamic network theory recognizes changes over time, which may lead to significant change.

2.2   Policy Diffusion Theory

The diffusion theory of the policy investigates the diffusion of an innovative policy and significant attributes in its diffusion [11–13]. The leader and laggard model of diffusion recognizes leaders and followers in policy adoption [11, 14]. Leaders, or pioneers, are risk-takers who are willing to test an untested policy and followers are those who are risk averse and adopt the policy after evaluating the success of once-untested policy from leaders [11, 14, 15]. The leader-laggard model assumes that if the government is socio-economically developed, the government is willing to take the risk and become a pioneer in adopting an untested policy [11, 14, 15].

3 Research Design

Dynamic network theory and diffusion theory postulate that the decision on policy adoption and implementation is influenced not only by socio-economic attributes of cities, but also by the network, to which the city belongs. Therefore, this study aims to investigate the contribution of socio-economic and social network attributes on the diffusion of the environmental regulatory policy by influencing the preference of the policy adopted and implemented.

In light of theories discussed above, the objective of this study is to examine following major objectives:

(a) To identify significant socio-economic attributes influencing preferences for degree of the environmental regulatory policy adopted and implemented.
(b) To prove the significance of social network attribute on influencing preference for the degree of the environmental regulatory policy adopted and implemented.

To satisfy objectives, this study hypothesizes that:

(a) H1: Wealth and education level of cities would influence the degree of policy implementation.
(b) H0: Wealth and education level of cities would not influence the degree of policy implementation.
(c) H1: The importance of the role a city played in terms of diffusion of policy in networks would influence the degree of policy implementation.
(d) H0: The importance of the role a city played in terms of diffusion of policy in networks would not influence the degree of policy implementation.

To satisfy objectives, this study uses Pooled OLS to investigate the influence of both a social network variables of eigenvector centrality measure and socio-economic variables of education level, median individual income, population density and proximity to the sea on degree of policy adopted and implemented.

The key dependent variable in this study is degree of policy adopted. The degree of policy adopted is measured into three categories as 1 being the least strict policy adopted which is 10-cent charges per reusable plastic bags, 2 being the moderately strict policy adopted which is 25-cent charges per reusable plastic bags and 3 being the strictest policy which is the total ban on plastic bags. Independent variables are eigenvector centrality, educated population, median individual income, population density and proximity to the sea. This study incorporated eigenvector centrality measure as one key explanatory variables to represent the importance of the role each city played in the diffusion of the policy adoption and implementation. It measures how well each node is connected with other well connected nodes in the network, measuring the central-peripheral position of each city in the network. In this paper, eigenvector centrality is measured as a percentage and higher value indicates a more important role in the diffusion of the policy. Educated population is measured as a percentage, reflecting the proportion of population with bachelor degree or higher. Higher indices of educated population represents that the population in the city is more educated. Population density is measured as the number of people per square mile in each city. As dynamic network theory highlights the importance of the change of network over time, all variables reflect yearly socio-economic and social network attributes from 2008 to 2017. However, proximity to the sea is fixed and measured dichotomously, as 1 represents the city with less than 5 miles distance from the sea and 0 represents a city with more than 5 miles distance from the sea.

In sum, the model specification is as follows;

$$
\begin{aligned}
\text{Degree of Policy Adopted} &= \alpha + \beta_1 \cdot \text{Eigenvector Centrality} \\
&+ \beta_2 \cdot \text{Educated Population} + \beta_3 \cdot \text{Median Individual Income} \\
&+ \beta_4 \cdot \text{Population Density} + \beta_5 \cdot \text{Proximity to the Sea} + \varepsilon
\end{aligned}
\tag{1}
$$

4   Data Description

The pooled OLS regression model is used to identify the influence of social network and socio-economic attributes. For the pooled OLS regression to precisely capture the influence of social network and socio-economic attributes of each micro-level agent, cities, this study constructed a dataset by using indices available online. The key dependent variable was collected from California against Waste. Socio-economic attributes, including median individual income, education level, population density and geographic proximity to the sea, were collected from US Census Bureau, which is widely regarded as a credible source [16, 17]. Eigenvector centrality measures, which is another key explanatory variable, were collected from previous work by Kim [4, 5]. Previous work by Kim generated and analyzed the network of 109 cities who adopted the plastic bag regulatory policy from 2008 to 2017 in California with 3 different form of the policy as distinction between subgroups and measured the changes of the influence of each subgroups to the diffusion of the policy from 2008 to 2017.

The completed data set includes 872 pooled observations, which consist of 109 cities who adopted a single-use plastic bags ban policy in years 2008, 2011, 2012, 2013, 2014, 2015, 2016, and 2017. Each observation has the degree of policy implementation, eigenvector centrality measures, percent of population with bachelor degree or higher, median individual income, population density, and geographic attributes, whether they are located close to the sea or not.

5   Findings

It was found that policy diffusion was the most active after 2012 [4, 5]. It was the least strict policy, which is a 10-cent charge per reusable/plastic bag, increased the most. From the perspective of size, the number of participating cities increased by 9, 6, 23, 19, and 10 from the years 2012 to 2016 respectively [4, 5]. Therefore, this study seeks to identify which socio-economic attributes influenced the degree of policy implementation before and after 2012.

Table 1. Sensitivity analysis.

VARIABLES	Model 1: all variable	Model 2: before 2013	Model 3: after 2012
	Policy	Policy	Policy
Eigenvector Centrality	−1.155***	−1.175***	−1.387***
	(0.0506)	(0.0423)	(0.257)
Educated population	−0.0136**	−0.00788**	−0.0632**
	(0.00617)	(0.00358)	(0.0276)
Median individual income	−0.0000008		
	(0.000001)		
Population density	−0.00001***	−0.00001**	−0.00004
	(0.000006)	(0.000005)	(0.00003)
Proximity to the sea	−0.0846**		
	(0.0430)		
Constant	2.463***	2.342***	3.137***
	(0.120)	(0.0542)	(0.200)
Observations R-squared	429	432	25
	0.650	0.673	0.681

Robust standard errors in parentheses
*** $p < 0.01$, ** $p < 0.05$, * $p < 0.1$

Pooled OLS Result. Eigenvector centrality educated population and population density are significantly influencing the degree of policy implementation.[1]

Table 1 reports pooled OLS results. Model 1 shows that eigenvector centrality, educated population, which is a percentage of bachelor degree or higher, population density and geographic proximity to the Pacific Ocean are statistically significant in influencing the preference of degree of policy adopted and implemented. Economic indices of median individual income is not significant. Model 2 and 3 show different effect of variables before and after 2013. Throughout the model, the magnitude of the

---

[1] Since the data suffers the heteroscedasticity, every model uses robust standard error.

estimated effect of eigenvector centrality measure is unchanged, but the educated population has greater magnitude of the estimated effect on the degree preference of policy adopted and implemented after 2013 and population density is insignificant after 2013.

As eigenvector centrality of a city is closer to 1, a less strict of degree of policy implementation the city prefers. Also 1 percentage point increase in the percentage of population with bachelor degree or higher reduces the policy strictness degree of implementation by 0.01 percent. So as population density increase by 1 would lead to the decrease in the policy strictness degree of implementation by 0.00001 percent. Median individual income and geographic attributes are not statistically significant.

Eigenvector centrality measure, which shows how important the city was in terms of policy diffusion, and percentage of educated population were statistically significant both before and after 2012. However, the magnitude of the estimated effect of educated population is about ten times greater after 2012 than it was before 2012 from -0.008 to -0.06. However, population density was statistically significant only before 2012 and insignificant after 2012.

Based on the regression results, it can be inferred that both social network attribute and sociological attributes significantly influence the degree preference of environmental regulatory policy in California. The more important role the city played and the more educated the city population is the less strict of a policy implementation the city prefer regardless activeness of diffusion of environmental regulatory policy. However, before the active diffusion, the more densely populated city is, the less strict of a policy implementation the city prefers, but after the active diffusion, the population density is insignificantly influencing the degree of policy implementation.

This study tested general models of leader and laggard model and dynamic network theory and provided congruent regression results, which empirically support the main argument of the theory. The leader and laggard model posits that wealthier and more educated cities are more likely to influence the adoption and implementation of untested policy than those who are poorer and less educated [11]. As dynamic network theory highlights the influence of network itself on stakeholder's preference and decision, this study provides congruent findings with dynamic network theory as well [7, 9, 10]. As outcomes support, education level of the city, which is the proportion of the population with bachelor degree or higher in this study, is proven to be significantly influencing the degree of policy implementation.

6   Conclusion

This study aims to provide a persuasive argument on the influence of social network and socio-economic attributes on the degree of policy implementation of single-use plastic bags ban policy in micro-level actors, cities, in California before the state referendum. By utilizing, operationalizing, and calibrating data available online, this study conducted pooled regression analysis in order to identify and measure significant social network and socio-economic attributes which influence the degree of policy implementation.

By analyzing the influence of social network and socio-economic attributes on degree of policy implementation, this study provided quantitative evidence on the diffusion of the single-use plastic bags ban policy. To some degree, this study illuminates the role of social network and sociological attributes on the diffusion of policy adoption based on activeness of policy diffusion. In order to encourage policy diffusion, it is preferred to focus on more densely populated cities with populations that are more educated. However, after the active diffusion of policy adoption and implementation, it is recommended to focus on cities with populations that are more educated. In addition, more active policy diffusion can be expected by recommending less strict regulation to those cities who play a more important role in diffusion.

This study also suggests that a persuasive argument on policy diffusion can be derived by employing both social network analysis and quantitative analysis. Simple indices could provide comprehensive interpretation and analysis on social phenomena, but by incorporating social network analysis and quantitative analysis together, more narrative and deep of an understanding of social phenomena can be obtained. Although this study focused only on the diffusion of environmental regulatory policy, the same methodological approaches can be applied in other policy studies to further strengthen both theoretically and empirically.

References

1. Rogers, P.: Voters approve plastic bag ban: What's happens next?, 11 November 2016. http://www.mercurynews.com/2016/11/10/voters-approve-plastic-bag-ban-whats-happens-next/. Accessed 09 Oct 2017
2. Calefati, J.: California bag ban: Voters to weigh industry's fate at the ballot box, 11 October 2016.   http://www.mercurynews.com/2016/09/16/california-bag-ban-voters-to-weigh-industrys-fate-at-the-ballot-box/. Accessed 09 Oct 2017
3. Californians Against Waste: Single-Use Bags Ordinances in CA (2016). https://static1.squarespace.com/static/54d3a62be4b068e9347ca880/t/583f1f57e4fcb5d84205b330/1480531800415/LocalBagsOrdinances1Pager_072815.pdf
4. Kim, S.: Analyzing the single-use plastic bags ban policy in California with social network model and diffusion model. In: International Conference on Applied Human Factors and Ergonomics. Springer, Cham (2018)
5. Hanneman, R.A., Riddle, M.: Introduction to Social Network Methods (2005)
6. Scott, J.: Social network analysis. Sage US Census Bureau. (2017). https://www.census.gov/quickfacts/fact/table/TX,NV,WA,CAOR/PST045216. Accessed 22 Nov 2017
7. Westaby, J.D.: Dynamic Network Theory: How Social Networks Influence Goal Pursuit. American Psychological Association, Washington, D.C. (2012)
8. Valente, T.W.: Network models of the diffusion of innovations. Comput. Math. Organ. Theory **2**(2), 163–164 (1996)
9. Jackson, M.O., Yariv, L.: Diffusion of behavior and equilibrium properties in network games. Am. Econ. Rev. **97**(2), 92–98 (2007)
10. Westaby, J.D., Pfaff, D.L., Redding, N.: Psychology and social networks: a dynamic network theory perspective. Am. Psychol. **69**(3), 269 (2014)
11. Berry, F.S., Berry, W.D.: Innovation and diffusion models in policy research. In: Theories of the Policy Process, vol. 169 (1999)

12. Volden, C., Ting, M.M., Carpenter, D.P.: A formal model of learning and policy dif-fusion. Am. Polit. Sci. Rev. **102**(3), 319–332 (2008)

13. Boehmke, F.J., Witmer, R.: Disentangling diffusion: the effects of social learning and economic competition on state policy innovation and expansion. Polit. Res. Q. **57**(1), 39–51 (2004)

14. Bennett, C.J.: What is policy convergence and what causes it? Br. J. Polit. Sci. **21**(2), 215–233 (1991)

15. Crain, R.L.: Fluoridation: the diffusion of an innovation among cities. Soc. Forces **44**(4), 467–476 (1966)

16. US Census Bureau (2017). https://www.census.gov/quickfacts/fact/table/TX,NV,WA,CA, OR/PST045216. Accessed 22 Nov 2017

17. Smith, T.F., Waterman, M.S.: Identification of common molecular subsequences. J. Mol. Biol. **147**, 195–197 (1981)

Virtual and Augmented Reality

Optimizing Occupational Safety Through 3-D Simulation and Immersive Virtual Reality

Ebo Kwegyir-Afful[1(✉)], Maria Lindholm[2], Sara Tilabi[1],
Sulaymon Tajudeen[1], and Jussi Kantola[1]

[1] University of Vaasa, Wolffintie 34, 65200 Vaasa, Finland
{ebo.kwegyir-afful, sara.tilabi, sulaymon.tajudeen,
jussi.kantola}@uwasa.fi
[2] University of Oulu, Pentti Kaiteran Katu 1, 90014 Oulu, Finland
maria.lindholm@oulu.fi

Abstract. This paper evaluates the effectiveness of computer simulation and the immersive virtual reality (IVR) technology for occupational risk assessment improvement. It achieves this by conducting a risk assessment on a 3-D simulation of a Lithium-Ion battery (LIB) manufacturing factory. This is necessary since calls for the enhancement of occupational risk assessments continue to dominate safety improvement measures in manufacturing context. Meanwhile, industries such as aviation, mining and healthcare employ advanced versions of IVR for risks awareness with successes. However, applications for safety in manufacturing context is only at the infancy although it utilizes IVR profitably for product and production optimization issues. The study involved 19 participants who performed the assessment with the aid of a safety checklist followed by open-ended semi-structured questions and interviews. Results indicates an outstanding utilization capability of IVR for risk assessment. Furthermore, the assessment pinpoints specific safety issues in the factory that requires attention and improvement.

Keywords: Risk assessment · Hazard identification · 3-D simulation ·
Immersive visual reality · Manufacturing industry

1 Introduction

This study integrates the potentials of the Immersive Virtual Reality (IVR) technology for enhancing occupational safety through a 3-D simulation of a lithium-Ion battery (LIB) factory. Specifically, the study aims to evaluate the extent that this technology can be employed to conduct an effective industrial risk assessment procedure that is key to occupational health and safety of manufacturing factories [1]. The Occupational Health and Safety Assessment Series standard (OHSAS 18001:2007) defines risk as the combination of the probability of occurrences and results of a predetermined hazardous event [2]. Risk assessment is therefore defined in the document as the process of calculating risk magnitude and deciding if the risk is tolerable. Besides, evidence indicates that safety measures are best implemented during the planning stages of a facility. Moreover, traditional means of analyzing risks at the planning stages is

© Springer Nature Switzerland AG 2020
D. N. Cassenti (Ed.): AHFE 2019, AISC 958, pp. 97–107, 2020.
https://doi.org/10.1007/978-3-030-20148-7_10

handicapped due to its inability to envisage details of production processes adequately [3, 4]. Implications are that, the safety of detailed manufacturing processes cannot be critically assessed until after construction and production. For this reason, the IVR technology which has the capability to simulate manufacturing equipment, the environment, robotic manipulations, products and production processes in real-time is currently gaining industrial acceptance [4, 5].

Although this technology was initially developed for computer games, today it has evolved as a viable industrial tool in example; engineering, construction, telecommunications, military and healthcare to optimize operations, product design and processes. Moreover, it is employed for safety improvements in some instances [5]. Generally, industrial applications of the technology for safety proves that IVR is the best currently known method for safety training, hazard identification and accident reconstruction [6].

However, there is scanty evidence to support its successes in manufacturing for safety improvement despite these impressive strides in provided by the virtual technology. For example, in a 154-article review relevant to Virtual Environment (VE) applications in manufacturing from 1992 to 2014, only 4 addressed applications of the technology for human factors [7]. One touched on workers safety training [8], another on safe working environment for the disabled [9] and two on simulations of human models for risk assessments [10]. Furthermore, future research directions of some of these applications seek to apply the technology pragmatically for safety analysis in emerging manufacturing fields. For these reasons, this paper formulates the following research questions:

RQ1: Is it possible to conduct a risk assessment in a 3-D model of a manufacturing factory?
RQ2: How close is a 3-D simulation to a factory in terms of layout, equipment and manufacturing procedures?
RQ3: What extent can IVR identify hazards and risks of a 3-D simulation in manufacturing context?
RQ4: Does a 3-D simulation of a manufacturing edifice and analysis through IVR provide better means of evaluation compared to traditional means of risk assessments at the planning stages?

Hence this paper is structured as follows to address these questions; in the next section, the methodology describes the structure, data and empirical framework of the research follows. Subsequently, literature based on applications of this technology for occupational safety optimization in manufacturing is reviewed to provide a theoretical background. Thereafter, presentation of results follows which elaborates and discusses the research findings. Finally, the study concludes in retrospect with the limitations and suggestions for further research.

2   Methodology

Figure 1 describes the design of the research which starts with the afore mentioned problem statement about the need to strengthen occupational risk assessments in manufacturing through simulation and IVR.

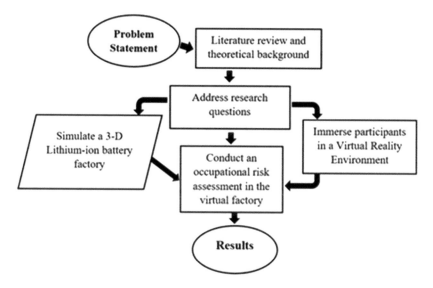

Fig. 1. The design of the research

The research follows both qualitative and quantitative research methods that includes experiments and surveys with yes or no answers. Three open-ended questions proceed this to complement the data. None of the participants had any adverse issues with an IVE nor mental disability. The mean age of participants was 34 (SD = 8.1) and had 8.5 (SD = 6.2) years of work experience. 47.4% of the respondents were from industry and manufacturing while others were from research and education institutions. The purpose and procedure of the research, particularly the experiment was first explained to participants individually. Afterwards, the researchers provided detailed tutorials for the navigation and control of the LIB virtual factory through utilization of the stereoscopic HMD device. Emphasis on the safe use of the system such as the proper adjustment of the head set for clarity and visibility follows. Subsequently, the researchers' highlighted the safety zone for the immersion. Thereafter, the signing of the informed consent form by both the researcher and each participant follows. After a thorough walk and inspection of the facility, example (Fig. 3), participants then completed the risk assessment form. This constituted 35 questions according to the workplace safety checklists (WSC) which conforms to the occupational safety and health (OSH) answer fact sheet. The safety checklist covers a broad range of health and safety issues that includes the factory environment, availability of First Aid kits, presence of safety signs and fixed guards for articulated robots and running machinery. Other issues are storage for chemicals and explosive materials, fire protection, warning systems and the possibility of electrocution. The first part of data gathered constitutes a quantitative analysis for the risk assessment based on disagree or agree questions and thematic categorization for the open-ended questions.

The Visual Components (VC) simulation software was used to build the 3-D virtual factory. This enables scrutiny of all processes for the manufacture of the 2170 LIB cell. The simulation processes consist of the raw material offloading stages through anode and cathode mixing, coating, calendaring and slitting. Others are cell winding, electrolyte filling, and cell welding. The formation cycling and packing into modules follows. Finally, the palletizing and shipment stages completes the simulation.

3 Theoretical Background

Several industrial applications of IVR simulations for safety exists. These are; training of employees in a simulation through IVR for safe operation [5, 11], Likewise, in risk assessments associated to human responses in emergency situations [12]. Furthermore, safety levels have been improved through the application of a Plant Simulator (PS) that constitutes process simulation and accident simulation for normal as well as abnormal accident scenarios [7, 13]. Moreover, the simulation-based training (SBT) software enhances human-robot collaboration (HRC) safety training. Meanwhile, virtual prototypes are also suitable for human factors and ergonomics at factory design stages [14]. Thus, through the application of 3-D simulation and scrutiny with IVR, a rise of safety levels in some high-risk sectors such as in the construction industry has become possible [15]. Similarly, the mining industry records successes for safety improvement [16]. Theoretically, applications of the technology for risk assessment involves (a) analyzing and evaluating the risks associated to a specific hazard: Termed risk analysis and/or risk evaluation. (b) Determining the most appropriate methods to eliminate the identified hazards and (c), controlling the risks when the hazard posed is impossible to eliminate. Termed: risk control. Accordingly, this research utilizes this sequence while participants experience immersion in the 3-D virtual factory.

3.1 Occupational Risk Assessment

The primary and key technique to achieve optimum workplace safety is an active and vibrant occupational risk assessment (ORA) procedure [17]. In manufacturing cycles, hazard-based qualitative risk assessment with risk management implies locating and identifying jobs, operations and procedures that have the tendency to increase the likelihood of exposure to injury, damage or even fatalities [18]. OHS management system therefore clarifies risk according to time horizons and severity such as imminent and serious risks to prioritize and structure control and intervention mechanisms. A workplace survey is vital in achieving a serene occupational safety environment. This implies identifying and assessing all health and accident risks at the workplace with suggestions for improvement. The design of this paper hopes to achieve that. Section 5 on Act 701/2006 of the Finnish Occupational Safety and Health Enforcement and Cooperation Act for Workplaces emphasizes and enforces requirement for frequent and efficient inspections to uphold safety standards [19].

3.2 Technology and Simulated Environment

According to Wang et al. [20], there are four independent definitions of the reality-virtuality (RV) simulation technology. These are augmented reality (AR), augmented virtuality (AV) pure real presence, and pure virtual presence (VR). Currently, the pure virtual presence termed Virtual Reality has gained industrial attention and acceptance [21]. Primarily, this is because in the immersion, VR possesses better visibility, accessibility and much more capable for analyzing complete virtual environments. In addition, it generates complete virtual images with the head mounted display (HMD), sensory handheld controllers and base stations. A Virtual reality environment is a 3-D real time graphical environment that makes it possible to visualize and interact with simulated models in a virtual environment [14]. One is immersed (termed immersive VR) in the environment when using the technology to interact with seemingly real or physical situations. The purpose of the application of IVR technology is to transfer the user of the headset from the current natural location to the simulation in the virtual realm. This makes it possible for critical scrutiny of the factory for hazards at the planning stages.

In order to run the 3-D simulation and immersion needed in the virtual realm, a Windows 10 computer with the following specifications and accessories was utilized: An Intel core i7 processor having a speed of 3.6 GHz Dual-Core and a random-access memory (RAM) of 32 GB. Importantly, it is installed with an NVIDIA GPU (GTX 1070 GeForce gaming graphics card. For the immersion, an HTC VIVE equipment incorporating a head-mounted display (HMD) coupled to hand-held controllers connects to advanced gesture controls for navigation. Two base stations track movements to produce the virtual environment of the 3-D simulated models. Figure 3 shows a participant immersed in the virtual environment of the LIB manufacturing factory.

3.3   Lithium Ion Battery (LIB) Manufacturing Factory

The choice of a Lithium Ion Battery (LIB) manufacturing factory for this simulation is due to growing concerns of global warming and the current interests and growth for

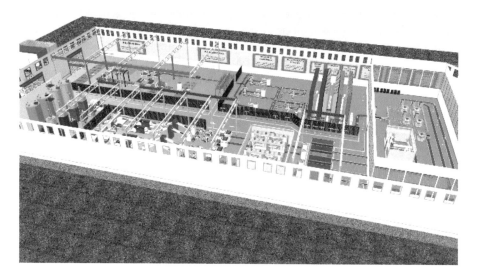

Fig. 2. The 3-D model of the utilized LIB factory

Fig. 3. Risk assessment of the 3-D LIB factory through IVR.

green energy. The facility incorporates advanced automation, robotics and state-of-the-art technology (Fig. 2) to produce the LIB that powers electric vehicles (EV).

4   Results and Discussions

Table 1 presents results of the risk assessment that answers RQ1 which seeks to investigate the possibility of conducting a risk assessment in a 3-D simulation through IVR. While answering the question in relation to statement A in Table 2 about the possibility of conducting risk assessments in the simulation, 52.63% of participants agreed and 47.37% strongly agreed. Accordingly, the mean value of 4.47 (SD = 0.50) for the statement A (Table 3) indicates proximity of respondent's perception. Furthermore, the answers obtained for statement C also in Table 2 of the capability of IVR in analyzing the safety of production attests to RQ1 in the affirmative (Fig. 4).

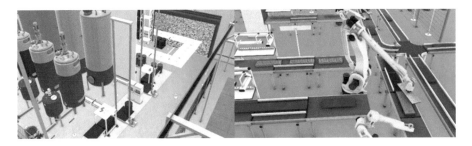

Fig. 4. Inside view of the factory showing mixer tanks and articulated robots at work

Table 1. Combined risk assessment results

	Risk assessment results (in percentages)	YES	NO	No answers
1	Are floor surfaces free of water, oil or other fluids?	95	5	0
2	Are passageways clearly marked?	68	32	0
3	Are walkways/doorways clear of boxes, cords and litter?	68	32	0
4	Are stairways clear of boxes, equipment and other obstructions?	95	5	0
5	Are stairs and handrails in good condition?	100	0	0
6	Will personnel be working above where others may pass?	63	37	0
7	Will personnel be working below others?	53	47	0
8	Are covers/guardrails in place around pits, tanks and ditches?	95	5	0
9	Is the level of light adequate for safe and comfortable of work?	100	0	0
10	Are work items that are regularly used within easy reach?	89	11	0
11	Is there sufficient access to machines/equipment?	100	0	0
12	Are appropriate manual handling aids readily available?	74	21	5
13	Are all machine parts adequately guarded?	74	16	11
14	Are warnings appropriate for any hazardous areas?	68	32	0
15	Are all hazardous products stored appropriately?	74	21	5
16	Is stored material stable and secure?	84	11	5
17	Are items placed neatly and securely on shelves?	79	21	0
18	Can items on high shelves be easily reached?	79	16	5
19	Are the elevated platforms properly and handrails secured?	89	11	0
20	Are switchboards in a safe operating condition and secured?	89	11	0
21	Are machine guards in place on all operating equipment?	84	16	0
22	Are emergency stop buttons clearly visible and operational?	53	47	0
23	Are chemical and hazardous substances stored safely?	89	5	5
24	Are hazardous products stored away from heat sources?	74	21	5
25	Is there a possibility of electrocution?	26	74	0
26	Is there adequate ventilation or an exhaust system?	84	16	0
27	Is ventilation equipment working effectively?	79	21	0
28	Are First Aid Kits easily accessible and prominent areas?	84	16	0
29	Is the location of the First Aid Kit clearly identified?	74	26	0
30	Are exits and exit routes equipped with emergency lighting?	79	21	0
31	Are exits and exit routes accessible?	84	16	0

(*continued*)

Table 1. (*continued*)

Risk assessment results (in percentages)		YES	NO	No answers
32	Are there signs and arrows indicating the direction to exits?	68	32	0
33	Are locations of fire alarms/firefighting clearly identified?	89	11	0
34	Are extinguishers properly mounted and easily accessible?	84	16	0
35	Are there enough extinguishers present?	58	37	5

Concerning the simulation's proximity to a real factory (statement D in Table 2), 63.16% agreed while 26.32% strongly agreed and 10.54% neither agreed nor disagreed. Similarly, 63.16% agreed while 31.58% strongly agreed and 5.26% disagreed to the proximity of the simulation to actual manufacturing processes (statement E in Table 2). The mean value for the statement D is 4.16 (SD = 0.59) and for the statement E it is 4.12 (SD = 0.69). These responses complement RQ2 that indeed, the simulation is quite close to a real factory.

Table 2. Results of IVR's suitability for risk assessment of the 3-D simulation

Perceptions of 3-D simulation and IVR for risk assessment (in percentages)	Strongly disagree	Disagree	Neither agree nor disagree	Agree	Strongly agree
A. Possibility to conduct a risk assessment	0	0	0	52.63	47.37
B. Ergonomics of the HMD	0	0	21.05	21.05	36.84
C. Capability to analyse production safety	0	0	0	63.16	31.58
D. Proximity to a real factory	0	0	10.54	63.16	26.32
E. Proximity to actual processes	0	5.26	0	63.16	31.58
F. Preference to IVR than traditional methods	0	5.26	10.54	42.10	42.10

Although much of the response showed a high level of correlation, there were however some questions that had YES and NO spit answers. For example, in question 22, that asks about the visibility and operational performance of emergency stop buttons. The split was because some participants concentrated on a YES answer for the visibility while others chose a NO answer based of its operation. Actually, there were emergency stop buttons, but they were not operational in the simulation. Hence, the split answers.

While analyzing the extent to which the IVR can identify hazards and risk factors present in the plan as RQ3 asks, statement C in Table 2 provides 63.16% in agreement and 31.58% strongly in agreement to the capability of IVR to analyze safety of production in the real factory. Secondly, the mean value of 4.21 (SD = 0.69) obtained from question F of Table 2 shows that IVR is more suitable for risk assessment than traditional means of assessment. Consequently, results from Tables 2 and 3 shows that IVR is truly a more appropriate and effective means of assessing risks at the planning stages compared to traditional methods such as text and 2-D plan and answers the RQ4.

Table 3. Results of IVR's suitability for risk assessment of the 3-D simulation

Questions related to the simulation and IVR walk (Strongly disagree = 1, disagree = 2, neither/nor agree = 3, agree = 4, strongly agree = 5)	Mean	Standard deviation
A. It is possible to conduct a risk assessment in a 3-D simulated factory	4.47	0.50
B. Ergonomics of the HMD for the assessment	4.16	0.74
C. Capability to analyse production safety	4.21	0.69
D. Proximity of the simulation to a real factory	4.16	0.59
E. Proximity of the simulation to actual manufacturing processes	4.21	0.69
F. Preference to IVR for risk assessment than traditional methods	4.21	0.83

During the risk assessment, participants identified diverse hazards present in the factory that requires attention. Two mentioned the potential for falling as the main hazard while four indicated gas leakage and another four participants also observed that the presents of fast-moving robots and machinery were the main hazards present in the simulation. Besides, there were nine hazardous issues that were reported only once (identified as other in Fig. 5a).

(a)   (b)

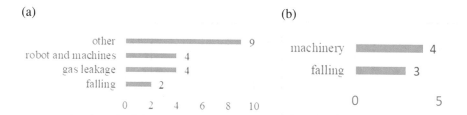

Fig. 5. a. Hazards identified in the LIB factory, b. Probable accidents

However, the most probable accidents identified in the simulation were related to machinery and falling (Fig. 5b) which are actually prevalent amongst the high causes of accidents in the industry [22].

5   Conclusions

This paper has employed an IVR technology for occupational risk assessment to a 3-D model of a LIB manufacturing factory. The risk assessment constituted the 35 WSC questions administered by 19 participants individually while immersed in the virtual realm. Overall, results of the research indicate that IVR is highly capable to improve work place safety through a more active occupational risk assessment that constitutes hazard identification and a safety walk of the facility even in the plan. Furthermore, the assessment provided ample suggestions for instituting and implementing the necessary control measures. Generally, participants overwhelmingly indicated preference to IVR risk assessment of the factory at the planning stages to traditional methods of risk assessment. The exercise has demonstrated abundantly that indeed, a full-scale occupational risk assessment procedure can occur in a 3-D simulation of a manufacturing factory with the aid of an IVR technology. Despite these results, the research encountered a few limitations necessary for recognition. Firstly, the exercise concentrated exclusively on VR and the simulation was equally limited to LIB production. Therefore, in the future, the researchers intend conducting the simulation and IVR in other manufacturing factories to compare and validate the results presented in this paper. Likewise, we hope to involve the safety, health and environment (SHE) managers within the sector for the risk assessment. Secondly, the overall simulation process froze repeatedly due to simultaneous multiple simulations of the production process. As such, the exercise sometimes took more time than initially anticipated although most participants enjoyed the walk and scrutiny in the virtual realm. While IVR's utilization for safety in manufacturing context is only at the initial stages, this paper demonstrates both empirically and pragmatically its utilization potentials in manufacturing to fill this research gap.

References

1. Kokangül, A., Polat, U., Dağsuyu, C.: A new approximation for risk assessment using the AHP and fine kinney methodologies. Saf. Sci. **91**, 24–32 (2017)
2. Kapp, E.A.: PW 2499 using ISO 45001 to improve safety & heath in the global supply chain (2018)
3. Wright, A., et al.: The influence of a full-time, immersive simulation-based clinical placement on physiotherapy student confidence during clinical practice. Adv. Simul. **3**(1), 3 (2018)
4. Yuen, K.K., Choi, S.H., Yang, X.B.: A full-immersive CAVE-based VR simulation system of forklift truck operations for safety training. Comput.-Aided Des. Appl. **7**(2), 235–245 (2010)
5. Nedel, L., Costa de Souza, V., Menin, A., Oliveira, J.: Using immersive virtual reality to reduce work accidents in developing countries. IEEE Comput. Graph. Appl. **36**(2), 36–46 (2016). https://doi.org/10.1109/MCG.2016.19
6. Michalos, G., et al.: Workplace analysis and design using virtual reality techniques. CIRP Ann. **1**, 141–144 (2018)
7. Choi, S., Jung, K., Noh, S.D.: Virtual reality applications in manufacturing industries: past research, present findings, & future directions. Concurr. Eng. **23**(1), 40–63 (2015)

8. Budziszewski, P., Grabowski, A., Milanowicz, M., Jankowski, J., Dzwiarek, M.: Designing a workplace for workers with motion disability with computer simulation and virtual reality techniques. Int. J. Disabil. Hum. Dev. **10**(4), 355–358 (2011)

9. Or, C.K., Duffy, V.G., Cheung, C.C.: Perception of safe robot idle time in virtual reality and real industrial environments. Int. J. Ind. Ergon. **39**(5), 807–812 (2009). https://doi.org/10.1016/j.ergon.2009.01.003

10. Qiu, S., Fan, X., Wu, D., Qichang, H., Zhou, D.: Virtual human modeling for interactive assembly and disassembly operation in virtual reality environment. Int. J. Adv. Manuf. Technol. **9**(12), 2355–2372 (2013)

11. Leder, J., Horlitz, T., Puschmann, P., Wittstock, V., Schütz, A.: Personality variables in risk perception, learning and risky choice after safety training: data of two empirical intervention studies contrasting immersive VR and PowerPoint. Data Brief **20**, 2017–2019 (2018)

12. Norazahar, N., Smith, J., Khan, F., Veitch, B.: The use of a virtual environment in managing risks associated with human responses in emergency situations on offshore installations. Ocean. Eng. **147**, 621–628 (208). https://doi.org/10.1016/j.oceaneng.2017.09.044

13. Nazir, S., Manca, D.: How a plant simulator can improve industrial safety. Process Saf. Prog. **34**(3), 237–243 (2015)

14. Aromaa, S., Väänänen, K.: Suitability of virtual prototypes to support human factors/ergonomics evaluation during the design. Appl. Ergon. **56**, 11–18 (2016)

15. Aven, T., Flage, R.: Risk assessment with broad uncertainty and knowledge characterisation: an illustrating case study. Knowl. Risk Assess. Manag. **3**, 1–26 (2017)

16. Van Wyk, E., De Villiers, R.: Virtual reality training applications for the mining industry. In: proceedings of the 6th international conference on computer graphics, virtual reality, visualization and interaction in Africa, pp. 53–63 (2009)

17. Pinto, A., Nunes, I.L., Ribeiro, R.A.: Occupational risk assessment in construction industry–overview and reflection. Saf. Sci. **49**(5), 616–624 (2011)

18. Anttonen, H., Pääkkönen, R.: Risk assessment in Finland: theory and practice. Saf. Health Work **1**(1), 1–10 (2010)

19. Ministry of Social Affairs and Health: Occupational Safety and Health in Finland (2016). ISBN 978-952-00-3780-2

20. Wang, W., Zhang, W., Feng, W.: The research of maintainability analysis based on immersive virtual maintenance technology. In: Advances in Intelligent Systems and Computing, pp. 573–582 (2018)

21. Chittaro, L., Corbett, C.L., McLean, G.A., Zangrando, N.: Safety knowledge transfer through mobile virtual reality: a study of aviation life preserver donning. Saf. Sci. **102**, 159–168 (2018)

22. Kjellen, U., Albrechtsen, E.: Prevention of Accidents and Unwanted Occurrences: Theory, Methods, and Tools in Safety Management. Leikannut Taylor & Francis Group. CRC Press, Boca Raton (2017)

Same Task, Different Place: Developing Novel Simulation Environments with Equivalent Task Difficulties

Benjamin T. Files[1(✉)], Ashley H. Oiknine[2,3], Jerald Thomas[4],
Peter Khooshabeh[1,3], Anne M. Sinatra[5], and Kimberly A. Pollard[1]

[1] US Army Research Laboratory, Los Angeles, USA
{benjamin.t.files.civ, peter.khooshabehadeh2.civ,
kimberly.a.pollard.civ}@mail.mil
[2] DCS Corporation, Los Angeles, USA
aoiknine@dcscorp.com
[3] Psychological and Brain Sciences, University of California at Santa Barbara,
Santa Barbara, USA
[4] Computer Science, University of Minnesota, Minneapolis, USA
thoma891@d.umn.edu
[5] Natick Soldier Research, Development & Engineering Center – Simulation
& Training Technology Center, Orlando, USA
anne.m.sinatra.civ@mail.mil

Abstract. We introduce a novel framework for creating and evaluating multiple virtual reality environments (VEs) that are naturalistic and similar in navigational complexity. We developed this framework in support of a spatial-learning study using a within-subjects design. We generated three interior environments and used graph-theoretic methods to ensure similar complexity. We then developed a scavenger-hunt task that ensured participants would visit all parts of the environments. Here, we describe VE development and a user study evaluating the relative task difficulty in the environments. Our results showed that our techniques were generally successful: the average time to complete the task was similar across environments. Some participants took longer to complete the task in one of the environments, indicating room for refinement of our framework. The methods described here should be of use for future studies using VEs, especially in within-subjects design.

Keywords: Virtual environments · Graph-theoretic measures ·
Within-subjects designs · Task development · Floorplan design ·
Bayesian modelling

1 Introduction

Virtual, augmented, and mixed reality technologies, and the virtual environments (VEs) that run on them, have become increasingly popular tools for research, education, and commercial applications. While some applications, such as gaming, may employ unnatural or fanciful VEs, naturalistic virtual spaces are valued in military

D. N. Cassenti (Ed.): AHFE 2019, AISC 958, pp. 108–119, 2020.
https://doi.org/10.1007/978-3-030-20148-7_11

training, architectural modeling, ergonomics, and spatial cognition research. Using VEs for these purposes has the potential to reduce costs and improve flexibility compared to real environments or physical models. Much research has aimed at determining what variables influence the effectiveness of VEs for these purposes [1, 2], or what variables influence spatial cognition [1], ergonomics [3], or task performance [4] in VEs when used to model real world experiences. Such variables may include different task framing, different levels of teamwork, different narrative, different display technologies or display parameters, or a multitude of other possibilities.

Research in these areas commonly proceeds by using a single test environment and changing the variable of interest across subjects. For example, each subject might be assigned to one of three different screen refresh rates or one of three different room temperatures. However, because individuals respond to VEs differently, it is often desirable to control for individual differences by using a within-subjects research design. This introduces new challenges, chief among them the need to manage order effects. Counterbalancing is essential, but even with this precaution, order effects may overwhelm the effects of the test variable if the participant becomes too familiar with the same VE over repeated exposure. In this paper, we describe our development process for generating a set of three different difficulty-balanced naturalistic VEs for use in within-subjects research designs. We then describe the validation of the equivalent task difficulty of the VEs and provide some recommendations.

A need for within-subjects design arose in our research on immersive technologies in spatial learning. We wished to examine the effects of three different levels of immersive display technologies—desktop monitor, mid-grade head-mounted display (HMD), and fully occlusive HMD—on spatial learning in VEs. To control for effects from individual differences, we favored a within-subjects design. This necessitated the development of three unique VEs and a common task, ideally of equivalent difficulty, that we could deploy using the three display technologies. The task we selected was an ordered scavenger hunt task. Within each VE, we attached a numbered flag to each of eight scavenger hunt items, and participants sought the items in serial order. Participants could see the flags of items that were not the current target, allowing them to note the location of a future hunt item as they sought their current target.

In designing the VEs and task, we considered three main challenges to achieving equivalent difficulty. The first challenge was constructing three different floorplans with equivalent navigability. The second challenge was the placement of scavenger hunt items. The third challenge was ordering the scavenger hunt items. In what follows, we describe the processes and principles we employed to address these challenges. Subsequently, we show evidence of the success (*i.e.*, difficulty equivalence) of our design as well as some possibilities for improvements.

2   Approach

The goal of our broader study was to use a within-subjects design to evaluate the effects of different immersive technologies in the context of a spatial navigation task. This goal required the creation of three different VEs that were nonetheless equivalent in navigational difficulty. Within these equivalent VEs, we required a task that ensured participants engaged with and explored the VEs. Here, we lay out our approach for designing the floorplans of the VEs and the task to perform therein.

2.1 Floorplan Design

The floorplan design process began with a 13-room template. The template for the three VEs was a floorplan of an actual home, although no VE used the unaltered original. We separated the template into four blocks that contained three or four rooms each. When selecting the blocks, we focused on constructing versatile bounding edges for ease of generating configurations. We rearranged and rotated these four blocks to generate new layouts aiming for compactness, completeness, and naturalistic fidelity. Compact layouts have a roughly square footprint; non-compact layouts might be much wider than they are tall or have rooms protruding into space. Complete layouts have no voids of un-reachable space that were discontinuous with the outer perimeter. Naturalistic fidelity reflects efforts to ensure that the configuration did not violate any norms expected in an environment layout (*e.g.,* a long, narrow room would typically be a hall leading to another room). Using the same blocks ensured that each VE would have the same surface area, number of rooms (13), and room dimensions.

The principles guiding placement of the doors between rooms were traversability, equality, and naturalistic fidelity. The VE had to be completely traversable so that participants could access all of the rooms in the VE. The process began by adding doorways to each edge that joined rooms. No doors were added to the perimeter. We reviewed naturalistic fidelity for each room to evaluate the practicality of door placement. We subsequently adjusted the number of doorways to equate the quantity of doorways across VEs. After this initial layout and door placement, each VE contained 13 doors.

2.2 Graph Representations of Floorplans

After this initial layout was set, we used graph-theoretic summaries to examine the navigational complexity of the VEs [5, 6]. A broader set of tools that include graph-theoretical elements are called Space Syntax, and have been used for similar ends in previous work [7]. We represented each of the three VEs with a binary undirected graph, in which nodes represented rooms and edges represented doors. We selected this representation because the goal was to understand the navigational difficulty/memorability of the VE by reflecting the number of choices the participant faced for where to go next rather than how much distance the participant would need to cover to get through a given door.

These graphs were analyzed using graph-theoretic summaries implemented in the Brain Connectivity Toolbox [8]. First, we counted the number of connected edges of each node for the initial layout, which is called the node degree. Node degree is a measure of centrality, meaning that nodes with higher degree are likely more important to the graph. In this setting, a node with higher degree has more doors, so it is likely to entail more navigational options and be involved in more routes compared with a node of lower degree. Other measures of centrality might have better captured the navigability of these VEs, but an advantage of degree distribution is that it is relatively easy to adjust by adding and deleting edges. We added, subtracted, and moved doors so that each VE had the same degree distribution: four nodes of degree 1, five nodes of degree 2, three nodes of degree 3, and one node of degree 5, while maintaining a naturalistic arrangement of doorways. After this process, each VE had 14 total doors.

After these adjustments, other graph summaries were computed: characteristic path length, which is the average length of the shortest path between two points in the graph; nodal eccentricity, which equals the shortest path length from a given node to the node that is farthest from it; graph radius which is the minimum of nodal eccentricity; and graph diameter, which is the maximum of nodal eccentricity. Characteristic path length, graph radius, and graph diameter for each VE appear in Table 1. Although these measures are not identical, they differed by no more than one unit, and there was no obvious way to equalize them without losing equality of degree distribution.

Table 1. Graph measures of the environments.

Environment	Characteristic path length	Graph radius	Graph diameter
Home	3.06	4	7
Office	2.78	3	6
School	3.09	3	6

After layout was determined, we built the VEs in a 3D game engine, finalized the theme of each room, and populated the VEs with objects consistent with each room's theme with careful consideration not to repeat objects across VEs. The object selection process and the development of transfer questions about those objects are detailed elsewhere [9]. Next, we faced the second challenge of where to put hunt items in the VE.

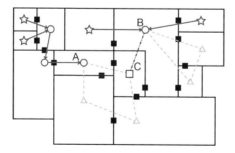

Fig. 1. Target object placement heuristics. Dead-end rooms (*stars*) all required target objects to ensure that they would be visited. Must-visit status propagates (*arrows*) from must-visit rooms with only one connection to a target-optional room into that room (*circles*). All routes from point A to point B pass through a common node (*square*, point C). Parallel chains (*dashed lines*) necessitated targets be added to rooms (*triangles*) to give them must-visit status. Rooms are connected by open doors (*black squares*).

2.3   Hunt Item Placement

One of the main goals of having a scavenger hunt task was to ensure participants fully explored the VEs. Therefore, the main criterion for hunt item placement was that finding all the items required visiting each room in the VE at least once. This criterion lead to a few heuristics (Fig. 1) that governed which rooms needed to have hunt items in order for each room to have must-visit status (i.e. the room must be visited to complete the scavenger hunt). All dead-end rooms (nodes of degree 1) needed to have a hunt object, because dead-end rooms are never a necessary step in a path to any room other than that dead-end room. Placing a hunt item in each dead-end room gave them must-visit status. Following this, all rooms directly connected to a dead-end room obtained must-visit status, because one would have to pass through that room to get to the dead-end room. This must-visit status propagates iteratively; at each step any must-visit room that is connected to exactly one room without must-visit status grants that room must-visit status. Following dead-end propagation, any room that was included in all possible paths between any two must-visit rooms also obtained must-visit status. This process covered most of the rooms in our VEs.

The remaining rooms formed parallel chains of rooms (*i.e.*, two different paths can get from point A to point B without including common nodes other than A and B). With parallel chains of rooms, a participant might always choose one of the two routes and miss seeing the rooms along the parallel route. When the dead-end propagation procedure did not already render each room in a parallel chain as must-visit, that room needed a hunt item to ensure the participant visited each step in the chain rather than possibly turning around instead of completing the chain. The dead-end propagation followed by parallel chain resolution dictated the minimum number and placement of hunt items.

Following these heuristics, the required number of hunt items for each VE was seven, eight, and eight for the home, office, and school VEs, respectively. To keep the number of hunt items equal, each hunt consisted of eight items. We could have added additional hunt items, but that might have required longer experimental sessions to complete.

2.4   Hunt Item Order

Having selected the number and location of the hunt items, we needed to specify an order for participants to seek the hunt items. For each VE, we selected a starting position, first hunt item, and last hunt item based on the scientific aims of the main study. For each possible hunt order respecting these constraints, the minimum path length (i.e. smallest number of nodes visited or revisited) to complete the hunt was calculated based on the graph representations of the VEs. The median minimum path lengths were 25, 23, and 27 for home, office, and school VEs, respectively. Therefore, we selected only hunt item orders with minimum path length of 25. Having a consistent minimum path length aimed to keep the variance in search time similar across VEs. There was no expectation that participants would complete the hunt following the shortest path length, but some proportion would have done so due to lucky guessing; we expected that a consistent floor value on the scavenger hunt path would minimize variability across VEs due to this kind of luck.

There remained more than 50 options in each VE that satisfied all criteria including the minimum path length of 25, so we applied an additional constraint that the distribution of minimum path lengths between subsequent hunt items should be the same. Only one distribution of minimum path lengths occurred in all VEs: one with path length 1, two each of lengths 2, 3, and 4, and one with path length 6. The final layout and scavenger hunt items and order appear in Fig. 2.

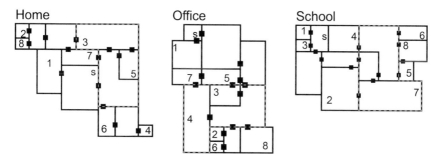

Fig. 2. Environment floorplans and scavenger hunt order. Participants began at the starting point (*s*) and explored the environment looking for the scavenger hunt items in the order indicated (*numerals*). Rooms are connected by open doorways (*solid black squares*). The floorplans are composed of four blocks of rooms taken from an original extant floorplan. One such block is highlighted (*dashed outline*).

3 Assessing Task Difficulty

3.1 Data Collection

Participants took part in three experimental sessions, each separated by at least two weeks to minimize any carryover effects. In each session, a participant filled out questionnaires (details and results discussed elsewhere), and was then set up with one of the three display technologies and a game controller. This was followed by a brief training session and practice to familiarize them with the game controller and display technology and to give them exposure to the types of spatial tasks they would encounter and questions that would later be asked. Participants then entered one of the three naturalistic VEs and were asked to complete a scavenger hunt, following on screen instructions. Participants spent a total of 15 min in each VE. Scavenger hunt items were marked with numbered flags, and participants were instructed to find each number in order and to press a button to mark each one as found. The intention behind the scavenger hunt approach was to make sure the participant was motivated to traverse the entire VE and see all the rooms. After the participant found all the scavenger hunt objects, they were instructed to freely explore the VE for the rest of their 15 min. The experiment software automatically recorded the participant's virtual location and scavenger hunt progress at 100 ms intervals. After exiting the VE and removing the display technology, the participants answered spatial memory and navigation questions to demonstrate spatial learning of the VE. The nature of these questions, their development, and results of these measures will be reported elsewhere. Participants returned after a break of at least two weeks and then performed the same tasks with a new display technology and new VE. The technology, VE, and the order of presentation were counterbalanced.

Fifty participants (37 F, 13 M) completed the scavenger hunt in each of the three sessions of the experiment. Data from two participants were excluded, because they did not finish the scavenger hunt by the time limit in the home VE. The mean age was 21 years (range 19−29). All participants were recruited through various online platforms that were associated with the community at the University of California, Santa Barbara including the Psychology Department's paid participant subject pool. Inclusion criteria were at least 18 years of age, normal hearing, color vision, and normal vision (or corrected-to-normal provided that participants would wear contact lenses to the experiment). Exclusion criteria included heart problems, pregnancy, easily getting motion sick, or 11 or more hours of experience with virtual reality equipment. The voluntary, fully informed, written consent of participants in this research was obtained as required by Title 32, Part 219 of the CFR and Army Regulation 70-25. All human subjects testing was approved by the Institutional Review Board of the U.S. Army Research Laboratory.

3.2   Model Fitting

The primary question of interest is whether the three different VEs and their respective hunt orders and locations had similar effects on task difficulty. Time to complete the hunt served as a measure for the overall difficulty of the scavenger hunt task. To answer this question, we fit two Bayesian models to the data. One model had separate effects for each of the three VEs, and the other had one common effect for all three VEs. In both models, we modeled participant effects as random additive effects. Time (all units in seconds) to complete the scavenger hunt for subject i in VE j was modeled with a truncated generalized Student's t distribution:

$$t_{i,j} \sim Student's\ t\left(v, \mu_{i,j}, \sigma\right)T[0, 900].$$ (1)

$$\mu_{i,j} = s_i + e_j.$$ (2)

Priors were $v \sim \Gamma\left(2, \frac{1}{10}\right)$, $\sigma \sim \mathcal{N}^+(0, 100)$, $s_i \sim \mathcal{N}(0, 200)T[-900, 900]$, and $e_j \sim \mathcal{N}(450, 200)T[0, 900]$. We selected a generalized Student's t distribution so the model would be robust to extreme values. Truncation at 0 and 900 s reflects that the scavenger hunt ended if the participant had not found the last item after 900 s. The common-mean model was similar to the three-mean model:

$$t_{i,j} \sim Student's\ t(v, \mu_i, \sigma)T[0, 900].$$ (3)

$$\mu_i = s_i + e_{common}.$$ (4)

Priors were $e_{common} \sim \mathcal{N}(450, 200)T[0, 900]$ and the same priors from the three-mean model for v, σ, and s_i. The only distinction between the two models is that in the three-mean model, e_j represents one parameter per VE, while e_{common} represents one parameter for all three VEs. The priors for these models are vaguely informative, in that they indicate that solutions with values near the middle of the time range are more likely than values near the extrema, but the large standard deviations represent low certainty about the expected locations of effects.

These models were implemented in Stan [10] using the RStan interface [11]. The default No U-Turn Sampler was used in 12 independent chains each using 2,000 warmup and 2,000 post-warmup iterations. Convergence diagnostics were used with the criteria that effective sampling ratio was over 0.1 and $\hat{R}$ was less than 1.1 for all parameters, and all chains ran with no divergences.

Model comparisons were done by computing expected log predictive density using leave-one-participant-out cross-validation [12]. We re-fit the models with N-1 participants' data and computed predictive density of the left-out data at each iteration. Repeated N times, leaving a different participant out each time, this provides the expected log predictive density (ELPD) of a new participant's results. We calculated model ELPD difference and report the mean and standard error of within-subjects differences.

4   Results

Mean completion times (in seconds) and bias-corrected, accelerated bootstrapped 95% within-subjects confidence intervals were 399.4 [355.8, 453.2], 400.2 [362.8, 437.3], and 433.0 [399.8, 467.9] for home, office and school VEs, respectively (Fig. 3). Consistent with the overlapping confidence intervals, a repeated-measures ANOVA using the Greenhouse-Geisser correction for non-sphericity indicated no strong incompatibility between the data and a null hypothesis of no difference in VE means, $F(1.75, 85.5) = 1.06$, $\varepsilon = .87$, $p = 0.342$, $\eta^2 = .021$.

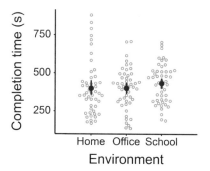

Fig. 3.   Scavenger hunt completion times in three virtual environments. Individual results (*light circles*) are summarized with the mean (*large dark circles*) and bootstrapped 95% within-subjects confidence intervals (*dark whiskers*).

In the Bayesian model fitting, all convergence criteria were satisfied. The main variables of interest, the VE means in the three-VE model and the common mean in the common-mean model appear in Fig. 4. In the three-VE model, posterior distribution mean and central 90% uncertainty intervals, conditional on the priors and data, were 375.3 [321.8, 429.1], 399.1 [346.9, 451.0], and 440.1 [388.0, 491.5] for home, office, and school VEs, respectively. In the common-mean model, the posterior distribution for the common mean was 407.6 [358.5 456.6].

Participant effects were similar across the two models. Posterior summaries for each participant appear in Fig. 5. For the three-mean model, mean participant effect posteriors ranged from −212.1 to 168.0 with an inter-quartile range of 115.4. In the common-mean model, mean participant effect posteriors ranged from −209.3 to 172.1 with an inter-quartile range of 110.9.

Posterior distributions for degrees of freedom v and standard deviation σ for the generalized Student's t distribution were obtained for both models. In the three-VE model, mean and central 90% uncertainty intervals for v were 5.7 [2.3, 13.7] and for σ were 92.8 [70.9, 117.7]. In the common-mean model, they were 12.2 [3.6, 31.7] and 109.7 [88.3, 130.2]. These values are similar, although the lower degrees of freedom in the three-VE model, indicating heavier tails than in the common-mean model, suggests that extreme values may be more apparent in the three-VE model.

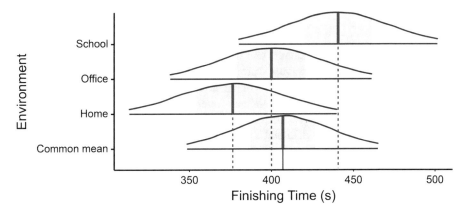

Fig. 4. Posterior distributions on independent environment effects in the three-mean model overlap substantially with the common environment effect in the common-mean model. The posterior distributions are summarized with mean (*heavy vertical bar*), central 50% uncertainty intervals (*shaded region*), and density estimates covering the central 90% uncertainty intervals (*lines*).

The ELPD was −974.6, SE 11.6 for the three-VE model and −976.9, SE 10.3 for the common-mean model. The ELPD difference between the three-VE model and the common-mean model was 2.4, SE 3.0, reflecting a predictive advantage of the three-VE model of less than one standard error. This shows that between-subjects variability in ELPD is larger than the systematic difference between the two models.

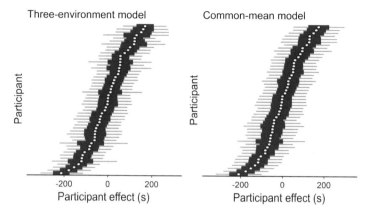

Fig. 5. Participant effect posteriors were similar for the three-environment and common-mean models. Participant data (N = 50) is sorted by posterior mean (*light central dot*). Central 50% (*dark horizontal bar*) and 90% (*whiskers*) uncertainty intervals summarize the posterior distributions.

5 Discussion

In general, the approach we followed was successful. An initial frequentist estimate showed no strong incompatibility between the data and a hypothesis of no VE-specific effects. Bayesian estimates of independent VE effects all had means falling within 65 s of each other, and the posterior distributions of the three VE effects overlapped substantially with the common VE in the common-mean model. Fitting with the common-mean model resulted in a negligible loss in predictive power relative to the three-mean model, indicating that the common-mean model was not appreciably worse at accounting for the data than the three-VE model. Taken together, these results support the claim that the difficulty of the task was similar across the three different VEs.

Of course, the difficulties were not identical and the three-VE model did estimate different, albeit very similar, VE effects. With this in mind, future analyses should attempt to control for environment effects, even though the evidence available indicates that the environment effects here are small.

The variabilities in the participant effects for both models were large relative to the VE effects. This variability occurred despite the participant sample representing a narrow demographic range. Studies recruiting from a more general population could expect even higher between-participant variability. This observation reinforces the value of within-subjects designs in VE studies, because these substantial between-participant differences can be isolated from the experimental effects of interest.

5.1 Extreme Values

The robust error term in the Bayesian models prevented extreme values from having a substantial impact on the effect distributions. However, extreme values were concentrated in the Home VE. Moreover, two participants who failed to complete the scavenger hunt in a session both failed to complete the Home VE. This suggests that our design process failed to account for something in the Home VE that, on rare occasion, lead to relatively slow searches. Without further experimentation we can only speculate on what might have caused these slow searches, but the Home VE did have a larger graph radius and diameter than the other two VEs. This relatively larger graph might have afforded more opportunities for costly navigation errors, leading to the small number of long hunt completion times. Another possible explanation is that specific hunt items might have been harder to spot within a room, or that some other factor accounts for these extreme values.

6   Conclusions

The approach described here to designing the VE layouts and the specific design of the scavenger hunt lead to a generally successful outcome. Typical completion times for the scavenger hunt in all VEs were within a range of 65 s, which is a considerably smaller range than the variability between individuals, regardless of VE.

Graph-theoretic measures are an important tool for understanding spatial layouts. Free, open source tools for computing such measures are available. Although advanced measures of environment graphs might be more informative, an advantage of the relatively simple measures we used was that they lend themselves to easy adjustments with a fair amount of leeway to account for other experimental or design constraints.

Our overall approach was not entirely sufficient to overcome all design challenges. In particular, although all our VEs had similar typical completion times, one VE induced a small number of participants to take an unusually long time to complete the hunt.

The greatest manual effort was in ensuring the hunt item placement satisfied the completeness criterion. The outlined heuristics were sufficient for our VEs, but they might not cover all possible environment layouts. Automated solutions would be needed when designing a relatively large number of environments or generating them algorithmically.

All these VEs were interiors. Different design principles may be needed for less path-constrained exterior and mixed indoor-outdoor VEs. In particular, applying graph theory when nodes and edges are not conveniently defined by rooms and doors (or streets and well-defined intersections) might be challenging. Past work suggests that space syntax tools are a viable complement to the graph-theoretic approach [7]. Isovists [13] can give an indication of environment inter-complexity for large scale spaces and even indoor environments such as ours.

Designing naturalistic VEs for use in within-subjects designs raises the problem of how to design VEs that are different but equivalent. Here, we laid out our approach to this conundrum and showed that the approach seems to have worked well. It is our hope this approach will be helpful for future experimental designs, unlocking the potential advantages of within-subjects designs for VE research and application.

Acknowledgements. This work was funded by the US Army Research Laboratory's Human sciences campaign. The views and conclusions contained in this document are those of the authors and should not be interpreted as representing the official policies, either expressed or implied, of the Army Research Laboratory or U.S. Government. The U.S. Government is authorized to reproduce and distribute reprints for Government purposes notwithstanding any copyright notation herein. The authors thank Debbie Patton, Mark Ericson, and the entire Training Effectiveness/Immersion group for their comments and suggestions on this work. Bianca Dalangin helped conduct the user study.

References

1. Witmer, B.G., Bailey, J.H., Knerr, B.W., Parsons, K.C.: Virtual spaces and real world places: transfer of route knowledge. Int. J. Hum.-Comput. Stud. **45**, 413–428 (1996)
2. Cummings, J.J., Bailenson, J.N.: How immersive is enough? a meta-analysis of the effect of immersive technology on user presence. Media Psychol. **19**, 272–309 (2016)
3. Grajewski, D., Górski, F., Zawadzki, P., Hamrol, A.: Application of virtual reality techniques in design of ergonomic manufacturing workplaces. Proc. Comput. Sci. **25**, 289–301 (2013)
4. Zimmons, P., Panter, A.: The influence of rendering quality on presence and task performance in a virtual environment. In: Proceedings of IEEE Virtual Reality 2003, pp. 293–294 (2003)
5. Steadman, P.: Graph theoretic representation of architectural arrangement. Archit. Res. Teach. **2**, 161–172 (1973)
6. Roth, J., Hashimshony, R.: Algorithms in graph theory and their use for solving problems in architectural design. Comput.-Aided Des. **20**, 373–381 (1988)
7. Levy, R.M., O'Brien, M.G., Aorich, A.: Predicting the behavior of game players - space syntax and urban planning theory as a predictive tool in game design. In: 2009 15th International Conference on Virtual Systems and Multimedia, pp. 203–208 (2009)
8. Rubinov, M., Sporns, O.: Complex network measures of brain connectivity: uses and interpretations. NeuroImage **52**, 1059–1069 (2010)
9. Sinatra, A.M., et al.: Development of cognitive transfer tasks for virtual environments and applications for adaptive instructional systems. In: Lecture Notes in Computer Science. Springer, Orlando (2019, forthcoming)
10. Carpenter, B., et al.: Stan : a probabilistic programming language. J. Stat. Softw. **76**, 1–32 (2017)
11. Stan Development Team: RStan: the R interface to Stan (2018)
12. Vehtari, A., Gelman, A., Gabry, J.: Practical Bayesian model evaluation using leave-one-out cross-validation and WAIC. Stat. Comput. **27**, 1413–1432 (2017)
13. Benedikt, M.L.: To take hold of space: isovists and isovist fields. Environ. Plan. B Plan. Des. **6**, 47–65 (1979)

3D User Interface for a Multi-user Augmented Reality Mission Planning Application

Simon Su[⊠], Vincent Perry, Heather Roy, Katherine Gamble, and Sue Kase

US Army Research Laboratory, 120 Aberdeen Blvd, Aberdeen Proving Ground, Adelphi, MD 1005, USA
{simon.m.su.civ, vincent.p.perry7.ctr, heather.e.roy2.civ, katherine.r.gamble2.civ, sue.e.kase.civ}@mail.mil

Abstract. In this paper, we present a 3D User Interface design for a shared augmented reality setup to support strategic, tactical, and training platforms. An accurate 3D holographic object registration is necessary to enable a shared augmented reality experience using multiple Microsoft HoloLens devices. We developed a sensor data fusion framework which uses both external positional sensor data and Microsoft HoloLens to reduce augmented reality registration errors in our shared augmented reality application. Our shared augmented reality 3D User Interface design implemented in the application, together with our sensor data fusion framework, enables the use of multiple Microsoft HoloLens to support a mission planning scenario.

Keywords: Multi-user augmented reality · 3D User Interface

1 Introduction

Researchers at the United States Army Research Lab have been using multiple Microsoft HoloLens devices to explore augmented reality technology for supporting a collaborative mission planning scenario [1]. The shared augmented reality spaces provide the users with a common virtual space for collaboration. However, it is common that people struggle to share unique information when collaborating to solve problems [2] and the effectiveness of team communication while using new technology diminishes [3]. Using capabilities unique to augmented reality in combination with 3D User Interface, we hope to present a shared augmented reality setup capable of supporting collaboration.

Our collaborative augmented reality application in team mission planning is using augmented reality technology to further enhance the traditional face-to-face communication where we use nonverbal information to supplement verbal language [4]. A face-to-face communication has the live feedback translated through the body language and facial expressions that provide essential clues to verify an effective communication. The use of Microsoft HoloLens in team collaboration retains all the advantages of face-to-face communication and at the same time, enhances the collaboration with superior ability to deliver any necessary visualization. In designing the

D. N. Cassenti (Ed.): AHFE 2019, AISC 958, pp. 120–131, 2020.
https://doi.org/10.1007/978-3-030-20148-7_12

3D User Interface for our shared augmented reality application, we took into consideration both face-to-face user interaction and the limitation of the Microsoft HoloLens device.

In the remainder of the paper, the related work section discusses other work in using immersive technology to support collaboration. Section 3 describes our shared augmented reality application using multiple Microsoft HoloLens devices. Section 4 describes a typical use case of our application and the results of our preliminary study on our collaborative augmented reality 3D User Interface design is presented in Sect. 4. In Sect. 5, we discuss our ongoing effort to expand the capability of our collaborative augmented reality visualization into a hybrid 2D and 3D visualization framework, concluding in Sect. 6.

2 Related Work

Our research into using shared augmented reality as a collaboration medium is not possible without the recently available augmented reality devices like the Microsoft HoloLens [5]. As a result, there are not a lot of publications on shared augmented reality research using the newly available augmented reality devices in the market. However, Leicht et al. [6], Shachaf et al. [7] and Hasler et al. [8] studied the effectiveness of team collaboration in virtual environment. Wagner et al. looked at shared reality meetings using marker based augmented reality [9] and Regenbrecht et al. investigated the interaction aspect of the marker based collaborative augmented reality environment [10]. Billinghurst et al. also looked at augmented reality interfaces for collaborative computing [11]. Piumsomboon et al. explored using both virtual reality and augmented reality systems for remote collaboration [12].

In our approach, we use data from an external tracking system to improve the registration accuracy of holographic objects in the shared augmented reality environment using Microsoft HoloLens devices. Oda et al. used 3D referencing techniques for physical objects to improve registration accuracy in a shared augmented reality [13].

3 Shared Augmented Reality Using Multiple Microsoft HoloLens

3.1 Hardware and Software Environment

The main focus of our shared augmented reality application is initial object registration and multi-user interaction in order to demonstrate a synchronized view among all users connected from separate HoloLens devices. The outer boxes of Fig. 1 shows the hardware components of our shared environment, with the inner boxes showing the software applications that run. For our shared augmented reality application, we use Unity to develop a client/server application, with the server being hosted on a dedicated server machine and the client applications deployed and run on individual Microsoft HoloLens devices. For the external sensor data, we use an OptiTrack Motion Capture system that uses the Motive software to convert the data captured by our 11 Infra-Red cameras setup into 3D positional and orientation data, and broadcast that data over the network. The additional 3D positional and orientation data essentially provides the individual Microsoft HoloLens with its location information within a global coordinate system.

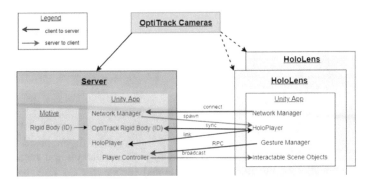

Fig. 1. Shared AR architecture.

To run our application, the Motive motion capture software must first be running. Markers need to be attached to the top of the HoloLens devices, each with their own unique configuration. Once the markers are attached to the HoloLens devices, the user must manually create a Rigid Body for each HoloLens marker configuration, providing a one-to-one correspondence from Motive to Unity. The OptiTrack cameras stream data from the Rigid Body attached to the HoloLens in the real world to the Unity server application. While the client/server application is running, the Motive software is continually capturing position and orientation data of the markers attached to the top of each HoloLens. As seen in the bottom left blue box of Fig. 1, both the Motive software and Unity server application run on the same dedicated server machine. The Unity server application streams in the Rigid Body data from the Motive camera software via a Unity-OptiTrack plugin. In the Unity server application, the real-world coordinate system captured by Motive is overlaid on top of the virtual Unity coordinate system that the HoloLens sees. The server essentially hosts the augmented reality world, knowing where the physical HoloLens devices are located in the real-world coordinate system with the Motive camera data, as well as the holographic scene in relation to where the physical HoloLens devices are.

The synchronization of Unity's virtual coordinate system with the Motive software's real-world coordinate system allows the server to always contain the exact coordinates of all physical and virtual objects for the shared augmented reality application. The server application displays the HoloLens devices as colored cubes based off of their real-world position and orientation streamed in from the Motive software and shows their relative position to the holographic model that each HoloLens is to be viewing. Once the server is up and running and the HoloLens devices are being tracked, each HoloLens device may launch the client application. Figure 1 shows how the HoloLens client connects to the server application and synchronizes with its corresponding Rigid Body in the real world. When the client application is launched, the HoloLens device connects to the server application. Because there is a one-to-one correspondence of each unique Rigid Body marker configuration attached to each HoloLens device, the server application can associate the connected HoloLens device to a Rigid Body being captured by Motive. The server then relays the position and orientation data of that Rigid Body collected by the Motive software to the client device such that the HoloLens syncs to its real-world position and orientation. Since the server also knows where the holographic scene should exist with respect to the real-world coordinate system, the object is registered within the HoloLens device in the appropriate position in the real-world coordinate system defined by the server. The use of the external motion capture sensors helps the HoloLens devices to initially register the holographic scene in a pre-defined real-world coordinate system instead of an unknown, dynamically mapped coordinate system.

3.2   System Architecture Design

While the Unity-OptiTrack plugin is used for synchronization of the HoloLens with its location in the physical environment, Unity Networking is used for multi-user shared interaction with the holographic environment between the HoloLens clients and server. In order for messages to be passed using a client/server model, the Command and ClientRPC attributes of the Unity Networking framework are used. The ClientRPC attribute allows the server to invoke the designated method on all connected clients whereas the Command attribute allows clients to send a command to invoke the designated method on the server. All of the Command and ClientRPC attribute methods are referenced in one script attached to the player prefab for each client that connects. The Command methods are implemented in the script that resides on the server, whereas the ClientRPC methods are implemented in the script that resides on the clients.

The blue and green arrows in Fig. 1 shows the client/server interaction between the HoloLens app and server app. When a client first runs the application, the client side sends a message to the server that a new player is connecting. The server in turn spawns a new player into the environment and the client connects. The server matches the end of the IP address of this new client to a streaming ID from the Motive software and communicates back to the client where it is physically located in the global coordinate system. Based off of where the HoloLens is in regard to the OptiTrack tracking area, the server will give the HoloLens client its position such that the scene will always appear at the same location in regards to the physical environment of the room. This reduces the error of the initial registration of the HoloLens when connecting to the shared environment. From there, the server grants the client the authority to color its cube marker a distinct color, visible by other clients above the user's head as shown in the top left of Fig. 1, then becomes static and waits for user input.

Once the environment is set up and the client applications have connected to the server, all HoloLens clients will be observing the same shared augmented reality environment. The last two links between the HoloLens and server of Fig. 1 shows how interaction occurs between a HoloLens user and all connected clients. The interaction sharing comes directly from the Gesture Manager implemented within each HoloLens client. When a user taps in the environment, the Gesture Manager registers that tap at the current gaze location. Instead of responding to the user's tap, the message is sent to the server application. Then, the server broadcasts this tap message and location out to all HoloLens clients so that each client application remains the same. In this case, all clients will have the same update happen in their scene and the scenario will remain shared.

The use of a Gesture Manager for registering interaction events and then sending messages in this fashion best suits a client/server interaction model. The newer HoloToolkit InputModule uses an Input Manager to detect gesture input where each tapped object can respond to that gesture directly. Although it provides a level of abstraction for native application input development, it is not ideal for a client/server model that requires shared interaction among multiple clients. For a client/server model, all interaction should be sent to the server before being recognized by the individual clients. Encompassing all interaction into one method of Gesture Manager is much more efficient and extensible than having input data dispersed among various scripts attached to every clickable object in the scene to then be sent to the server. The server receives the tapped event data from one of the clients and broadcasts out the RPC to all clients to execute the tap event. Each client will execute the same response to a tapped event delivered by the server. This allows for synchronization of all clients since each tap event is first sent to the server and then distributed to all clients, removing the case where a client's tap only registers within that user's HoloLens application.

Fig. 2. Shared Augmented Reality Application running showing users views of the virtual environment and their physical space. Top left is showing the 3D User Interface.

3.3   Shared Microsoft HoloLens Application

As shown in Fig. 2, our application is designed as a mission planning scenario for multiple users to collaborate in a shared mixed reality environment. The main motivation for our application is for users to be able to plan and discuss key mission objectives and paths using augmented reality technology such that collaborators can view a 3D virtual scene, but also interact with each other face to face. The purpose of using augmented reality technology for viewing of a 3D scene is to enable users to have both audible and visual feedback with their collaborators sharing the scene, allowing users to talk and interact with each other in a natural environment. The augmented reality technology enables this natural interaction among multiple users that is lost in a virtual reality environment. However, the AR technology still enables users to view a 3D virtual scene to plan a mission, which provides added capability to multiple users standing around a 2D map to plan a mission. The mission planning application provides an environment for users to plan an extraction scenario of a High Value Target (HVT) from an overrun building, with landing occurring via helicopter and extraction occurring via boat. Once the HVT has been rescued, the team of soldiers is to exit the building, then make their way to the extraction location. The team is to meet the boat at the extraction point for the scenario to complete.

There is a server application that runs on the same machine that hosts the OptiTrack's Motive software, and a client application that runs on the HoloLens devices. In order to run the application, the server must first be running and connected to the Motive software data streaming. Then, the client application running on each independent HoloLens may be started and the client application will connect to the server over the network. When the client connects, the user will see the predefined scene in a shared environment. All users will see the scene in the same position and orientation at a tabletop height in the real-world environment, in the center of the OptiTrack cameras' capture area. Underneath the scene is a virtual cylinder that rests on the ground as if the scene were sitting on a real-world object like a table. This allows users to physically point out objects in the scene and discuss with each other naturally as if they were standing around a table in the real world. The predefined scene is set with the team at a start location, an end location, hostiles, and a target.

To implement a shared mission planning scenario for multiple HoloLens users, the application is intended for multiple users to connect to the same environment and collaboratively plan a rescue mission through interaction with the mixed-reality scene. The mission planning involves constructing a path to rescue a high value target from a building and advance to a pre-specified location where a boat will meet for extraction, all while avoiding being sighted by hostiles. The different planning maneuvers are initiated via menu buttons located above the building in the middle of the scene and can be performed by any of the users in the scene at any time, making communication and collaboration imperative to success. While the menu buttons are shared among all users, the menu will remain facing each individual user as the user physically walks around the scene. The main menu consists of three buttons: Start Plan, Reset Plan, and Execute Mission. When Start Plan is tapped, a pop-up menu shows a layer of 5 buttons to plan the scenario. These buttons allow the users to place path markers, delete path markers, place a marker to initiate the boat moving to the extraction location, delete the boat marker, or end the planning phase altogether and close the pop-up menu. This menu can be seen in the top left of Fig. 2.

When the place path marker button is selected, any user can tap on the terrain and a marker will be placed at whatever position their gaze cursor interacts the terrain. This marker will be instantiated and appended to an internal list, and a new line will be rendered between the previous marker and the new one. The path will always start at the same starting location which is labeled as an arrow next to the helicopter in the scene. Every mission will begin from this location. To delete the previous placed marker, the delete path marker button may be selected to delete the marker from the scene. To place the boat marker, any user can select the place boat marker button and then tap anywhere on the terrain. However, the boat will only begin moving if the team passes over this boat marker, so it should be placed somewhere along the path that the team will take. To remove the marker, the remove boat marker button can be selected. Only one boat marker can ever be present in the scene at any time. When the users are finished with the planning phase, the end plan button can be selected to return to the main menu. From there, users can either reenter the planning phase by selecting the start plan button again, can reset the entire planned scenario by selecting the reset plan button, or choose to have the scenario run by selecting the execute mission button. If the reset plan button is selected, all path markers will be deleted and the team, boat, and hostiles will be placed back at their starting locations. If the execute mission button is selected, the team will begin moving along the path and the scenario will play out from start to finish.

Users can plan, reset, and execute the mission for as many iterations as they would like. Users are also timed on how long it takes to plan the rescue mission, how long it takes to execute the mission, and total time spent in the application itself. Points are deducted for passing through walls, being seen by hostile radars, and failing to complete mission objectives (capture HVT, meet the boat for extraction).

Throughout the mission planning, the scene will always look the same for all users. This includes the path as it is created by the users, as well as the speed at which the team, hostiles, and boat move when the mission is executing. The menu buttons will also be synchronized for all users, and actively selected buttons will persist in a green state, while non-selected buttons will be colored black. Any user can select buttons at any time, making communication among users imperative. Because the application is network-based, it is possible for a user to disconnect from the server. When this occurs, the user can simply relaunch the application and reconnect to the server, then click the reset plan menu button. This will coordinate the disconnected client back with the users still in the scene instead of all users having to disconnect and restart the entire application.

3.4   Scenario Planning Use Case

As shown in Fig. 2, a typical use case for our shared mission planning scenario application is for two users to put on a HoloLens device and start the client application. Once connected, both users will see the main scene in front of them, and both of them will be seeing the scene in approximately the same location in the physical environment. The users begin to inspect the layout of the scene, such as where the building is located and where the target is within the building, where the start and end locations are located on the map, where the hostiles are located, and where the buttons are to perform actions in the scene. The mission planning begins once one user gazes at and clicks the "Start Plan" button as shown in the upper left corner of Fig. 2.

Once this button is selected, the hostiles will begin moving around in the scene. The users will take special note of the speed at which the hostiles are moving, as well as the areas covered by their patrol. The next step is to construct a path for the team, initiated by either user selecting the "Place Path Markers" button. Once this button is selected, the button will turn green and the users will discuss the best path to take and take turns placing markers along the path by gazing at a specific location in the scene and clicking. A user may place a marker at an undesirable location and decide that they do not want that to be part of the path anymore. Either of the users can gaze at and select the "Remove Path Marker" button, which removes the most recently placed marker from the path. They may then re-select the "Place Path Markers" button to continue placing markers.

Both users will take turns placing markers and discussing the best location to plan the next path marker. The users will take special care to plan a path that enters safely into the building and passes the blue arrow, representing the HVT in the scene. The path will then continue safely out of the building and to the extraction point designated by a black arrow. The last marker placed in the scene will be carefully placed as close to the black extraction arrow as possible. Once the users are happy with the path they constructed, the next plan of action is to determine where to place the boat marker.

One of the users will select the "Place Boat Marker" button from the menu. They will then find a location along the path that they think will coordinate the time at which the team and boat will meet at the extraction location. This can be done by carefully observing the speed at which the hostiles are moving about the scene, since the team and boat move at the same speed. Thus, the users will try to calculate the correct distance from the last marker placed, and click to place the boat marker at that location. After careful examination, they decide that the boat marker should be placed farther along the path at a location closer to the extraction location. One of the users can select the "Remove Path Marker" button to remove the marker, then re-select the "Place Path Marker" button to place it at the newly desired location. Once both users agree that they have constructed a desirable path, the "End Plan" button is selected.

This takes the button menu back to the main menu, where the users can select to start planning, reset the plan, or execute the mission. Since the users are happy with their path, they do not want to reset the plan. Instead, they want to execute the mission. One of the users watches the path of the hostiles as the other user focuses on the "Execute Mission" button. The users will know when to select "Execute Mission" based off of the speed and location of hostiles. When the hostiles are at a certain spot in their patrols that prevents them from seeing the team traverse the path, the user audibly voices to the other user to click the button. The other user will select the "Execute Mission" button and then both users can watch as the mission is executed.

4   3D User Interface Preliminary Pilot Study

The purpose of this informal study was to receive feedback regarding how well our setup and application enabled a shared, collaborative, augmented reality experience. It was primarily used to help us gain feedback on the design decisions we made to enable the sharing capability among users. We ran 8 groups of 2 users each through the shared mission planning scenario, for a total of 16 participants. Prior to running the scenario, each pair of participants was briefed on the layout of the scene, the UI buttons and shared interaction capability, as well as the goals and objectives of the scenario to plan a mission. Each pair of participants ran through the scenario, performing actions very similar to the use case described. Once the users were finished conducting the study, we asked them to fill out a simple questionnaire in order to get their feedback.

All questions were to be answered using a Likert scale from 1 to 5, with 1 being the most negative answer and 5 being the most positive. We asked the users some general questions first such as how experienced they were with augmented reality technology and whether or not they liked the experience. Using the general questions, we could analyze the influence their experience ratings had on more relevant questions later. The general consensus was mixed for prior experience using the technology, but every participant gave a 4 or 5 for how well they liked the experience.

Other questions asked regarded the stability of the gaze, stability of the scene, selection interaction capability, and the user interface menu. While these questions were useful for our own feedback, we were most interested in the shared capability of our application. This included the scene registration and how closely it appeared to the other user's scene, as well as how permitting the HoloLens device is for normal collaboration. The first figure shows the ratings of the 16 participants for how well they thought the scene appeared to their partner's scene while conducting the scenario.

As can be seen from Fig. 3, only one participant gave a neutral rating of 3 on the Likert scale from "Not close" to "Very close." The rest of the participants thought that the scene appeared pretty or very close, for an average rating of 4.625 out of 5. This is very reassuring to know that our shared environment provides multiple users with a very close augmented reality scene in which to collaborate. The second figure shows the ratings of the 16 participants regarding how capable they felt collaborating with their partner both visually and audibly.

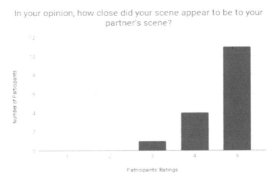

Fig. 3. User ratings on a Likert scale from 1 to 5 of how close the scene appeared to his/her partner's scene

As can be seen from Fig. 4, only two participants gave a rating of 3 on the Likert scale from "Not capable" to "Very capable," with an average rating of 4.5. In terms of usability, we do not want the HoloLens technology to limit the natural communication among individuals, especially when performing a critical task such as mission planning. Virtual reality technology limits a user's ability to physically see their

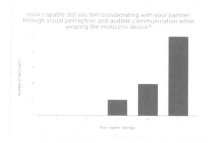

Fig. 4. User ratings on a Likert scale from 1 to 5 of how capable the user felt collaborating with his/her partner both audibly and visually while wearing the HoloLens device.

collaborators and provides a barrier for a user to know be who may be talking to them due to a lack of visual cues. With augmented reality technology, there is no limiting factor in communication considering a user can still see their collaborators in the real-world through the clear visor and can use those visual cues to talk and share ideas without a hint of doubt. In addition, the augmented reality technology provides a way to collaborate and interact in a 3-dimensional virtual environment that more closely emulates a real-world scenario instead of a 3D projection on a 2D monitor or piece of paper. Only one user gave a 3 for how likely they would be to choose augmented reality to conduct mission planning, with every other participant giving a rating of 4 or 5. This is insightful feedback, from which we may conclude that using a shared augmented reality environment can provide many benefits to conducting a collaborative mission planning exercise.

In addition to the questionnaire, we informally asked the participants if they had any other feedback in regards to the scenario and sharing capability. One piece of feedback we received was that it would be beneficial for the gaze cursor to change color in order to provide more visual feedback beyond just the menu buttons changing color. One idea would be to have the user's gaze cursor turn blue when the users are either placing the path or boat marker to remain consistent with the color of the team and target to capture. When removing markers, the cursor could then turn red. This would enable the users to know what their partner selected based off of just the color of the cursor instead of having to look up at the button menu each time.

Another piece of feedback that we received from multiple participants was the difficultly in knowing exactly where their partner wanted to place a marker. While the general area can be described through communication, the only hope of knowing the precise location would be through physically pointing at a location in the scene from one user's point of view. Due to the imperfections in the registration of the scene, as well as the holographic view obstructing the view of what is in the real world behind it, it is hard to be confident on the precise location that a partner points to as they see it in their view. To enable a more assertive collaborative experience, feedback on precise location is imperative. Some ideas received were to be able to see their partner's gaze in the scene so they know where they are looking, or to be able to place a removable marker indicator that all users can see.

5   Discussion and Future Work

In the future, we plan to incorporate the HoloLens into our reconfigurable visual computing architecture [14]. This will allow HoloLens users to not only communicate in the shared augmented reality environment, but also with a 2D environment elsewhere. To accomplish this, a web socket connection will need to be established between the HoloLens server application and the Event Translation Router. Currently, when a HoloLens user interacts with the environment, the interaction is first sent to the HoloLens server to broadcast to all HoloLens clients. With the websocket established, the HoloLens server will be responsible for not only broadcasting the interaction to all HoloLens clients, but also sending the interaction over a websocket to the Event Translation Router. In turn, the Event Translation Router will need to implement the mapping of incoming messages from the HoloLens server to relevant RPC calls on the Application Server. This will allow the 2D application(s) to receive the HoloLens user's interaction update from the Application Server and update the 2D visualization accordingly.

In addition, based on our preliminary pilot study findings, we have developed a full usability study to investigate the effectiveness of our collaborative augmented reality application to support team mission planning. In addition to the technology aspect, we are also very interested in the human factor aspect of the team mission planning that we explore in great depth. At this time, we are well into our experimental efforts and initial analysis of the full usability study appears to be inline with our pilot study's assessments.

6   Conclusion

We investigated using shared augmented reality technology, coupled with traditional face-to-face communication, to facilitate team mission planning. Using additional sensor data to improve holographic object registration in the virtual environment, we were able to use icon based menu system with user interaction provided by Microsoft HoloLens to facilitate the collaboration. Our preliminary user study shows the users' acceptance of using augmented reality technology and its ability to facilitate face-to-face communication for collaboration. With future advancements to the Microsoft HoloLens and other augmented reality technology, more users may be willing to engage in using augmented reality technology for collaborative visualization and interaction. With our shared augmented reality environment in place, we can extend our application to share other environments beyond that of a mission planning scenario.

Acknowledgments. This work was supported in part by the DOD High Performance Computing Modernization Program at The Army Research Laboratory (ARL), Department of Defense Supercomputing Resource Center (DSRC).

References

1. Su, S., et al.: Sensor data fusion framework to improve holographic object registration accuracy for a shared augmented reality mission planning scenario. In: Chen, J.Y., Frago-meni, G. (eds.) Virtual. Augmented and Mixed Reality: Interaction, Navigation, Visualiza-tion, Embodiment, and Simulation, pp. 202–214. Springer International Publishing, Cham (2018)
2. Stasser, G., Titus, W.: Pooling of unshared information in group decision making: biased information sampling during discussion. J. Pers. Soc. Psychol. **48**(6), 1467–1478 (1985). https://doi.org/10.1037/0022-3514.48.6.1467
3. Kiesler, S., Siegel, J., McGuire, T.W.: Social psychological aspects of computer-mediated communication. Am. Psychol. **39**(10), 1123–1134 (1984). https://doi.org/10.1037/0003-066X.39.10.1123
4. Hosobori, A., Kakehi, Y.: Eyefeel & Eyechime: a face to face communication environment by augmenting eye gaze information. In: Proceedings of the 5th Augmented Human International Conference, AH 2014, pp. 7:1–7:4. ACM, New York (2014). https://doi.org/10.1145/2582051.2582058
5. Furlan, R.: The future of augmented reality: Hololens - Microsoft's AR headset shines despite rough edges. IEEE Spectr. **53**(6), 21 (2016). https://doi.org/10.1109/MSPEC.2016.7473143
6. Leicht, R.M., Maldovan, K., Messner, J.I.: A framework to analyze the effectiveness of team interactions in virtual environments. In: Proceedings of the 7th International Conference on Construction Applications of Virtual Reality. University Park, PA, USA, 2007
7. Shachaf, P., Hara, N.: Team effectiveness in virtual environments: an ecological approach. In: Teaching and Learning with Virtual Teams, pp. 83–108. Idea Group Publishing (2005)
8. Hasler, B.S., Buecheler, T., Pfeifer, R.: Collaborative work in 3D virtual environments: a research agenda and operational framework. In: Ozok, A.A., Zaphiris, P. (eds.) Online Communities and Social Computing, pp. 23–32. Springer, Heidelberg (2009)
9. Wagner, M.T., Regenbrecht, H.T.: Shared reality meeting - a collaborative augmented reality environment. In: The First IEEE International Workshop Augmented Reality Toolkit, p. 2, September 2002. https://doi.org/10.1109/art.2002.1106970
10. Regenbrecht, H.T., Wagner, M.T.: Interaction in a collaborative augmented reality environment. In: CHI 2002 Extended Abstracts on Human Factors in Computing Systems, CHI EA 2002, pp. 504–505. ACM, New York (2002). https://doi.org/10.1145/506443.506451
11. Billinghurst, M., et al.: Mixing realities in shared space: an augmented reality interface for collaborative computing. In: 2000 IEEE International Conference on Multimedia and Expo, ICME 2000. Proceedings. Latest Advances in the Fast Changing Worldof Multimedia (Cat. No. 00TH8532), vol. 3, pp. 1641–1644, July 2000
12. Piumsomboon, T., Dey, A., Ens, B., Lee, G., Billinghurst, M.: [POSTER] CoVAR: mixed-platform remote collaborative augmented and virtual realities system with shared collaboration cues. In: 2017 IEEE International Symposium on Mixed and Augmented Reality (ISMAR-Adjunct), pp. 218–219, October 2017. https://doi.org/10.1109/ismar-adjunct.2017.72
13. Oda, O., Feiner, S.: 3D referencing techniques for physical objects in shared augmented reality. In 2012 IEEE International Symposium on Mixed and Augmented Reality (ISMAR), pp. 207–215, Nov 2012. https://doi.org/10.1109/ismar.2012.6402558
14. Su, S., et al.: Reconfigurable visual computing architecture for extreme-scale visual analytics. In: Proceedings of the SPIE 10652, Disruptive Technologies in Information Sciences, 106520M, 9 May 2018. https://doi.org/10.1117/12.2303887

Human Factors and Simulation in Transportation

Pupillary Response and EMG Predict Upcoming Responses to Collision Avoidance Warning

Xiaonan Yang[✉] and Jung Hyup Kim

Department of Industrial and Manufacturing Systems Engineering,
University of Missouri, Columbia, USA
xyr29@mail.missouri.edu, kijung@missouri.edu

Abstract. Nowadays, an advanced driver assistance system (ADAS) has become popular in a new automobile. Even though cutting-edge technologies in ADAS provide high accuracy collision avoidance warnings, the distraction-related crashes caused by collision avoidance warnings are likely to become an emerging problem. Hence, it is necessary to understand how a driver responds to collision warning and how to reduce driver distraction caused by ADAS. Recent studies have found that physiological data, such as pupillary responses and electromyography (EMG), can forecast the human's physical responses. Therefore, in this study, pupil and EMG signals were applied to predict drivers' physical responses to collision warning. Logistic regression was applied to predict if drivers would like to give a physical response or not. The findings of the current study will contribute to improving the safety feature of collision warning systems and help to design the advanced driver assistance system with better device-user interaction.

Keywords: Collision avoidance · Pupillary responses · Electromyography · Logistic regression · Support vector machines

1 Introduction

In the past few years, in-car advanced driver assistance systems (ADAS) has become ubiquitous. Several researchers have proved a collision warning provided by ADAS which can reduce the crash rate and improve the driving safety [1–4]. However, Parasuraman [5] found that the present analysis of the current warning system would make it hard to achieve the expected level of reduction in crash rate after considering human interactions, because drivers have some aversion to the ADAS warnings if they are exposed to frequent false alarms. The high rate of a false alarm can cause distracted driving [6] and reduce the effectiveness of warning in ADAS [7, 8]. Therefore, motivated by the benefit of driving safety by using ADAS [9, 10] but accounts for the high risks caused by false alarms [7, 11] in ADAS, it is necessary to advocate a better design of ADAS that can adjust the warning functions based on drivers' behavior in order to decrease the number of false alarms and increase the effectiveness of the collision warning. However, to achieve this goal, detecting and predicting the driver response to collision warning is essential.

© Springer Nature Switzerland AG 2020
D. N. Cassenti (Ed.): AHFE 2019, AISC 958, pp. 135–143, 2020.
https://doi.org/10.1007/978-3-030-20148-7_13

So how the ADAS systems elicit drivers' upcoming response to the collision warning? Recently, many research studies used physiological measures to advance our understanding of the relations between human biometric signals and physical reactions, such as pupillary response, electroencephalography, electromyography signals, etc. Peysakhovich [12] stated the dilation of pupil could reveal the upcoming choice of the human about half a second before the actions in the dynamic decision-making process. Also, other studies have supported that the patterns of pupillary response could reflect human perceptual decisions [13, 14]. Therefore, in this study, we assumed the pupil dilation changes could be one of the indicators of physical responses once drivers give a physical reaction to the collision warning. Also, electromyography (EMG) is an electrodiagnostic medicine technique for evaluating and recording the electrical activity produced by skeletal muscles. As it quantifies muscle force levels, it has been widely used to classify human postures, with decoding user intent and indicating patients' physical movement in healthcare [15]. So it was introduced as another indicator of physical movement in this study. Therefore, the research objective is to predict drivers' upcoming physical reactions to collision avoidance warnings by using the ratio of pupil diameter changes and EMG responses as two physiological measures. The response variable Y is a binary variable: "Yes" represents there is a physical reaction to collision warning (like steering or braking), "No" means drivers ignored the collision warning as no clear physical responses were detected. Time series features such as mean absolute value and waveform length were applied to pupil and EMG data, it has been proved in many researches [16, 17].

The goal of this analysis is to detect a driver's responses corresponding to collision warnings by using pupillary response and EMG data features within a driving simulator. For this classification problem, logistic regression was tried as the first step of prediction. The motivation for the use of logistic regression in this study is that it is one of the most commonly used standard statistic methods for building prediction models for a binary response. Logistic regression models have been widely applied in customer behaviors [18], disease detection [19] and bankruptcy prediction [20]. It also has been extended for physical activity with physiological data by many authors [21]. Logistic regression not only performs as well as the other machine learning technique in two-class classification, which perfectly fit with the purpose of this study, but can also naturally generalize the underlying probabilistic view of prediction. Furthermore, it provides model coefficients as indicators of predictors' feature importance. Classification rate was obtained from a 5-fold cross-validation as results. It shows 94% classification rate when the driver ignored the warning. Even though some environment limitations and high variance of human physiological data, this traditional statistical learning prediction model still shows potentially good outcomes. To our limited knowledge, this is the first study applying both pupillary and EMG signal at the same time to predict human physical responses.

Indeed, if we can successfully predict the drivers' responses before their physical actions were made, the outcome of this study will help engineers to develop an algorism for a highly advanced ADAS which can assess how drivers behave differently to the warning. For example, the smart ADAS device might generate a follow-up warning if the device catches a driver whom seems to ignore a critical warning. On the other hand, the ADAS device could mute some of the functions itself if a driver is fully aware of the situation with following up physical actions, and everything is under control. In doing so, what we expect is greater safety benefits because future ADAS will be able to predict drivers' upcoming actions which could be delivered to avoid a crash. The number of false alarms can be decreased as increasing of user preference for warning systems with considering drivers' interactions. In the future, the Intelligent Driver Assistance System, with functions of collecting data that related to the road environment, also be able to sense, analyze, predict and react to a driver [22]. The finding of this study will improve the collision avoidance features in ADAS. Also, it could be used as a key component of the software algorithm in a next-generation smart vehicle that can identify not only potential surrounding hazards, but also driver's status in order to provide a better and safer driving experience.

2   Method

2.1   Apparatus

In this study, the experiment was conducted in a driving simulator environment by using OpenDS, which is an open-source driving simulation software. It provides multiple modules to redesign scenarios and create complex driving environments. Three looped driving scenarios were generated on the highway condition. A full set of Logitech G27 driving device was connected to the OpenDS with the steering wheel, brake, and accelerator. Tobii eye-tracking glasses were used to collect scan path, eye gaze points, and pupil diameter. During this experiment, two Myo armbands were placed on both left and right forearm to record the changes in electrical potential around the muscle due to the depolarization of the cell muscle during muscle contraction. There are eight channels of each armband. Also, a GoPro camera was placed at right-behind to record the driver's physical movements (Fig. 1).

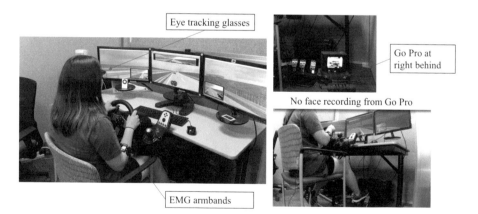

Fig. 1. Experiment environment

2.2   Participants

A total of ten university students (mean age: 22.8, standard deviation: 5.52) partici-pated in this experiment, including six males and four females. Their average driving experience was five years (mean year: 6.9, standard deviation: 5.49). Three participants had never experienced a vehicle accident. However, the other seven participants did experience a car accident determined to be 1.75 (standard deviation: 1.25) based on a severity level scale (scale 1 – minor accident and scale 5 – severe accident).

2.3   Procedure

The experiment was conducted with three scenarios. Each scenario took around 2–3 min which depends on drivers' actual speed. A three minutes break was provided between each scenario. After finishing all three scenarios, the participants completed a questionnaire to collect demographic information such as age, gender, driving expe-rience (in years) and whether the participant had ever been in a car accident. If they had, the severity level was determined based on a scale of 1 to 5.

3   Data Analysis

After the experiment, the pupil and EMG signal under each warning (one-second length) were picked for deep analysis, averaging the data from five seconds before a warning happened as baseline. When reviewing all GoPro recordings, drivers' responses to collision warnings were classified into "YES" or "NO" groups, which depend on drivers gave physical responses or not.

3.1   Important Features

After completing all data collection, time series features have been applied to pupil and EMG data. According to the research done by [16], the following time series features are recommended for pupil and EMG signal pattern recognition.

- Mean Absolute Value (MAV)
  Mean absolute value is an estimate of summation absolute value and measures contraction level of the pupil and EMG signals, MAV is an integral of the pupil and EMG.

$$MAV = \frac{1}{N}\sum_{k=1}^{N}|X_k| \tag{1}$$

Where X_k could be EMG and pupil data at k and N is a number of samples.

- Waveform Length (WL)
  Waveform length (WL) measures the complexity of the signal. It has been applied to pattern recognition of many physiological data [23]. It is defined as the

cumulative length of the signal waveform in a certain time segment, which can be used to indicate the degree variations of EMG and pupil changes. It is given as:

$$WL = \sum_{K=1}^{N} |X_{k-1} - X_k| \tag{2}$$

3.2 Logistic Regression for the Classification Problem

Binomial Logistic regression (LR) is one of the most widely used classification methods [24]. It comes from generalized linear regression which has a linear combination of predictors. Unlike discriminant function analysis, there is no assumption requiring that predictor variables and classes are distributed as a multivariate normal distribution with equal covariance. It also considers all the data points in the sample. As we only got a small data sample, logistic regression would be the best first try of building this classification prediction model. Let response Y be a binary variable to indicate drivers' physical response to collision warning, it has two classes:

$$Y = \begin{cases} 1, & \text{if driver gave a physical response(Yes)} \\ 0, & \text{no physical respose was detected(No)} \end{cases} \tag{3}$$

The probability that one observation x belongs to one of the two classes is [25]

$$y = \pi(x) = E(Y|x_1, \ldots, x_p) = \frac{e^{\alpha + \beta_1 x_1 + \ldots + \beta_p x_p}}{1 + e^{\alpha + \beta_1 x_1 + \ldots + \beta_p x_p}} \tag{4}$$

This models the probability of an event occurring in one of the two classes of a dichotomous criterion by using Logit transformation

$$logit(y) = log\left(\frac{y}{1-y}\right) = \alpha + \beta_1 x_1 + \ldots + \beta_p x_p \tag{5}$$

α is the intercept, x_p are a set of predictors, and β_p are the regression coefficient of each predictor. In this study, participants' age, gender and driving experience in year have been considered in the model. Also, the time series features that are mentioned before of pupil change ratio for both eyes and EMG signal for both arms are important predictors. The reason we have chosen logistic regression for this drivers' response classification problem is that it can easily handle continuous and categorical features to binary response. It fits as well as support vector machine for classification problem within two groups (drivers' physical response: Yes or No). The accuracy of the predicted classes can be summarized in a confusion matrix. If the estimated probability is greater than the pre-defined threshold value, this data point is classified into the success (Yes) group; otherwise, it is classified into the failure (No) group. The decision rule of the pre-defined threshold can be easily adjusted under a different circumstance. Right now, 0.5 is set as the pre-defined threshold. Figure 2 shows the process of data analysis.

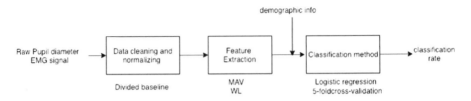

Fig. 2. The process of data analysis

4　Results

As each participant had three trails, the number of warnings were different and depended on how they drive in each scenario. Some data points were not qualified because the driver gave a physical response even before the warning. Therefore, only a small amount of data could be used, there may not be enough data for both training and testing. In this case, 5-fold cross-validation was applied. The numbers reported for classification rate are the averages over all 5 test sets. Independent variables, gender, age, MAV and WL feature of pupil change ratio and EMG signals were included in the final regression model with the drivers' physical response results of the validation sample summarized in the following table. The statistical significance of the coefficients was also tested in the model. From the output, none of the predictors show significant importance, so it is hard to tell which features are important to predict drivers' response from this database. From the results in Table 1, it is observed that the correct classification rate is 25% for Yes group and 94% in no response group where 2 were true positive and 17 were true negative. The misclassification rate was 75% with 6 observations with physical response misclassified as no physical response, while 6% with 1 observation had no physical response but the prediction said yes.

Table 1. Confusion matrix of validation in logistic regression model

Actual response	Predicted response	
	Yes (1)	No (0)
Yes (1)	2 (25%)	6 (75%)
No (0)	1(6%)	17(94%)

Not only the misclassification probability, but also the cost of misclassification in each class should be taken into account in order to obtain a model with the smallest expected misclassification costs for the real world application [26]. Table 2 summarizes the Type I and Type II errors of logistic regression model from testing and validation sample. It is apparent that the costs associated with Type I errors (driver physically responded to the collision warning but classified into no response group) and Type II errors (no physical response in actual is misclassified as responded) are significantly different. In this study, the misclassification costs associated with Type II errors could lead to a car crash once potential dangerous happened, which were much higher than those associated with Type I errors (double unnecessary warning).

Table 2. Type I and Type II errors of logistic regression model

Model	Testing sample		Validation sample	
	Type I	Type II	Type I	Type II
Logistic regression	7%	31%	6%	75%

5 Discussion and Conclusion

The purpose of this research is to predict the drivers' upcoming physical reactions to collision avoidance warning by using pupillary and EMG responses. A logistic regression-based approach for classifying drivers' physical response to a collision warning has been performed in this paper. Logistic regression focuses on maximizing the probability of the data to get the classification rate in each group. The logistic regression performance in terms of classification rate in the no response group is around 94%, which 25% for the yes group with physical response. From the results, the logistic regression model did a great job at selecting the true negative cases, where a driver had no physical responses to the collision warning. As this experiment was conducted in a driving simulator, drivers often chose to ignore the warning because they were not endangered compared to real driving. However, the high Type II errors need to be reduced, can decrease the possible high risks of car accident associated with it. This can be done by properly adjusting the pre-defined threshold for different situations. For example, if a driver is truly ignoring a collision warning for a dangerous situation, the cost of an initial misclassification is very high. However, if a driver does respond, the misclassification into no response is not so severe because the driver already realized the potential danger. However, for some companies with advanced road environment detection, they may prefer to design a collision device with less false or unnecessary warnings to increase user satisfaction. As the logistic regression only provides a linear decision boundary of two classification groups, which in some degree caused relative low prediction score, the other methods with different shapes of boundary could be applied, such as QDA, SVM, even nonparametric analysis like KNN. Definitely, the sample size needs to be increased in the future to cover more people in different age groups, some time series model like ARIMA model could also be tried for further analysis to study the pupil and EMG signal pattern based on the time to give a response.

One of the main limitations of pupillary response is that any changes in pupil diameter as a result of illumination level are the first order effects (0.1 mm to 8 mm) in contrast to second-order effects resulting from changes in the human cognitive and decision-making process that affects pupil size (0.05 mm to 0.1 mm) [23]. Hence, in order to block confounding variables in cognitive experiments involving pupillometry as a dependent variable, light conditions and stimulus brightness have to be controlled during this experiment. However, in real driving conditions, it is hard to maintain in a certain level of illumination. So in the future, a prediction model implementation in real life application with considering the environmental factors is necessary, such as brightness, noise level. Moreover, the effectiveness of collision warning varied in response to driving conditions [7]. Additionally, the road condition and surrounding objects movements should be considered in the model as well for a better prediction accuracy. Moreover, the different driving habits should be studied by detailing drivers' response into steering, brake, push the gas, etc. Physiological data could be directly affected by the way how drivers choose to respond the collision avoidance warning, and it could be changed as the driving time goes by. The longer time driving will also be helpful to study the drivers' response change trend to the collision warning.

The findings of this study could contribute to improving the safety feature of the collision warning system. All these safety technologies in ADAS could be used as a combination of hardware (sensors, cameras, and radar for the surrounding environment) and software (algorithm to predict drivers' response) to help vehicles identify certain safety risks and understand drivers' status so they can provide a better device-user interaction in avoiding a crash.

References

1. Lee, J.D., et al.: Collision warning timing, driver distraction, and driver response to imminent rear-end collisions in a high-fidelity driving simulator. Hum. Factors **44**(2), 314–334 (2002)
2. Hojjati-Emami, K., Dhillon, B., Jenab, K.: Reliability prediction for the vehicles equipped with advanced driver assistance systems (ADAS) and passive safety systems (PSS). Int. J. Ind. Eng. Comput. **3**(5), 731–742 (2012)
3. Haas, E.C., Van Erp, J.B.: Multimodal warnings to enhance risk communication and safety. Saf. Sci. **61**, 29–35 (2014)
4. Yang, X., Kim, J.H.: Acceptance and effectiveness of collision avoidance system in public transportation. In: International Conference of Design, User Experience, and Usability. Springer (2018)
5. Parasuraman, R., Hancock, P.A., Olofinboba, O.: Alarm effectiveness in driver-centred collision-warning systems. Ergonomics **40**(3), 390–399 (1997)
6. Horowitz, A.D., Dingus, T.A.: Warning signal design: a key human factors issue in an in-vehicle front-to-rear-end collision warning system. In: Proceedings of the Human Factors and Ergonomics Society Annual Meeting. SAGE Publications Sage CA, Los Angeles (1992)
7. Abe, G., Richardson, J.: Alarm timing, trust and driver expectation for forward collision warning systems. Appl. Ergon. **37**(5), 577–586 (2006)
8. Yang, X., Kim, J.H.: The effect of visual stimulus on advanced driver assistance systems in a real driving. In: Proceedings of IIE Annual Conference. Institute of Industrial and Systems Engineers (IISE) (2017)
9. Ben-Yaacov, A., Maltz, M., Shinar, D.: Effects of an in-vehicle collision avoidance warning system on short-and long-term driving performance. Hum. Factors **44**(2), 335–342 (2002)
10. Maltz, M., Shinar, D.: Imperfect in-vehicle collision avoidance warning systems can aid drivers. Hum. Factors **46**(2), 357–366 (2004)
11. Lees, M.N., Lee, J.D.: The influence of distraction and driving context on driver response to imperfect collision warning systems. Ergonomics **50**(8), 1264–1286 (2007)
12. Peysakhovich, V., et al.: Pupil dilation and eye movements can reveal upcoming choice in dynamic decision-making. In: Proceedings of the Human Factors and Ergonomics Society Annual Meeting. SAGE Publications (2015)

13. Einhäuser, W., et al.: Pupil dilation reflects perceptual selection and predicts subsequent stability in perceptual rivalry. Proc. Nat. Acad. Sci. **105**(5), 1704–1709 (2008)
14. Tang, R., et al.: Indicating severity of vehicle accidents using pupil diameter in a driving simulator environment. In: International Conference on Digital Human Modeling and Applications in Health, Safety, Ergonomics and Risk Management. Springer (2018)
15. Parker, P., Englehart, K., Hudgins, B.: Myoelectric signal processing for control of powered limb prostheses. J. Electromyogr. Kinesiol. **16**(6), 541–548 (2006)
16. Arief, Z., Sulistijono, I.A., Ardiansyah, R.A.: Comparison of five time series EMG features extractions using Myo Armband. In: 2015 International Electronics Symposium (IES). IEEE (2015)
17. Phinyomark, A., Phukpattaranont, P., Limsakul, C.: Feature reduction and selection for EMG signal classification. Expert Syst. Appl. **39**(8), 7420–7431 (2012)
18. Lee, T.-S., et al.: Mining the customer credit using classification and regression tree and multivariate adaptive regression splines. Comput. Stat. Data Anal. **50**(4), 1113–1130 (2006)
19. Liao, J., Chin, K.-V.: Logistic regression for disease classification using microarray data: model selection in a large p and small n case. Bioinformatics **23**(15), 1945–1951 (2007)
20. Premachandra, I., Bhabra, G.S., Sueyoshi, T.: DEA as a tool for bankruptcy assessment: a comparative study with logistic regression technique. Eur. J. Oper. Res. **193**(2), 412–424 (2009)
21. Tomioka, R., Aihara, K., Müller, K.-R.: Logistic regression for single trial EEG classification. In: Advances in Neural Information Processing Systems (2007)
22. Kannan, S., Thangavelu, A., Kalivaradhan, R.: An intelligent driver assistance system (I-DAS) for vehicle safety modelling using ontology approach. Int. J. UbiComp **1**(3), 15–29 (2010)
23. Lotte, F.: A new feature and associated optimal spatial filter for EEG signal classification: waveform length. In: International conference on pattern recognition (ICPR) (2012)
24. Dreiseitl, S., Ohno-Machado, L.: Logistic regression and artificial neural network classification models: a methodology review. J. Biomed. Inform. **35**(5–6), 352–359 (2002)
25. Hosmer Jr., D.W., Lemeshow, S., Sturdivant, R.X.: Applied Logistic Regression, vol. 398. Wiley, Hoboken (2013)
26. Härdle, W., Simar, L.: Applied Multivariate Statistical Analysis, vol. 22007. Springer, Heidelberg (2007)

Gender Differences Measured on Driving Performances in an Urban Simulated Environment

Chiara Ferrante[1](✉), Valerio Varladi[2], and Maria Rosaria De Blasiis[1]

[1] Engineering Department, Roma Tre University,
Via V. Volterra 62, 00146 Rome, Italy
chiara.ferrante@uniroma3.it
[2] RISE - Research and Innovation for Sustainable Environment, Ltd.,
Via G. Trevis 88, 00147 Rome, Italy

Abstract. For decades, researchers have highlighted significant differences in gender and age groups in terms of driving behaviour. There are several studies, concerning gender, about the different perceptions of risk, traffic accident involvement and risky driving. The present research is a part of a research project about risk perception between genders that aims to estimate and quantify gender differences in terms of driving behaviour. In particular, this paper aims to focus on differences between male and female drivers in performing stopping maneuver in an urban environment. As confirmed by several medical studies, the gender difference involves cognitive and psychophysiological differences, which have a significant impact on the risk perception assessment. In the past, several researches have investigated gender differences on driving behaviour through questionnaire, statistical analysis and psychological driving task, but today technological advances have allowed the development of new tools to study the drivers' behaviour. In fact, many studies in the field of road safety were recently conducted with the virtual reality driving simulator. In this instance, the analysis is carried out through the virtual reality driving simulator, situated in the LASS3 Virtual Reality Laboratory of University Research Centre for Road Safety, by implementing a simulated scenario of an urban condition with many sudden events (e.g. intersection, pedestrian crossing, merging vehicle into the traffic, etc.) that can lead to a stopping maneuver. The LASS3 virtual reality driving simulator collects all the kinematic and dynamic driving measures (e.g. speed, acceleration, position respect to the lane, pedal pressure, etc.) with a frequency of 0.1 s. A sample of 40 drivers were subjected to the driving test and the results of three indicators of risk perception are studied in order to confirm the results of the previous analysis. The selected indicators for the analysis are: time to collision (TTC), pressure on brake pedal (PB) and slip ratio (SR). As results show, male and female drivers have a different behaviour in performing stopping maneuver. Even if they have the same perception of the potential risk of the road environment (TTC values), they perform in a very different way the stopping maneuver: men more carefully than women (PB, SR).

Keywords: Virtual reality driving simulation · Risk assessment · Driver behaviour · Human factor

© Springer Nature Switzerland AG 2020
D. N. Cassenti (Ed.): AHFE 2019, AISC 958, pp. 144–156, 2020.
https://doi.org/10.1007/978-3-030-20148-7_14

1   Background

The topic of gender differences is a matter of great interest for researcher from sociology, behavioural sciences, psychology and engineering. These differences, which have biological reasons, affect the lifestyles, the stress levels, more in general all the aspect of the human behaviour. The focus of this research is the different driving behaviour between genders which could lead to risky maneuvers and consequently to different traffic accidents involvement.

According to several sociology' researches the stress levels are different between genders, women are affected by higher stress levels than men. In particular, [1] have represented the trend of stress levels along a day and has underlined a different trend between genders, almost opposite, with two main peaks for women's trend in two different moments of the day: the first one in the mid morning and the second one in the afternoon. Regarding the first peak, it could be due [2] to high stress levels in the work place mainly women suffer because of the difficulty in reconciling work tasks and private life. A recent Italian study [3] shows that the difference in stress levels influences all the various aspect of human behaviour and, in this regard, [4] confirms that these differences in lifestyles could affect their risk perception, also during the driving task.

Many researches have investigated gender differences on driving behaviour through questionnaire, statistical analysis and psychological driving task. A result of a survey [5] shows that women report lower levels of confidence in their driving skills than men. Therefore, while women have less confidence in their driving skills and tend to avoid driving at night, in bad weather, with a lot of traffic, on highways [6], men are more confident and sometimes show a more aggressive behaviour [7, 8]. In fact, a five-year study [9] shows that men are more inclined to violate the road laws. This behaviour can be explained by analyze the force control capabilities (FCCs), as indicator of motor performance [10]. In fact, differences in FCCs can influences driving performance and control capabilities. Results highlight that the gender has a significant influence on FCCs and in particular, male performance are significantly better than female performance in all the FCCs. Another drivers' aggressiveness indicator [11] is the following distance. In fact, as shown by the results, a driving behaviour characterized by a close following distance, typical of men, shows impatience and aggressiveness.

Being the driving behaviour directly related to the accident involvement [12], the number of accidents is highest among men than women. It could be also due to female drivers drive less than males and consequently are less likely to be involved in road accidents. Nevertheless, women are more involved mainly in minor accidents: the percentage of minor accidents was 45% for females and 39% for males.

In the recent years, many studies in the field of road safety were conducted with the use of driving simulator in virtual reality. This tool allow to correlate the dynamic and kinematic characteristics of the driving with the different design parameters of the simulated scenarios (traffic flow and road geometry) as well to characterize the driving behaviour. In this regard, a research conducted in a simulation environment [13] has focused on the different visual scanning behaviour between genders during simulated driving using an eye tracking system. Furthermore, an age and gender analysis [14], based on a statistical analysis of percentage of the fatal accidents on two-lane highways, is directly related to overtaking maneuvers. The results show interesting and significant differences in the overtaking behaviour of drivers depending on their age and gender. These differences are mainly in the overtaking maneuvers frequency, overtaking time duration, following distances, critical overtaking gaps, and desired driving speed. In particular, men are more confident in overtaking maneuvers than women, but women have a longer overtaking time duration, about 0.9 s more.

In this regard, another study [15] is developed by our research group in order to quantify gender difference on driving behaviour. The input is an accident statistical analysis that have highlighted the most frequent risky maneuver in urban environment (rear off) in order to implement the environment conditions in virtual reality and analyze the behavioural aspects that lead to a rear off. In particular, a scenario with an intersection is built and each stopping maneuvers for all the drivers sample is studied through three risk indicators: time to collision, pressure on brake pedal, slip ratio. As results show, female drivers, although perceive in early the potential risk in the road environment respect men do, perform a delay stopping maneuver as shown by the values of brake pedal pressure that consequently return in slip risk maneuvers. On the other hand, men are more able to calibrate the braking with acceptable values of brake pedal pressure. Following this aim and in the same research project, the LASS3 group conducts the present study in order to extend the previous analysis to a several stopping maneuvers.

2   Aim of Study

The present research is developed in the field of the cited studies with the aim to estimate gender differences in terms of driving behaviour, focusing on stopping maneuver. Respect to the previous research, the purpose is to extend the analysis of the virtual reality driving experiment to other case of studies thanks to other simulated events that leads different stopping maneuvers with different approaching speed and reaction time.

For this reason, another simulated scenario in virtual reality is built and other tests drive are conducted. The new scenarios reproduce a real urban environment with many influences and stimuli to the driving behaviour (intersection with and without priority, pedestrian that cross the main road with and without pedestrian crossing, merging vehicle into the traffic from right and left) in order to study many times the driving performances in stopping maneuver. Furthermore, the same data analysis procedure based on the study of the outcomes measures from driving simulator of the previous study is applied in order to give response to the objectives set out above.

3   Methodology

In this instance, the present research carries out a virtual reality driving experiment keeping the outcomes of the statistical accidents analysis of the previous study that highlights the rear off as the most risky maneuver in urban area. Therefore, starting from the road layout and the traffic flow conditions of an urban area, a simulated scenario is built in order to test a sample of drivers, defined and validated from a statistical point of view. From all the outcome measures, the three significant indicators in terms of risk perception are analyzed. Below will be shown in detail the different phases of the methodology (Fig. 1).

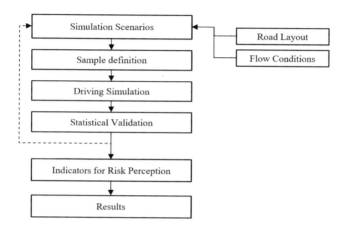

Fig. 1. Flow chart methodology

3.1   Simulation Scenarios

A simulated scenario is designed with the STISIM DRIVE® software in order to reproduce a typical condition of an urban area characterized by many sudden event, frequent intersections and some pedestrian crossing (Fig. 2). The layout is a typical urban road with two lanes of 3.5 m, ones for each direction. Regarding to the traffic flow, a value of Level of Service (LOS) has been determined and implemented in virtual reality scenario. This value correspond to a travel speed defined as a percentage of base Free-Flow Speed (FFS). With a travel speed of 30–35 km/h in an urban area charac-terized as described above and a FFS of 50 km/h, the corresponding LOS value is C.

Respect to the previous study, there is not difference in pavement condition and a value of real friction of a road pavement in good condition is imposed, equal to 0.6 (average of the two values imposed in the previous research: 0.8 for dry pavement and 0.4 for wet pavement).

3.2   Sample Definition

An homogeneous sample of subjects, recruited as volunteers from the Department of Engineering at the University Roma Tre, is composed by 40 subjects, 20 women and 20 men, 35 years old on average both for women than for men. None of the subjects had previous experience with driving simulator. Through Chauvenet statistical crite-rion, the number of participants has been verified from a statistical point of view and, and according to the criterion, no data has been rejected. In order to avoid biasing of results induced by driver attitude, experience in driving, age, stress phenomena, emotional state or neuro - cognitive status or by other subjective factors, the same driving conditions have been reproduced for each driver.

3.3   Driving Simulator

The driving simulator of the LASS3 (Fig. 2) is an advanced tool capable to investigate how different factors both external and internal to the driver can affect the driving risk perception. In particular, how traffic flow conditions combined with geometrical characteristics or the differences in age and gender can influence the drivers' behaviour [16] in terms of risk perception. The reliability of the tool has been fully validated, by several findings [17, 18], which also have demonstrated that the values carried out by simulation tests, are representative of a real drive.

From the hardware point of view, the simulator is a vehicle, a Toyota Auris, converted in a driving simulator by removing all unnecessary parts and integrating with the components that will communicate with the workstation computer, equipped with the software STISIM DRIVE®. Thanks to high-tech projectors, the images in front of the car and sideways are projected in order to cover a visual angle of 180°. Furthermore, the sound speakers are located in the hood of the car in order to emulate the acoustic environment at the best. Therefore, a good perception of the real conditions during the tests is provided. The outputs of the simulation tests are 45 parameters recorded with a frequency equal to 0,25 s that allow to estimate several kind of indicators.

Fig. 2. STISIM DRIVE® driving simulator of LASS3

3.4   Statistical Validation

A statistical validation of the outcomes measures of the driving tests is necessary to generalize the results. In this regard, according to findings in literature [19, 20] a one way analysis of variance (ANOVA) is applied in order to highlight the statistical significance of the indicators' values. The ANOVA test is used to research the statistically significant differences in the indicator values among the analyzed configurations. For each parameter, an analysis is performed to evaluate the effects due to the gender effects. The null hypothesis was that the average of the dependent variable is the same for the configuration being investigated. Rejecting the null hypothesis would mean that the independent variable (i.e. gender) influences the dependent variable. The results of ANOVA test will be show in the Sect. 4.1 ANOVA Results.

3.5   Indicators

According to the procedure of the previous research, three significant indicators in terms of risk perception assessment, focused on the stopping maneuver, are selected in order to estimate the differences in driving risk perception between genders.

Time to collision (TTC)

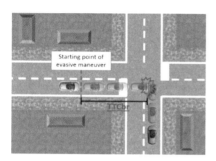

Fig. 3. Example of TTC

TTC is defined as "the time required for two vehicles to collide if they continue at their present speed and on the same path" [10]. Moreover, according with [21], TTC at the beginning of braking (TTCbr) represents the available maneuvering space when the stopping action starts (Fig. 3). The TTC allow to estimate the time since driver perceives the obstacle and releases until the obstacle itself (in the simulated scenario, it is represented for example by a vehicle at the intersection or a pedestrian that cross the road). Therefore, it deals with time that driver estimates to have before the collision with the obstacle. Obviously, in the calculation of this indicator, it is necessary to simplify the braking maneuver, for example, considering constant the speed value, that driver has before breaking down. High values of TTC indicates a good risk perception and prudent behaviour, on the contrary a low value indicates a poor risk perception and a greater risk exposure.

Pressure on the brake pedal

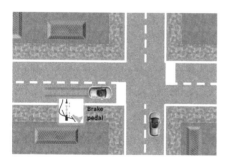

Fig. 4. Example of pressure on brake pedal

The pressure on the brake pedal (Fig. 4), measured in pounds, is an indicator that can give a significant information about how the drivers perform the stopping maneuver since perceive the obstacle [22]. If the driver slow down with a calibrated braking that does not require a high value of pressure on the brake pedal, it is possible to associate this behaviour to a prudent behaviour and a good risk perception. On the contrary, an high value could correspond to a delayed or wrong scan of the risk, that require an high pressure in order to stop before colliding with the obstacle. The value of pressure on brake pedal is measured when the driver start to braking, close to the obstacle.

Slip ratio

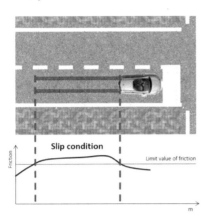

Slip condition

Limit value of friction

Friction

m

Fig. 5. Example of slip ratio

The last indicator is the slip ratio (Fig. 5), between the value of friction applied during braking and the value of friction offered by pavement (in this instance, 0.6). As shown by [23] the slip ratio is a widely diffused indicator that allows to investigate the risk conditions in acceleration and deceleration maneuvers. In this regard, the ratio is calculated by means of (1):

$$SL = \frac{Friction_{applied}}{Friction_{offered}} \qquad (1)$$

Values of SL greater than 1 represents the condition where the friction applied during a braking is grater then the friction offered by pavement with consequent loss of friction in the contact wheel– road. Slip ratios greater than 1 are likely to occur with a low risk perception condition when the driver did not perceive and estimate correctly the risk. On the contrary, values of SL less than 1 correspond to a calibrating braking in safe conditions, without slip.

4   Results and Discussion

The results of driving tests are analyzed by means of an Excel code in order to provide both a graphical and analytical information of indicators. For each indicator and for each drivers the x measure is calculated respect to the events that leaded to a stopping maneuver.

4.1   Anova Result

As mentioned before, a statistical analysis of data is develop in order to validate the results of the three risk perception indicators. In particular, the ANOVA test is applied on genders factors, for each indicators, using the F distribution, as shown in Table 1.

Table 1. ANOVA results

TTC (seconds)	Pressure on brake pedal (pounds)	Slip ratio
$F_{(1,210)} = 1.9$	$F_{(1,170)} = 40.6$	$F_{(1,165)} = 9.0$
$P = 0.20$	$P < 0.01$	$P < 0.01$

The results is satisfied because on the gender factors all the indicators is significant with high percentage of probability. The variation of values of pressure on brake pedal and the slip ratio between two groups depend on genders differences with 99% of probability (null hypothesis rejected with more than 99%), while for the TTC the difference between data is significant with a percentage of 80%.

Therefore, after checking a statistical significance of the results, below the discussion of the indicators.

4.2   Indicators Results

As opposed to several literature findings that attest differences in lifestyles could affect the risk perception in driving task between genders [1], the results of the TTC for male and female drivers are very similar. It means that both groups estimate the potential risks of the road environment in the same way and it could be due to a similar risk perception between genders that lead to react to stimuli around 6 s before the obstacle itself.

From a numerical point of view, the TTC average values are mf = 5.81 s; mm = 5.39 s and the standard deviation of both groups are comparable, a little bit grater for women (s = 2.74) than men (s = 1.65). This higher value for female drivers represents the higher variability of the female driving behaviour respect to male drivers, which will be confirmed in the following results (Fig. 6).

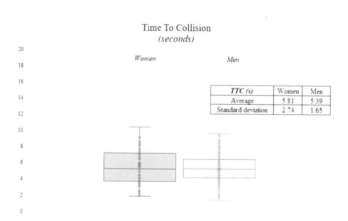

TTC (s)	Women	Men
Average	5.81	5.39
Standard deviation	2.74	1.65

Fig. 6. Time to collision

In the previous study, the TTC indicators showed (Table 2) that female drivers reacted in early respect to male do, paying attention to the pavement conditions. The differences are 0.7 s in the dry condition and 2.9 in wet condition.

Table 2. Results of [15] – TTC

TTC (seconds)	Female	Male
Dry condition	6.4	5.7
Wet condition	8.2	5.3

In this instance, without difference in imposed friction and assessing an higher numbers of different stopping maneuvers, the results seems matched in a quite average value between genders.

As well as the values of the TTC are similar between genders and consequently they have a similar risk perception and reaction, it is very different how they perform the stopping maneuver. The male drivers approach to a stopping maneuver with very lower values of pressure on brake pedal respect to female drivers. From a numerical point of view, the average of the pressure on brake pedal is m = 83.50 lb for women and m = 55.44 lb for men (33% of reduction), but the averages values, especially for female drivers, are affect by an high standard deviation of pressure values.

In fact the SD of female group is equal to s = 39.39, respect to s = 25.31 of the male one. This values distribution confirms the trend of the TTC in terms of variability of female driving behaviours (Fig. 7).

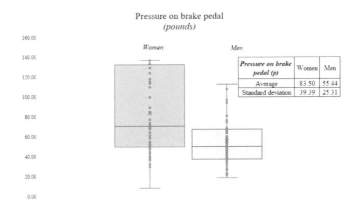

Fig. 7. Pressure on brake

Despite, women have different behaviours in the stopping maneuver performances within female sample, the average value of pressure on brake pedal means that female group perceive in a wrong way the needed space to perform a safe stopping maneuver. They realize in delay to brake, when the available space is not enough to perform a safe braking and they are lead to brake with higher values on brake pedal before the obstacle. On the other hand, men reach to estimate better the space available for the stopping maneuver and consequently they are able to calibrate braking pressure without avoiding the slip risk.

The same behaviour was found in the previous study (Table 3), where female drivers reached high values of pressure on brake pedal both in wet and dry condition respect to men do. The results showed an approximately increase of 32% in both pavement condition, similar to the value found in the present study.

Table 3. Results of [15] – Pressure on brake pedal

Pressure on brake pedal (pounds)	Female	Male
Dry condition	68	46
Wet condition	86	57

The last indicator results show the average of female maneuvers equal to slip threshold ($m = 1$). As mentioned before, a value of slip ratio greater than 1 represents the condition where the friction applied during a braking is grater then the friction offered by pavement.

Instead, the average of male group values is 0.95 and around the 20% of values are over the threshold and 80% below the threshold, in safety condition. The standard deviation of both groups are comparable ($sf = 0.11$; $sm = 0.18$), therefore there are not significant differences in driving behaviours within two groups (Fig. 8).

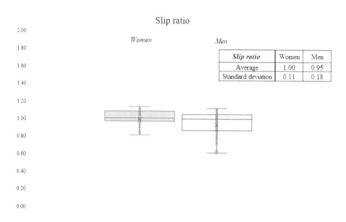

Fig. 8. Slip ratio

The results for this indicator are very significant. In fact, there is an evidence in the correlation between the pressure on the brake pedal and the likely to perform a risky maneuver in terms of slip. For female group the high values of pressure lead to perform all stopping maneuvers over the slip threshold, while for male group the calibrating braking assure a good percentage slip ratio values below the safety threshold.

As well as the pressure on brake pedal, the slip ratio values of the previous findings confirm the trend of the women to perform stopping maneuvers closer to limit threshold, than men do (Table 4). However, the previous results show values below the threshold both for male - female and wet - dry condition. In this regard, it is necessary to underline the case of study of the previous analysis, that is only an urban intersection. Increasing the cases of study of the performances of stopping maneuver the result is more exact and it highlights how the female behaviour is more inclined to perform risky braking.

Table 4. Results of [15] – Slip ratio

Slip ratio	Female	Male
Dry condition	0.58	0.44
Wet condition	0.83	0.56

The comparison shows that women are less careful in approach to an obstacle, despite a good perception in recognition of the object (time to collision), brake in delay when there is not enough time and space to stop in safety condition with result high value of pressure on brake pedal and lead to slip risk braking. On the contrary, men have a better risk perception than women have and estimate exactly the potential risks of the road environment, carrying out a calibrating braking that assure the 70% of times a safety stop.

5 Conclusion and Future Prospective

The present study is developed in the field of a project research aims to evaluate the differences of genders behaviour on the driving performances. It is well know that women and men have cognitive and psychophysiological differences, which have a significant impact on the risk perception assessment. This facet could be affect the driving behaviour and, in this regard, many studies is carried out in order to study how much this difference have an impact on driving performance.

In the previous study a statistical accidents analysis is developed in order to point out the most risky maneuver in an urban road environment. The result shows that the rear off is the most likely type of accident in an urban area due to many influences and stimuli that affect the driving behaviour in terms of attention and concentration. Therefore, the studied maneuver is the stopping, in the previous case referred to an intersection, in this study extending the case of study to different intersections and pedestrian crossings.

Starting from the standard procedure of a virtual reality driving experiment, the same analysis methodology is applied and three risk perception indicators are studied to understand the behaviour of women and men in the stopping maneuver performances.

Results show a different maneuver performing more than a different risk perception, in fact, the values of the time to collision demonstrate female and male drivers perceive and recognize the potential risk in the road environment in the same way, around 6 s before the obstacle itself. The space and the time to perform a safety maneuver is enough to calibrate the braking. Nevertheless, female drivers brake in the last part of this space with high values of pressure on brake pedal because they realize in delay that the available space is not sufficient to stop in safety condition. In addition, as shown by slip ratio results, the total of these maneuver exceed the safety threshold of the friction value and it exposes them to an high slip risk.

Contrariwise, men are more careful to spend in a correct way the available time and space since they perceive the obstacle, reaching it with a calibrating brake (low value of pressure on brake pedal) and pay attention to the friction offered by the pavement (slip ratio under the safety threshold).

Respect to the previous analysis, this result is more accurate because is validated on several stopping maneuvers condition, with different layout and different approaching speeds. Nevertheless, as mentioned, the present research is a part of a project. Therefore, the study could be extended to other aspects of driving, for example assessing the gender differences in overtaking maneuver with the same methodology to compare it, or studying a different road environment, or implementing an additional equipment (eye tracker, EEG, etc.) in order to see and estimate the differences both in psychophysiological and driving aspects at the same time.

References

1. Mariotti, R., Biapencheri, R., Dell'Orso, L.: Un approccio di genere alla salute. Atti del Convegno Da Esculapio a Igea. University Press, Pisa (2007)
2. Lavoro, A.E.: Stress legato all'attività lavorativa. Lussemburgo (2002)
3. Rosiello, A., Quarà, L.: Lo stress correlato al lavoro e le differenze di genere. Pianeta Lavoro e Tributi (2009)
4. Simon, F., Corbett, C.: Road traffic offending, stress, age, and accident history among male and female drivers. Ergonomics **39**, 757–780 (1996)
5. D'Ambrosio, L.A., Donorfio, L.K., Coughlin, J.F., Mohyde, M., Meyer, J.: Gender differences in self-regulation patterns and attitudes toward driving among older adults. J. Women Aging **20**, 265–282 (2008)
6. Rosenbloom, S., Herbel, S.: The safety and mobility patterns of older women. Public Works Manage. Policy **13**(4), 338–353 (2009)
7. Wickens, M., Mann, R., Stoduto, G., Anca, I., Smart, R.: Age group differences in self-reported aggressive driving perpetration and victimization. Transp. Res. Part F **14**(5), 400–412 (2011)
8. Hennessy, D.A., Wiesenthal, D.L., Wickens, C., Lustman, M.: The impact of gender and stress on traffic aggression: are we really that different? In: Focus on Aggression Research, pp. 157–174 (2004)
9. Fletcher, D.: A five year study of effects of fines, gender, race and age on illegal parking in spaces reserved for people with disabilities. Rehabil. Psychol. **40**, 203–210 (1995)
10. Hayward, J.: Near Miss Determination Through Use of a Scale of Danger. The Pennsylvania State University, Pennsylvania (1972)
11. Parker, D., Lajunen, T., Stradling, S.: Attitudinal predictors of interpersonally aggressive violations on the road. Transp. Res. Part F **1**, 11–24 (1998)
12. Laapotti, S., Keskinen, E., Rajalin, S.: Comparison of young male and female drivers' attitude and self-reported traffic behaviour in Finland in 1978 and 2001. J. Saf. Res. **34**(5), 579–587 (2003)
13. Pradhan, A.K., Li, K., Bingham, R., Simons-Morton, B.G., Ouimet, M.C., Shope, J.T.: Peer passenger influences on male adolescent drivers' visual scanning behavior during simulated driving. J. Adolesc. Health **54**, S42–S49 (2014)
14. Farah, H.: Age and gender differences in overtaking maneuvers on two-lane rural highways. Trasp. Res. Rec. 30–36 (2011)

15. De Blasiis, M., Ferrante, C., Veraldi, V., Moschini, L.: Risk perception assessment using a driving simulator: a gender analysis. Road Safety and Simulation – RSS Proceedings. The Hague (2017)
16. Underwood, G., Crundall, D., Chapman, P.: Driving simulator validation with hazard perception. Transp. Res. **14**(6), 435–446 (2011)
17. Bella, F.: Driving simulator for speed research on two-lane rural roads. Accid. Anal. Prev. **40**, 1078–1087 (2007)
18. Bella, F.: Validation of a driving simulator for work zone design. Transp. Res. Rec. J. Transp. Res. Board **1937**, 136–144 (2005)
19. Calvi, A., Benedetto, A., De Blasiis, M.: A driving simulator study of driver performance on deceleration lanes. Accid. Anal. Prev. **45**, 195–203 (2012)
20. Cuevas, A., Febrero, M., Fraiman, R.: An ANOVA test for functional data. Comput. Stat. Data Anal. **47**(1), 111–122 (2004)
21. Van der Horst, R.: A time-based analysis of road user behaviour in normal and critical encounters. Delft University of Technology (1990)
22. De Blasiis, M., Ferrante, C., Veraldi, V., Santilli, A.: Weaving lanes in different operating speed conditions: a driving simulator study. Advances in Transportation Studies: an international Journal, Section A 40 (2016)
23. Lee, C., Hedrick, K., Yi, K.: Real-time slip-based estimation of maximum tire–road friction coefficient. IEEE/ASME Trans. Mechatron. **9**(2), 454–458 (2004)

Train Intelligent Detection System Based on Convolutional Neural Network

Zining Yang[1(✉)], Virginia Cheung[1], Chunhai Gao[2],
and Qiang Zhang[2(✉)]

[1] Claremont Graduate University, Claremont, CA, USA
zining.yang@cgu.edu
[2] Traffic Control Technology Co.Ltd, Beijing, China
13601249262@163.com

Abstract. Autonomous driving train is an important component in the future rail transit system, as it can greatly improve the efficiency and safety of train operation. The most critical part in self-driving train is the active perception, due to the fact there's no active sensing system in the existing train control systems today, we hereby develop a "Train Intelligent Detection System" ("TIDS") to achieve reliable and real-time capable environment perception. The TIDS system consists three modules to simulate the real-world transit environment. The first module is to recognize the rail area by using semantic segmentation. In the second module, a convolutional neural network ("CNN") is used to identify the train in the image. We then use the third module to judge if the identified train affects normal train operation on the recognized rail area. The detection result of TIDS system has been tested in multiple real-world scenarios including but not limited to tunnel and turnouts environment, the results so far have been stable and positive.

Keywords: Rail safety · Rail area detection · Train detection

1 Introduction

The degree of automation in rail transit operation keeps on increasing, and we are getting more and more dependent on automation than ever before. In real world any of these software control systems will inevitably fail down from time to time unexpectedly, leaving us no choice but to relay entirely on manual drive mode in the case of main control system failure. In manual drive mode, the safety of the train fully depends on driver's driving skill and human eyesight. That's why manual drive can be considerably unreliable, more than 75% of world's major rail accidents happened when the drivers have to operate manually under tremendous pressure after the main system failure. If there can be any supporting control system to help drivers distinguish rail track and other abnormal obstacles on rail track, we can therefore expect great improvement of the safety level. Any safety improvement in train operation can give significant positive social impact, it can benefit our whole society with not only higher public safety and efficiency for passengers, but can also reduce potential economic loss from unexpected accidents caused by human errors.

© Springer Nature Switzerland AG 2020
D. N. Cassenti (Ed.): AHFE 2019, AISC 958, pp. 157–165, 2020.
https://doi.org/10.1007/978-3-030-20148-7_15

Despite remarkable theoretical progress achieved [1, 2] in the field of train active environmental perception in the past few years, many challenges remain due to the complexity of train environment, the first of which is the complex and dynamic rail. In the real train operating environment, the rail track is not all way straight, there are also curves, turnout and other scenes. The second challenge to overcome is the complexity of illumination. The light condition changes rapidly as the train moves, such as shadows, reflection, dark light and many other factors can all affect illumination. Fortunately, with the help of rapidly developed deep learning [3, 4] technics theses years, vision method becomes a potential solution for us.

We hereby developed a "Train Intelligent Detection System" ("TIDS") to overcome the aforementioned challenges, as a new application of Convolutional Neural Network ("CNN") model in rail transit industry to improve rail operation safety. This newly developed TIDS system is designed to perform beyond human capabilities to achieve train intelligent sensing and control, safety and reliability could be improved through numerous replicable operations. The system consists three modules to simulate the real-world transit environment. The first module is designed to identify the rail area through semantic segmentation. Based on the recognized rail area, CNN is used to identify the trains in the second module. Finally, whether the train detected in the second module can affect normal train operation is carried out elaborately in the third module. The structure of this system is shown in Fig. 1 below.

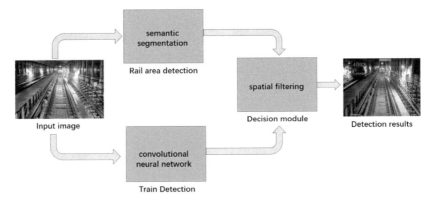

Fig. 1. Framework of the "TIDS" system

We benefit from this system for the following distinguish features. Based on the characteristics of each layer of the rail image, different characteristics on each layer are cascaded to ensure high efficiency of multi-level feature extraction and recognition in image processing. The system can also optimize driver's driving state simulation and learn different scenarios like curves, ramps, platforms, turnouts, and level crossings, to exert greater sensing and stability of the equipment to achieve high accuracy identification.

We start this paper with theoretical modelling first, followed by brief introduction of how rail area detection method is utilized in Sect. 2, methodological details of train detection in Sect. 3, determination process of whether the detected train can affect normal train operation in Sect. 4. We will then demonstrate our experiment results in Sect. 5 with final conclusion and discussion in Sect. 6.

2   Efficient Rail Area Detection

The core function of TIDS system is rail area detection, i.e. to determine the exact area of the rail tracks in the image efficiently and accurately. Accurate rail area detection can provide not only precise driving region of the train [5, 6], but can also delimit regions of interest for detecting other obstacles [7, 8]. In recent years, a major breakthrough has been made in the field of deep learning. This method is now being applied to many fields [9, 10], but its application in the field of railway is still lacking. The main difficulty and our focus to develop this TIDS system is to detect the rail area rapidly and accurately. We have found that semantic segmentation [8] could be a potentially good solution due to its strong ability to extract features of image, we thus applied this semantic segmentation method for rail area detection in our TIDS system, the structure of this network is shown in Fig. 2 below.

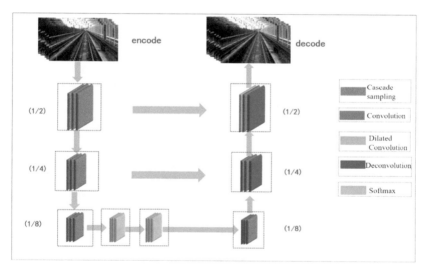

Fig. 2. Network architecture.

The efficient rail area detection network [8] is mainly divided into two parts: encoding part and decoding part. The encoding part is used to analyze the characteristics of the rail area, consisting a series of convolution layers and pooling layers. Dilated-cascade connection carried out the convolution operation in the feature maps of different resolutions, and the features of each layer are fused, the network makes full use of these information obtained from different sizes of the feature map. The characteristics between rail regions and the feature of a single rail are fully obtained by cascade sampling. In the following decoding part, the main purpose of decode segments is to upsample encoder's feature maps to match the original resolution. At the end of the network, a softmax layer is added, which is used to classify the pixels in the image thus to determine the rail areas. Furthermore, the residual connection [11] is applied in our network structure in order to obtain higher detection accuracy. The architecture of the network is shown in Table 1.

Table 1. The detailed layers architecture of the network.

Layer	Layer type	Number of feature map	Size of feature map
1	Cascade sampling	16	240 × 180
2–3	Convolution	64	240 × 180
4	Cascade sampling	128	120 × 90
5–7	Convolution	128	120 × 90
8	Cascade sampling	256	60 × 45
9–10	Convolution	256	60 × 45
11–12	Convolution (dilated 2)	256	60 × 45
13–14	Convolution (dilated 4)	256	60 × 45
15	Deconvolution	128	120 × 90
16–18	Convolution	128	120 × 90
19–21	Deconvolution	64	240 × 180
22–23	Convolution	64	240 × 180
24	Deconvolution	16	480 × 360
25	Softmax	2	480 × 360

3   Train Detection for TIDS

In the process of train operation, communication system malfunction can cause serious train collisions. Therefore, if we can detect other train or abnormal obstacle on rail track in advance, it can help us to prevent these kinds of potential collisions. For a better train identification result, we need to detect rail area first to provide an accurate boundary for further processing, then TIDS system continuous to work on detecting if there's any train within the defined rail area. Since accurate and fast train detection algorithm is the most critical part of the whole TIDS system, and we found that CNN performs well in the field of objection detection, thus, we applied a fast train recognition network for train detection in TIDS system, which has been applied on autonomous vehicle [12].

In order to meet the real-time application requirements of embedded platform, we take 18-layer residual net as base network. The full network of train detection is shown in Fig. 3. In our network, the deeper feature map can extract the global features of the train. On the other hand, the resolution of the deeper feature map is lower, which loses most of the original image information and make it difficult to detect small trains. Inspired by SSD [13], a fast object detection network, we choose several layers to predict default boxes of trains. We set the layers for prediction to res3b, res4b, res5b, Conv6_2, Conv7_2 and Conv8_2.

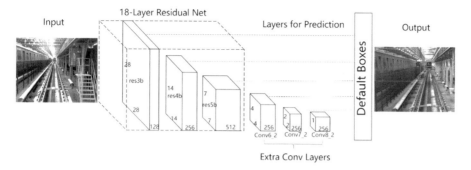

Fig. 3. The architecture of our detection network.

The scale of the default boxes progressively increases from the lowest feature map to the highest feature map. The formula to calculate the default boxes scale predicted by the k-th feature map is

$$s_k = s_{min} + \frac{s_{max} - s_{min}}{m - 1}(k - 1), \quad k \in [1, m]. \tag{1}$$

where s_{min} is the minimal value of the scale and s_{max} is the maximal value of the scale, m is the number of feature maps chosen for prediction. At the end of our network, we add three extra layers which is used for vehicle detection [12] to further improve our train detection accuracy, the detail of the extra layers is shown in Table 2.

Table 2. The detail of extra layers

Layer	Filter	Output size
Conv6_1	1 * 1 * 128	7 * 7 * 128
Conv6_2	3 * 3 * 256	4 * 4 * 256
Conv7_1	1 * 1 * 128	4 * 4 * 128
Conv7_2	3 * 3 * 256	2 * 2 * 256
Conv8_1	1 * 1 * 128	2 * 2 * 128
Conv8_2	2 * 2 * 256	1 * 1 * 256

4   Train Chosen Method

Based on the method proposed in module 2, we can detect trains as needed. However, in the actual operation of the train, not all trains detected affects train operation, like in the case of the train in adjacent rail as shown in Fig. 4. That's why we need to further determine whether the train detected can affect normal train operation.

Fig. 4. Train in adjacent rail.

We proposed to use spatial filtering algorithm to judge whether the train detected is dangerous as shown in Algorithm 1. The detailed procedure of the spatial filtering algorithm is summarized as follows:

Step 1: Calculate the lower boundary center point of the detected train, and this point is saved as point P.
Step 2: Determine whether the point P is in the rail area detected in module 2. If point P is in the rail area then the detected train is dangerous.
Step 3: Calculate the right and left lines of the rail area detected in the module 2. The right line l_r is expressed as l_r: $x = k_r * y + b_r$, and the left line l_l is expressed as l_l: $x = k_l * y + b_l$, where (x,y) represents the coordinates of the image and k_r, b_r, k_l, b_l represent the right and left lines parameters respectively.
Step 4: Judge whether point P is in the left of the line l_r, as well as in the right of the line l_l, if both are satisfied, the detected train is a dangerous object (Table 3).

Table 3. Spatial filtering algorithm

Algorithm 1: spatial filtering algorithm

Input: rail area, detected train bounding box
Output: whether the train detected is dangerous object

P: the lower boundary center point of the detected train
l_l: the left line of the rail area
l_r: the right line of the rail area
If **P not in the rail area** then
    if **P in the left of the l_r** and **P in the right of the l_l** then
      The train is dangerous object
    else
      The train is not dangerous object
else
      The train is dangerous object

---

5   Experiments

In order to make quantitative analysis of TIDS, we have collected a series of video frames in the real world and built a dataset. The dataset provides 3,000 images split up into a training, and test set (2,200 and 800 images, respectively), the resolution of the dataset is 1280 * 720.

To evaluate the train detection results which is in front of train, two important indicators are used to evaluate our algorithm including true positive rate (TPR) and false negatives rate (FNR). TPR is the proportion of the train detected correctly in all the train of all frames, and FNR is the proportion of the miss detected train in all trains of all frames. The formula of TPR and FNR is shown as follows:

$$TPR = \frac{\text{the train detected correctly}}{\text{the number of all target train}}. \tag{2}$$

$$FNR = \frac{\text{the miss detected trains}}{\text{the number of all target trains}}. \tag{3}$$

In addition, the detected train is correct only if the Intersection over Union (IoU) between the ground-truth bounding box and the detection result is greater than 0.5. The test result of our TIDS is shown in Table 4.

Table 4. Experiments result.

Metric	Result
TPR	98.8%
FNR	2.15%

To further visualize the effectiveness of TIDS system, rail area detection result is shown in Fig. 5, dangerous train detection result is shown in Fig. 6. From these figures, we can see the method we chose works well in different scenarios.

However, there are still some problems in some scenarios, e.g. when the train is far away from the camera, i.e. the train has very few pixels in the image, the detection result is not yet satisfied. Besides, when the train ahead turns on the train lights, the train detection result can be affected.

Fig. 5. Rail area detection results.

Fig. 6. Dangerous train detection results.

6   Conclusion

In this paper, we introduced an active rail transit collision avoidance system - TIDS. This TIDS system consists three main modules: rail area detection, train detection and the dangerous train chosen. The first module recognizes the rail area using the efficient rail area detection method. In the second module, CNN model is used to further identify the train based on the recognized rail area. We use the third module to define whether the detected train affects normal train operation. This TIDS system has been tested in lab for different scenarios, it has also been tested in real world, and then applied on operating metro lines in Beijing metro Yanfang line. Both the lab and on-site testing results have been so far stable and positive. To further improve this system, we are now working on image and lidar fusion, to combin the 2D image and 3D laser point cloud could offer much more information, for us to further leverage their different advantages to achieve pixel-level fusion of recognition and measurement.

References

1. Selver, M.A, Er, E.: Camera based driver support system for rail extraction using 2-D Gabor wavelet decompositions and morphological analysis. In: Proceedings of ICIRT, Birmingham, UK, pp. 270–275 (2016)
2. Wang, Z., Cai, B., Jia, C.: Geometry constraints-based visual rail track extraction. In: Proceedings of WCICA, Guilin, China, pp. 993–998 (2016)
3. Krizhevsky. A., Sutskever, I., Hinton, G.: ImageNet classification with deep convolutional neural networks. In: Proceedings of NIPS, Nevada, USA, pp. 1097–1105 (2017)
4. Szegedy. C., Liu, W., Jia, Y.: Going deeper with convolutions. In: Proceedings of CVPR, Boston, MA, pp. 1–9 (2015)
5. Nassu, B.T., Ukai, M.: A Vision-based approach for rail extraction and its application in a camera pan-tilt control system. IEEE Trans. Intell. Transp. Syst. **13**(4), 1763–1771 (2012)
6. Ruder, M, Mohler, N., Ahmed, F.: An obstacle detection system for automated trains. In: Proceedings of IEEE IV, Columbus, OH, USA, pp. 180–185 (2003)
7. Selver, M.A., Zoral, E.Y.: Predictive modeling for monocular vision based rail track extraction. In: Proceedings of IEEE CISP-BMEI 2017, Shanghai, China, pp. 1–6 (2017)
8. Wang, Z., Wu, X., Yu, G.: Efficient rail area detection using convolutional neural network. IEEE Access **6**, 77656–77664 (2018)
9. Li, X., Liu, Z., Luo, P.: Not all pixels are equal: difficulty-aware semantic segmentation via deep layer cascade. In: Proceedings of CVPR, Honolulu, HI, USA, pp. 6459–6468 (2017)
10. Ren, S., He, K., Ross, G.: Faster R-CNN: towards real-time object detection with region proposal networks. IEEE Trans. Pattern Anal. Mach. Intell. **39**(6), 1137–1149 (2017)
11. He, K., Zhang, X., Ren, S.: Deep residual learning for image recognition. In: Proceedings of CVPR, Las Vegas, NV, USA, pp. 770–778 (2016)
12. Hu, C., Wang, Y., Yu, G., Wang, Z.: Embedding CNN-based fast obstacles detection for autonomous vehicles. In: Proceedings of Intelligent and Connected Vehicles Symposium, Kunshan, China, pp. 1–11 (2018)
13. Liu, W., Anguelov, D., Erhan, D.: SSD: single shot multibox detector. In: Proceedings of ECCV, Amsterdam, The Netherlands, pp. 21–37 (2016)

Driving Simulator Validation of Surface Electromyography Controlled Driving Assistance for Bilateral Transhumeral Amputees

Edric John Nacpil[1(✉)], Tsutomu Kaizuka[1], and Kimihiko Nakano[1,2]

[1] Institute of Industrial Science, The University of Tokyo, 4-6-1 Komaba,
Meguro-ku, Tokyo 153-8505, Japan
{enacpil, tkaizuka, knakano}@iis.u-tokyo.ac.jp
[2] Interfaculty Initiative in Information Studies, The University of Tokyo,
7-3-1 Hongo, Bunkyo-ku, Tokyo 113-0033, Japan

Abstract. A driving assistance interface controlled by surface electromyography signals from the biceps brachii muscles has been developed to enable bilateral transhumeral amputees to accelerate, brake, and steer at low vehicle speeds. Driving simulator trials were conducted as a pilot study to validate the path following accuracy of the interface with respect to a conventional steering wheel and pedals interface. Human drivers used the interfaces to execute a circular 270° right turn with a radius of curvature equal to 3.6 m. The driving assistance interface and conventional interface had intertrial median lateral errors of 0.6 m and 0.5 m, respectively. A statistical analysis of the driving simulator data indicated that the driving assistance interface was comparable to the conventional interface. Based on the validated accuracy of the driving assistance interface, the investigators planned to further develop the interface to perform path following validation for an actual automobile.

Keywords: Automated driving · Surface electromyography ·
Human-machine interface · Assistive technology

1 Introduction

Bilateral transhumeral amputation resulting in residual upper limbs above each elbow can lead to challenges with daily activities such as tying shoe laces, eating a sandwich, and driving automobiles [1–3]. Drivers often learn to use their feet in place of their hands to accomplish driving tasks [4]. One legal challenge in Korea is the prohibition of driving with a transhumeral amputation, unless the automobile has been designed and constructed to accommodate the amputation [3]. In some countries, permission can be granted to install assistive technology in automobiles, such as joysticks that control steering and acceleration, although automobiles operated in Japan are required to retain functioning steering wheels and pedals for nondisabled drivers [4, 5]. Aside from joysticks, various gesture interfaces have been proposed for automobiles, including motion detection and strain gauge signals [6]. An interface employing surface

© Springer Nature Switzerland AG 2020
D. N. Cassenti (Ed.): AHFE 2019, AISC 958, pp. 166–175, 2020.
https://doi.org/10.1007/978-3-030-20148-7_16

electromyography (sEMG) signals from the forearm has also been proposed to recognize driver intentions through neural network technology [7]. However, there is a paucity in the literature concerning sEMG controlled automobile interfaces that assist bilateral transhumeral amputees who have no forearms. Since the biceps brachii muscles of transhumeral amputees are typically able to provide sEMG signals, a driving assistance interface was designed for controlling steering, braking, and acceleration with sEMG signals resulting from the isometric contraction of biceps brachii muscles [1]. The interface relied on sEMG signals resulting from the isometric contraction of the right arm biceps brachii to control the steering wheel angle (SWA), whereas braking and acceleration were controlled with sEMG signals from the isometric contraction of the left arm biceps brachii. The feet of the drivers were therefore free to do other tasks besides driving, since feet were frequently used as hands.

The current study pertains to the effect of the sEMG interface on the cornering ability of a vehicle on residential streets with a maximum speed limit of 30 km/h [8]. Since path following is a primary driving task during cornering, the path following accuracy of the interface was validated in a driving simulator with pre-programmed acceleration, braking, and steering in order to simulate the implementation of the interface in a partially automated vehicle [9]. The sEMG interface was validated because it was comparable in accuracy to a steering wheel and pedals interface.

In order to elaborate on the development of the sEMG interface, Sect. 2 pertains to the design of the interface, whereas Sect. 3 provides an overview of the equipment used to implement the interface for driving simulator trials. Section 4 details the experimental protocol for validating the interface with driving simulator data. Based on the results and discussion in Sect. 5, a conclusion is provided in Sect. 6.

2 Design of the Driving Assistance Interface

When the biceps brachii muscle contracts, the consequent sEMG signals could control devices such as prosthetic limbs and automobiles [10–12]. For the proposed driving assistance interface, sub-second isometric contractions from the left arm biceps brachii wirelessly controlled braking and acceleration, whereas the same type of isometric contractions from right arm biceps brachii wirelessly adjusted the SWA (Fig. 1) [13].

Since the driving assistance interface is designed to be easily implemented relative to some previously developed sEMG interfaces, a bipolar electrode configuration consisting of three dry electrodes per muscle is utilized instead of a sEMG pattern recognition configuration possibly consisting of dozens of electrodes [14]. Each armband in Fig. 1 measures the sEMG signal of a biceps brachii with a pair of bipolar electrodes on the muscle belly and a ground electrode on the lateral side of the muscle [15, 16]. In order to mitigate the effects of motion artifacts and electromagnetic interference on the measurement accuracy of sEMG signals, the driving assistance system recognizes signals above a threshold that is set to a percentage of the peak amplitude [17]. Since inter-subject sEMG amplitude variability affects the detection ability of sEMG measurement equipment, the threshold is calibrated for each driver to ensure that sEMG input is readily received by the driving assistance system [18].

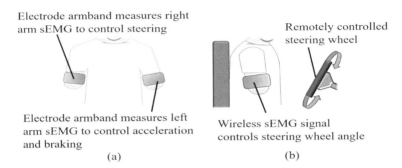

Electrode armband measures right arm sEMG to control steering

Remotely controlled steering wheel

Electrode armband measures left arm sEMG to control acceleration and braking

Wireless sEMG signal controls steering wheel angle

(a)   (b)

Fig. 1. (a) Driving assistance interface receives sEMG input from isometric contractions of biceps brachii muscles. (b) Interface input is transmitted wirelessly to perform driving tasks such as steering.

Acceleration and braking are controlled by isometric contraction of the left arm biceps brachii as described in Table 1. If the automobile is at a full stop, contraction initiates acceleration. A subsequent contraction prompts the computer to decelerate by adjusting the throttle. Further contraction during deceleration initiates braking. Regardless of whether the driver provides appropriate input for driving conditions, the extent to which acceleration, deceleration, and braking are applied is determined by an onboard vehicle automation computer based on information about the driving environment. This information can be obtained from computer vision sensors, GPS tracking, or other technologies that are utilized by an automated vehicle [19]. Hence, the driving assistance system is an instance of Level 4 automation as defined by SAE International [20].

Table 1. Relationship between vehicle motion and sEMG input from isometric contraction of left arm biceps brachii.

sEMG input	Vehicle motion output
Contraction when car at full stop	Initiate acceleration
Contraction during vehicle acceleration	Initiate deceleration
Contraction during deceleration	Initiate braking

Although a vehicle with the driving assistance system could execute turns autonomously, the driver has the option to intervene by inputting steering commands with sEMG signals of the right arm biceps brachii (Table 2). Two isometric contractions of the biceps brachii within a period of 1 s toggles between rightward and leftward steering wheel rotation. Turn signal indicator lights in a visible dashboard location, such as the instrument cluster, will flash to indicate the selected direction. In order to initiate a turn, one isometric contraction within a period of 1 s rotates the steering wheel in the selected direction. The initial angle of the steering wheel changes from the neutral position, i.e. a SWA of 0°, to a final SWA determined by the vehicle automation computer. One subsequent isometric contraction within a period of 1 s returns the steering wheel to the neutral position.

Table 2. Relationship between steering and sEMG input from isometric contraction of right arm biceps brachii.

sEMG input	Steering output
Two contractions within 1 s	Toggle between leftward and rightward steering direction.
One contraction within 1 s with steering wheel in neutral position	Rotate steering wheel in selected direction.
One contraction within 1 s with steering wheel rotated in selected direction	Rotate steering wheel to neutral position.

3 Experimental Equipment

In accordance with previous studies, the biceps brachii of nondisabled test subjects were used to approximate the isometric contraction of transhumeral amputees [21, 22]. The forearms of the test subjects were stabilized during isometric contraction with a pair of clamps (Fig. 2). Adhesive silver-silver chloride (Ag/AgCl) wet electrodes were mounted on the biceps brachii muscles, in accordance with SENIAM recommendations (Surface EMG for the Non-Invasive Assessment of Muscles) [23]. As a more affordable and readily available substitute for dry electrodes, the wet electrodes would facilitate interface design iteration and experimental replicability [24]. Signals measured by the electrodes controlled the virtual SWA of the driving simulator (Digital Battlespace 2™), rather than the SWA of a physical steering wheel, to ease the implementation of the interface.

The driving simulator was executed on a Windows platform laptop (Panasonic CF-LX6 laptop, 14 inch 1920 × 1080 resolution screen) (Fig. 2). Two custom-built

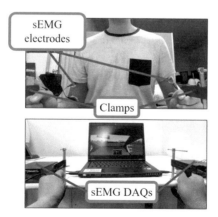

Fig. 2. Operation of driving simulator with driving assistance interface (top) and view of driving assistance interface from perspective of driver (bottom).

sEMG data acquisition devices (DAQs) were used to convert the signals into USB joystick signals that were recognized by the driving simulator. A custom circuit was used to amplify sEMG signals with a gain of 5000 and to filter sEMG signals outside the range of 2 Hz to 530 Hz. An Arduino Uno R3 microcontroller averaged and rectified the signals with a sampling rate of 10 kHz and a moving average window of 50 data points. The microcontroller employed the software and firmware package, UnoJoy!, to convert the processed signals into USB joystick commands that were sent to the laptop [25]. The game controller calibration program packaged with Windows 10 was used to determine the maximum input signal, whereas JoyToKey software on the laptop was used to set the input threshold from 10 to 30% of the peak input signal [26]. Calibrating the threshold for each test subject mitigated the effect of inter-subject sEMG amplitude variability on the input detection ability of the laptop. Since the driving simulator only accepted keyboard commands to control the steering wheel rate (SWR) of the virtual car, the JoyToKey software converted the joystick commands into keystrokes.

The SWR of the driving simulator was configured to approximate the SWR of 720 deg/s that is provided by some commercially available steering actuators [27]. Given that the maximum SWA of the virtual car was $65°$, dividing the maximum SWA by the SWR yields a period of 0.1 s for the car to transition between a longitudinal trajectory to a circular trajectory with a radius of curvature equal to the minimum turning radius of the car. Since this period is shorter than the previously observed minimum period of 0.268 s for human drivers, it was expected that the driving assistance interface would enable drivers to initiate turns faster than a steering wheel interface [28]. A force feedback game steering wheel (Driving Force™ GT) and pedals were compared to the driving assistance interface with respect to path following accuracy. The game steering wheel had a steering ratio of 1:1 and came with a set of pedals to control braking and acceleration.

4   Methodology

The subsections that follow pertain to the experimental protocol for the driving simulator trials. Sections 4.1 to 4.3 describe the design and execution of the trials, whereas Sect. 4.4 describes the analytical methods applied to data from the trials.

4.1   Test Subject Recruitment

Two nondisabled test subjects, consisting of one female and one male, ages 24 and 28 respectively, were recruited from personnel at The University of Tokyo. Both test subjects had previous experience with a driving simulator as well as an average of 5 years of actual driving experience. The test subjects provided written consent to participate in experimental trials approved by the Ethics Committee of the Interfaculty Initiative in Information Studies, Graduate School of Interdisciplinary Information Studies, The University of Tokyo (No. 14 in 2017).

4.2 Driving Scenario

All test subjects were instructed to complete a 270° turn with a radius matching the minimum turning radius of the virtual car in the driving simulator (Fig. 3).

Since the sEMG interface was intended to be operated at or below the speed limit of 30 km/h, the automated acceleration of the interface was configured in the driving simulator so that the vehicle would reach a constant speed of 20 km/h during the turn and a maximum speed of 25 km/h throughout the entire scenario. This type of acceleration simulates the ability of vehicles with Level 4 automation to control acceleration so that excessive speed is avoided in various driving scenarios [20].

In driving scenarios that allow an automobile with front steering to follow a circular path with a minimum constant speed as well as constant SWA and lateral position, path following accuracy could be maximized, if the front road wheels are steered to follow the circular path [29]. The tested driving scenario allows for these steady-state conditions to occur, since there is a circular path for the simulated vehicle to travel at minimum constant speed, while maintaining a constant lateral position and SWA.

It was expected that the sEMG interface could appropriately orient the front road wheels faster than the game steering wheel with a transition time of 0.1 s, as opposed to the estimated time of 0.268 s for the game steering wheel [28]. Since the shorter

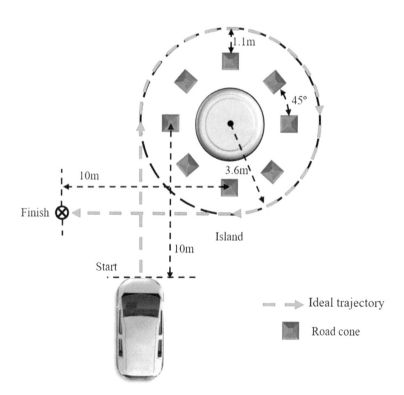

Fig. 3. Overhead view of driving scenario performed by test subjects in driving simulator.

transition time of the sEMG interface could enable the vehicle to reach steady-state sooner into the turn than the steering wheel and pedals interface, the sEMG interface was hypothesized to be associated with the highest path following accuracy between the two interfaces.

4.3 Test Subject Training and Driving Simulator Trials

Each test subject was trained to use the sEMG interface as well as the steering wheel and pedals interface. Training consisted of a presentation containing written directions and videos of an expert driver demonstrating the proper operation of each interface. Then, for the sEMG interface, test subjects were in a seated position with elbows at approximately 90° of flexion (Fig. 2). With the distance between the clamps of the sEMG interface being approximately equal to the shoulder width of a given test subject, the 270° turn had to be successfully performed twice. The same turn then had to be successfully performed twice with the steering wheel and pedals interface. No data was recorded throughout the training sessions.

For the initial experimental trial, a given test subject was allowed five attempts to complete the 270° turn with either the sEMG interface or the steering wheel and pedals interface. During the next trial, five attempts were allowed to complete the same turn with the remaining interface (Fig. 3). The test subjects were randomly assigned so that one test subject completed the first trial with the sEMG interface, whereas the other test subject completed the first trial with the steering wheel and pedals interface. Data from the first three successful attempts of each trial were used for analysis to ensure that enough data was available for processing.

4.4 Data Analysis

Path following accuracy was calculated in several steps. First, the minimum distance between the trajectory of the center of the front bumper and the edge of a given road cone was determined to be 1.1 m as shown in Fig. 3. The ideal trajectory was defined as the circumference of a circular path with a minimum distance of 1.1 m from the edge of any of the seven road cones along the trajectory. Lateral error for a given road cone was calculated by finding the absolute value of the difference between the minimum distance of the ideal trajectory from the cone and the minimum distance of the actual trajectory from the cone. This error calculation was performed seven times per trial, since there were seven road cones along the ideal trajectory. The path following accuracy for each interface was calculated as the intertrial median lateral error.

Since the Shapiro-Wilk test indicated that some of the data distributions for efficiency and accuracy were non-normal, the nonparametric Friedman test was employed to compare the sEMG interface to the steering wheel and pedals interface [30, 31]. If the p-value was less than 0.05, then it was determined that there was a statistically significant difference between the two interfaces. Otherwise, the interfaces would be considered as being comparable to each other. The sEMG interface would be validated, if it was at least comparable to the steering wheel and pedals interface.

5   Results and Discussion

The intertrial median lateral error for the 270° turn is 0.5 m for the steering wheel and pedals interface and 0.6 m for the sEMG interface (Fig. 4). The interfaces are comparable to each other because the Friedman test p-value is greater than 0.05. Therefore, the path following accuracy of the sEMG interface is validated.

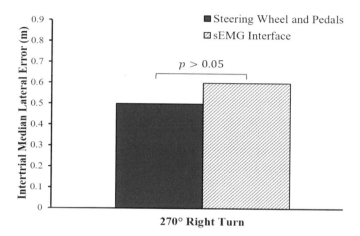

Fig. 4. Comparison of sEMG interface to steering wheel and pedals interface with respect to path following accuracy, i.e. intertrial median lateral error.

As discussed in Sect. 4.2, the sEMG interface was expected to be the most accurate interface because it could steer the front road wheels faster than the game steering wheel. However, this advantage may have been mitigated by understeer during the turn. Regardless of which interface is being used, longitudinal acceleration causes centrifugal acceleration on the vehicle, thereby resulting in understeer [29]. In the case of the steering wheel and pedal interface, understeer correction could be applied through the release of the accelerator pedal to lower centrifugal acceleration, but accelerator pedal correction is not possible when operating the sEMG interface. Furthermore, even though it was possible to decelerate with the sEMG interface in the event of understeer, the test subjects may have mostly relied on the driving assistance interface to determine the amount of acceleration during the turn. Although the sEMG interface would allow the vehicle to reach steady state conditions sooner, its inability to correct for understeer automatically may have contributed to its path following accuracy being comparable, rather than superior, to the steering wheel and pedal interface. On the other hand, it is possible that the test subjects would learn to compensate for understeer by using the sEMG interface to decelerate or initiate the turn earlier. Driving simulator trials with more attempts could be analyzed to determine whether any learning effects would increase path following accuracy.

One potential method for improving the path following accuracy of the sEMG interface is to realize automatic acceleration that adjusts to lessen the effect of understeer. The operation of the interface could also be improved by converting the wet sEMG electrode configuration into an wireless armband of dry electrodes that increases mounting efficiency, prevents electrode wires from hindering driver movement, and eliminates the need to clean residual adhesive and conductive gel from wet electrodes [16].

6 Conclusion

A driving assistance interface for bilateral transhumeral amputees was validated in the context of a circular 270° turn. Relative to a game steering wheel and pedals, the driving assistance interface had comparable path following accuracy. Since the results of this pilot study support sEMG controlled driving assistance as a substitute for a conventional steering wheel and pedal interface, further development of the driving assistance interface has been planned. One objective for future work is to further develop the automatic acceleration of the driving assistance interface to address understeer. Future trials involving an actual automobile, an increased number of attempts from each subject, subsequent interface iterations, a larger number of test subjects, and a variety of driving conditions would generate more data to determine the feasibility of the driving assistance interface.

Acknowledgments. This study was funded in part by the Otsuka Toshimi Scholarship Foundation.

References

1. Pierrie, S.N., Gaston, R.G., Loeffler, B.J.: Current concepts in upper-extremity amputation. J. Hand Surg. Am. **43**, 657–667 (2018)
2. Sullivan, R.A., Celikyol, F.: Post hospital follow-up of three bilateral upper-limb amputees. Orthot. Prosthet. **28**, 33–40 (1974)
3. Jang, C.H., Yang, H.S., Yang, H.E., Lee, S.Y., Kwon, J.W., Yun, B.D., Choi, J.Y., Kim, S. N., Jeong, H.W.: A survey on activities of daily living and occupations of upper extremity amputees. Ann. Rehabil. Med. **35**, 907–921 (2011)
4. Davidson, J.H., Jones, L.E., Cornet, J., Cittarelli, T.: Management of the multiple limb amputee. Disabil. Rehabil. **24**, 688–699 (2002)
5. Östlund, J.: Joystick-controlled cars for drivers with severe disabilities. Technical report, Swedish National Road Administration (1999)
6. Murata, Y., Yoshida, K., Suzuki, K., Takahashi, D.: Proposal of an automobile driving interface using gesture operation for disabled people. In: The Sixth International Conference on Advances in Computer-Human Interactions (2013)
7. Kwak, J., Jeon, T.W., Park, H., Kim, S., An, K.: Development of an EMG-based car interface using artificial neural networks for the physically handicapped. Korea IT Serv. J. **7**, 149–164 (2008)
8. Dinh, D.D., Kubota, H.: Profile-speed data-based models to estimate operating speeds for urban residential streets with a 30 km/h speed limit. IATSS Res. **36**, 115–122 (2013)
9. Ackermann, J.: Robust control prevents car skidding. IEEE Control Syst. Mag. **17**, 23–31 (1997)

10. Xu, Y., Zhang, D., Wang, Y., Feng, J., Xu, W.: Two ways to improve myoelectric control for a transhumeral amputee after targeted muscle reinnervation: a case study. J. Neuroeng. Rehabil. **15**(1), 37 (2018)
11. Pulliam, C.L., Lambrecht, J.M., Kirsch, R.F.: Electromyogram-based neural network control of transhumeral prostheses. J. Rehabil. Res. Dev. **48**, 739–754 (2011)
12. Nacpil, E.J., Zheng, R., Kaizuka, T., Nakano, K.: Implementation of a sEMG-machine interface for steering a virtual car in a driving simulator. In: AHFE 2017 International Conference on Human Factors in Simulation and Modeling (2018)
13. Yamazaki, Y., Suzuki, M., Mano, T.: Pulse control during rapid isometric contractions of the elbow joint. Brain Res. Bull. **34**, 519–531 (1994)
14. Ando, K., Nagata, K., Kitagawa, D., Shibata, N., Yamada, M., Magatani, K.: Development of the input equipment for a computer using surface EMG. In: IEEE Engineering in Medicine and Biology Society (2006)
15. Neilson, P.D.: Frequency-response characteristics of the tonic stretch reflexes of biceps brachii muscle in intact man. Med. Biol. Eng. **10**, 460–472 (1972)
16. Raez, M.B.I., Hussain, M.S., Mohd-Yasin, F.: Techniques of EMG signal analysis: detection, processing, classification and applications. Biol. Proced. Online **8**, 11–35 (2006)
17. De Luca, C.J.: The use of surface electromyography in biomechanics. J. Appl. Biomech. **13**, 135–163 (1997)
18. Bekey, G.A., Chang, C.-W., Perry, J., Hoffer, M.M.: Pattern recognition of multiple EMG signals applied to the description of human gait. Proc. IEEE **65**, 674–681 (1977)
19. Kukkala, V.K., Tunnell, J., Pasricha, S., Bradley, T.: Advanced driver-assistance systems: a path toward autonomous vehicles. IEEE Consum. Electron. Mag. **7**, 18–25 (2018)
20. J3016. Technical report, SAE International (2014)
21. Alshammary, N.A., Dalley, S.A., Goldfarb, M.: Assessment of a multigrasp myoelectric control approach for use by transhumeral amputees. In: 2012 Annual International Conference of the IEEE Engineering in Medicine and Biology Society (2012)
22. Ruhunage, I., Perera, C.J., Nisal, K., Subodha, J., Lalitharatne, T.D.: EMG signal controlled transhumerai prosthetic with EEG-SSVEP based approach for hand open/close. In: 2017 IEEE International Conference on Systems, Man, and Cybernetics (SMC) (2017)
23. Hermens, H., Freriks, B.: The state of the art on sensors and sensor placement procedures for surface electromyography: a proposal for sensor placement procedures. In: SENIAM, Enschede (1997)
24. Searle, A., Kirkup, L.: A direct comparison of wet, dry and insulating bioelectric recording electrodes. Physiol. Meas. **21**, 271–283 (2000)
25. Chatham, A., Walmink, W., Mueller, F.: UnoJoy!: a library for rapid video game prototyping using Arduino. In: CHI 2013 Extended Abstracts on Human Factors in Computing Systems (2013)
26. JoyToKey. https://joytokey.net/en/
27. Forkenbrock, D., Elsasser, D.: An assessment of human driver steering capability. Techinical report, National Highway Traffic Safety Administration (2005)
28. Pandis, P., Prinold, J.A.I., Bull, A.M.J.: Shoulder muscle forces during driving: sudden steering can load the rotator cuff beyond its repair limit. Clin. Biomech. **30**, 839–846 (2015)
29. ISO 4138:2012(E). Technical report, International Standards Office (2012)
30. Shapiro, S.S., Wilk, M.B.: An analysis of variance test for normality (complete samples). Biometrika **52**, 591–611 (1965)
31. Friedman, M.: The use of ranks to avoid the assumption of normality implicit in the analysis of variance. J. Am. Stat. Assoc. **32**, 675–701 (1937)

Evaluating Rear-End Vehicle Accident Using Pupillary Analysis in a Driving Simulator Environment

Rui Tang$^{(\boxtimes)}$ and Jung Hyup Kim

Department of Industrial and Manufacturing Systems Engineering,
University of Missouri, Columbia, USA
rtgv4@mail.missouri.edu, kijung@missouri.edu

Abstract. This study analyzed the driver's pupil diameter changes involved in rear-end car accidents. The purpose of this research is to understand the relationship between driving behaviors and changes in pupil diameter. The diameter of pupil changes not only in response to light, but also in response to other factors involving both cognitive and emotional activities. In this study, we evaluated rear-end vehicle accidents from drivers' psychological perspective by using pupillary analysis. We hypothesized that the patterns of pupil diameter changes could serve as an indicator to evaluate the driver's behaviors responding to a dangerous situation. An eye-tracking technology was used to collect pupil diameter and records the events of car accidents.

Keywords: Pupil diameter · Driving simulator · Eye tracking · Driving safety

1 Introduction

Through the analysis of vehicle accidents, we can advance our understanding related to the causes of the accidents and develop the ways to prevent them. Driving is a highly complexed dynamic task that requires continual integration of human perception, cognition, and motor response. For that reason, most of car accidents are caused by human errors. According to Gong and Yang [1], traffic accidents relevant to drivers' misjudgment and miss-operation take up about 80%-85% of all accidents. Therefore, it is vital to study how drivers respond when they are in the course of vehicle accidents. However, most of the research related to drivers' behavior analysis are focused on the driver's different physical responses in various pre-accident conditions. Only few articles investigated the pattern of driver's physiological data (i.e., eye movement, EEG, EMG, blood pressure, and skin conductance). Among them, many studies show that eye-tracking data could reveal hidden relations between driver's behavior and vehicle accident [2–6]. Hence, in the present study, the pupillary response data was used to evaluate different driving behaviors corresponding to the vehicle accidents. Also, according to Sivak [7], 80%–90% of driving information comes from visual cues in driving. Therefore, in this study, we hypothesized that the pattern of pupil diameter changes could be used as an indicator to evaluate the driver's behaviors responding to a rear-end vehicle accident.

© Springer Nature Switzerland AG 2020
D. N. Cassenti (Ed.): AHFE 2019, AISC 958, pp. 176–186, 2020.
https://doi.org/10.1007/978-3-030-20148-7_17

2   Method

2.1   Apparatus

In this study, the experiment was conducted in a driving simulator environment. OpenDS simulation engine (see Fig. 1) was used to create different experimental conditions. It is a cross-platform and open-source driving simulation software [8]. In this study, some of driving modules were redesigned and modified to make realistic driving environments. Logitech G27, which is shown in the middle of Fig. 1 was used as a steering wheel, brake, and accelerator [9]. They were connected to the OpenDS driving simulator.

For the eye-tracking device, Tobii Pro Glasses 2 (see Fig. 1) was used to collect the data related to the driver's eye movements and recorded all the details of the surrounding environments. It is a wearable device with a frame rate of 100 Hz.

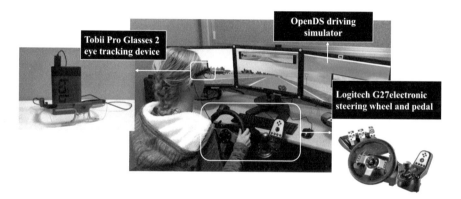

Fig. 1. Apparatus of the experiment

2.2   Participants

A total of forty-three college students (mean age: 21.28, standard deviation: 1.67) participated in this experiment (male: 38 and female: 5). Their average driving experience was five years (mean: 5.09, standard deviation: 1.36).

2.3   Experimental Setup

Based on the previous research done by McCartt et al. [10], one hazard event was selected and regenerated for the current study. It has been noted that drivers have a higher risk of having a crash when they are driving on freeway interchanges as opposed to other sections of the freeway. McCartt et al. [10] examined more than one thousand crashes occurred on high-traffic urban ramps in Northern Virginia and summarized the types and characteristics of ramp-related crashes. One of their results showed that crashes were most common on the on-ramp and off-ramp sections, and this type of accident increased when vehicles exited the interstate, or when there were curved sections on the ramps. Based on their study findings, we developed driving scenarios with different visibility conditions (i.e., sunny weather and foggy weather). The design of traffic condition (such as the number of vehicles in a high way, speed limits) of each scenario was based on the previous study done by Yang and Kim [5]. Also, both of the scenarios include the hazard event, which was triggered when participants drove onto the ramp and tried to merge onto the highway. In this hazard event, there was a car at the point B (see Fig. 2) in front of the participant subject that suddenly reduced speed during the on-ramp entrance to the highway. Because it requires abrupt and short-term visual searching and decision making to execute a merge onto the highway, the possibility of a vehicle crash is pretty high, especially in the experiment scenario which required drivers to divide their attention. During entrance merging, a driver must check the upcoming car (the point A in Fig. 2) in the target lane in order to find an opportunity to merge. If the lead car abruptly reduces speed, that requires a driver to divide his or her attention and make a decision and response quickly and simultaneously, which significantly increases the risk of a vehicle crash. Since the driving scenario is looped, drivers can experience the event multiple times; however, in order to prevent drivers from remembering the event, the event appeared randomly in experiments.

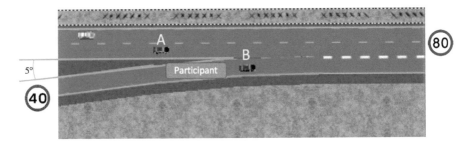

Fig. 2.   The hazard event occurs when participants drive onto the ramp try to enter the highway

2.4   Procedure

At the beginning of the experiment, the participants were required to complete a questionnaire for collecting demographic information (such as age, gender, driving experience) of each participant. After finishing the questionnaire, there was a five-minute orientation provided the tasks to the participants. To be familiar with the driving simulator, each participant was assigned to complete a training scenario in the OpenDS simulation. Then, the eye tracking glasses were set up and calibrated by the participants looking at the calibration card. During the experiment, the participants need to drive in two scenarios with the same hazard event but different weather (sunny, foggy) conditions. Each scenario took 30 min to complete. Between these two scenarios, ten minutes break was given. They also finished NASA-TLX questionnaires when they completed each scenario. After finishing the second scenario and completed the NASA-TLX surveys, the experiment ended. It took about 90 min for the whole experiment (see Fig. 3).

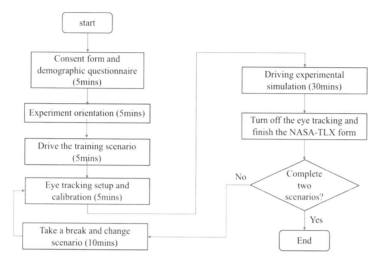

Fig. 3. The procedure of driving simulator experiment

3 Data Analysis

After the experiment, one hundred and seventy-four accident data were collected (including accident video records and driver's pupil diameter changes) to analyze the relationship between the pattern of pupil dilation and vehicle accidents. During the analysis, we found that after the drivers became aware of the danger, the drivers' maneuver and behaviors also affected their pupillary responses. Also, since the number of severe traffic accident, which was obtained from the scenario with a good visibility condition was very small, they are not included in the analysis of classification and comparison. Only the scenario with bad visibility condition was used for the data analysis.

To examine the pupillary responses in each vehicle accident, we set the beginning and the ending moment of the accident video data. The beginning moment was defined as driver's first fixation point to the accident vehicle, and the moment of vehicle collision is considered as the endpoint of the accident video data. By comparing the changes in the driver's pupil size at the moment of the vehicle accident to driver's reactions corresponding to the danger, it would be possible to find the relationship between the driver's psychological effort and the maneuver to avoid the accident.

After comparing the scatterplots of the pupil diameter to corresponding a video record of the vehicle accident, we found that the pupil diameter changes showed differences depending on the accident video record. Figure 4(a)–(c), presents three types of pupil change traces related to vehicle accident data points. Figure 4(a): pupils continued to dilate after noticing the lead vehicle until the moment of collision. Figure 4(b): pupil diameter size increased at first, but decreased before the endpoint. Figure 4(c): pupil dilated early, then constricted, and finally dilated.

According to the video record data, two patterns (types A and B) were found based on the drivers' driving maneuvers. In the type A, the emergency mode, which is a behavior where a driver hits the brake once they noticed the front car is suddenly slowing down, was always happened in the video. On the other hand, in the type B, a driver suspended an emergency mode for a moment, then the driver resumed it to avoid the accident. The car accidents that do not meet the features of types A and B were classified as other accidents. Figure 4(a) and (b) show the pupil change trace that corresponds to the types A and B, respectively. Figure 4(c) represents an example of other types.

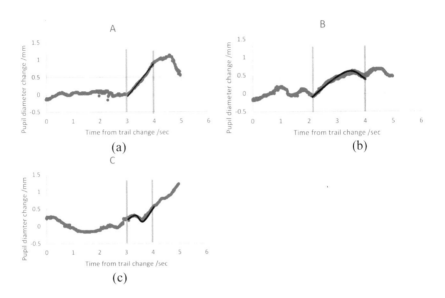

Fig. 4. Three tendencies of pupil diameter change among vehicle accidents data points

4   Results

After completing the data analysis, there was 40 cases of type A and 19 cases of type B. The remaining 15 cases were classified as other types. Figure 5 shows the pupillary responses of 40 cases in type A. In the type A, the drivers' maneuvers kept their focus on the front car while pressing the brake until the moment of a crash. As can be seen, the pupil diameter size increased in a directly linear line after the vertical orange line, which indicates the point in time when drivers first noticed the front car before the crash.

Figure 6 shows one case of type A as an example. The pictures on the left are the screenshots of an accident video record corresponding to the time, which is highlighted in the graph on the right. The red circle is the eye gaze point of the participant. We found that when the driver first noticed a front car (Fig. 6, 1), the pupil began to dilate and kept dilating until the end of the crash (Fig. 6, 2). During this period, the red circle was still on the front car, and the brake pressed.

Figure 7 shows that the change of pupil diameter in type B case slightly decreased after an initial increase. According to the accident video records, we found that when the participants started an emergency mode, the pupil size was increased linearly until the drivers moved their eye gaze points off the front car or until the front car stopped pressing a brake pedal. Then the pupil size started to decrease until the moment of the car crash.

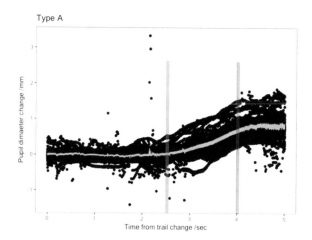

Fig. 5. The pupillary response of type A case

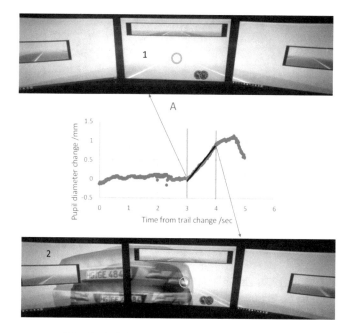

Fig. 6. Individual car accident case for type A

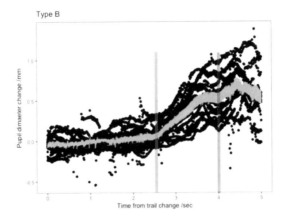

Fig. 7. The pupillary response of B type of vehicle accidents

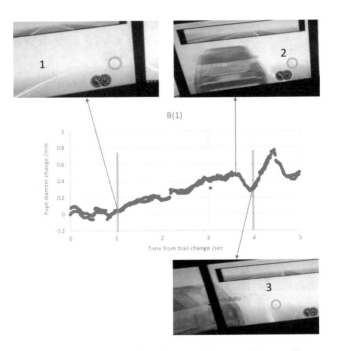

Fig. 8. The first individual car accident case for type B

There are two different driver's actions followed by the suspension of an emergency mode in type B. Figure 8 shows the first driver's action for this. As can be seen from the image 1 on the left of Fig. 8, the red circle is on the front car; this is the first time that the driver noticed the vehicle. Meanwhile, the driver hit a brake and remained to focusing on the car until the time corresponding to the picture 2 in Fig. 8. During this period, the driver's pupil size linearly increased. However, from the image 2, the eye gaze point was moved to the outside of the car, which indicates that the driver suspended an emergency mode. After that, drivers were tried to change a lane to avoid an accident. During that moment, the pupil size was constricted until the collision moment (see the image 3 in Fig. 8).

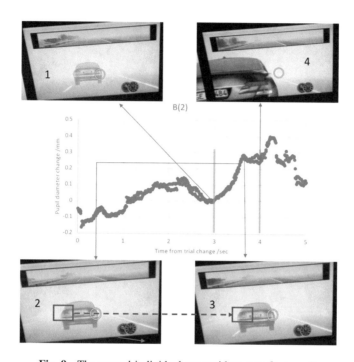

Fig. 9. The second individual car accident case for type B

Figure 9 shows the second driver action for the type B case. In this case, the image 1 shows the moment when a driver noticed the front vehicle; he hit the brake once he noticed a front car and then kept looking at the car. As can be seen from the graph in Fig. 9, pupils started to dilate at the moment of image 1, and then constricted at the moment of image 3. Although the driver did not change his driving behavior, the brake signal from the front car influenced the driver. The graph shows that there was a decrease in the pupil size at the moment of image 3. The last image 4 is the moment of the car crash.

Figure 10 shows that there was no common pupil change pattern to explain the vehicle accidents. One reason could be a small sample size of the other accidents. Another reason is that it was difficult to explain the cause of the pupil change only through video analysis because we did not find any change in the surrounding environment or driver behavior in the video.

Figure 11 shows one example video of the other type accidents. Although a driver noticed the front car from the picture 1, which is on the left part of Fig. 11, the video record shows that the driver did not reduce a vehicle speed. According to the video record, the driver started to press the brake at the moment of the picture 2, and from the

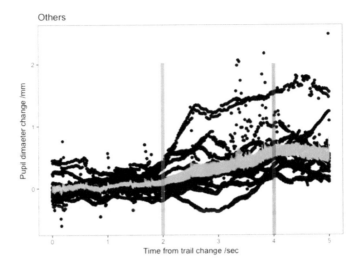

Fig. 10. The pupillary response of another type of vehicle accidents

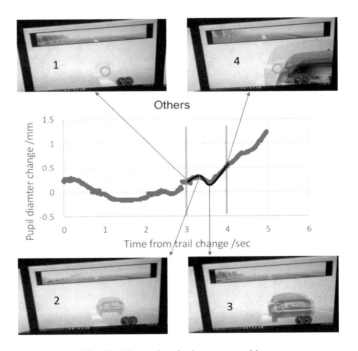

Fig. 11. Example of other type accidents

graph of pupil change, the pupil started to constrict as well. However, though the driver pressed a brake, the speed panel did not show any changes in speed. Therefore, the pupil dilated again by the time of picture 3. These pupil changes could be related to a brake latency, which resulted in an unclear pattern of pupil dilation.

5   Discussion

The present study demonstrates that the pattern of pupil dilation can provide information about how a driver responds to potential danger. The objective of this study is evaluating rear-end vehicle accident by using pupillary analysis. We hypothesize that the pupil dilation pattern can be used as an indicator to evaluate the driver's behaviors responding to a dangerous situation. Based on our results, two types of patterns, called type A and type B, were observed before the drivers made accidents. These two groups revealed different patterns of pupil changes. For type A, the drivers recognized potential danger on the road, and a series of maneuvers were executed to respond to the danger. During the situation, the pupil size continually expanded while the drivers maintained a high degree of attention. For type B, the drivers did not maintain a high degree of vigilance when they noted the hazardous event. Instead, they relaxed their vigilance shortly before the car accident occurred. The pupil size initially increased but declined right before the moment of the crash. In the type A, after the driver detected a front vehicle, he or she responded with a series of reactions (e.g., pressing the brake and keeping constant attention on the front car) to prevent a car accident. On the other hand, in the type B, the pupil size was decreased after an initially sustained increment during the time period (see Fig. 7). By comparing the accident videos, the increment of pupil size was monitored, and the driver's behavior was similar to the type A. They maintained close attention on the car in front of them and continued to apply the brakes. However, after a while the pupils began to shrink, correspondingly, the driver began to check the surrounding situation instead of focusing on the car that decelerated. This change of the driver's behavior does not mean that the situation is no longer dangerous because the driver collided with the front vehicle shortly after he or she shifted his gaze. The driver might have presumed that he or she could manage the situation.

6   Conclusion and Future Work

In conclusion, through the comparison of pupil changes and video data, this study summed up two pupil patterns corresponding to driving behaviors when they encounter the hazard event. The outcomes of this study will help engineers to develop an innovative way to predict driver's behavior through the driver's pupil dilation pattern.

For the limitations of the current study, some of the pupil dilation data were considered as noise due to the latency issues in a driving simulator during the data analysis. Therefore, we need to solve the latency problems and reduce the noise for the next experiment. Moreover, the participants in this study were all college students, in the future, different age groups people should be involved in the experiment. Finally, other factors that might affect pupil diameter changes in driving, such as different level of driving experience, drivers' ability of perception, mental workload, and fatigue, should be considered in our future work.

References

1. Gong, J., Yang, W.: Driver pre-accident operation mode study based on vehicle-vehicle traffic accidents. In: 2011 International Conference on Electric Information and Control Engineering (ICEICE), pp. 1357–1361. IEEE (2011)
2. Tang, R., Kim, J.H., Parker, R., Jeong, Y.J.: Indicating severity of vehicle accidents using pupil diameter in a driving simulator environment. In: International Conference on Digital Human Modeling and Applications in Health, Safety, Ergonomics and Risk Management, pp. 647–656. Springer (2018)
3. Praveen, J.: Biometrics wave poised to transform future driving. Biom. Technol. Today **2017**(5), 5–8 (2017)
4. Puspasari, M.A., Iridiastadi, H., Sutalaksana, I.Z.: Oculomotor indicator pattern for measuring fatigue in long duration of driving: case study in Indonesian road safety. J. Traffic Logist. Eng. **5**(1), 26–29 (2017)
5. Yang, X., Kim, J.H.: The effect of visual stimulus on advanced driver assistance systems in a real driving. In: IIE Annual Conference, Proceedings, pp. 1544–1549. Institute of Industrial and Systems Engineers (IISE) (2007)
6. Kim, J.H.: Effectiveness of collision avoidance technology (2016). https://www.mem-ins.com/
7. Sivak, M.: The information that drivers use: is it indeed 90% visual? Perception **25**(9), 1081–1089 (1996)
8. Math, R., Mahr, A., Moniri, M.M., Müller, C.: OpenDS: a new open-source driving simulator for research. In: GMM-Fachbericht-AmE 2013 (2013)
9. Bian, D., et al.: A novel virtual reality driving environment for autism intervention. In: International Conference on Universal Access in Human-Computer Interaction, pp. 474–483. Springer (2013)
10. McCartt, A.T., Northrup, V.S., Retting, R.A.: Types and characteristics of ramp-related motor vehicle crashes on urban interstate roadways in Northern Virginia. J. Saf. Res. **35**(1), 107–114 (2004)

Study on Operation Simulation and Evaluation Method in the Ship Limited Space

Zhang Yumei[1](✉), Jiang Lei[1], Wang Songshan[2], and Ding Li[3]

[1] China Ship Development and Design Center, Wuhan 430064, China
zhangyumei821202@sina.com
[2] Army Engineering University, Shijiazhuang 050003, China
[3] Beihang University, Beijing 100191, China

Abstract. There are many different types of equipments, complex pipeline and cable in the ship. So the overall space resource is poor. The crew is easy to encounter accessibility and ergonomics obstacles, when he is engaged in maintaining equipments in the limited space. The paper gives out an operation simulation and evaluation method in limited space base on Siemens Jack secondary development. Firstly the virtual human arm reachable domain and view frustum are proposed respectively to inspect the encounter accessibility and ergonomics obstacles by using the envelope generation method based on point-by-point scanning algorithm. Then the Task Analysis Toolkit (TAT for short) indicator weights based on the therbligs are confirmed by expert investigation. The calculation formula and rating rule of each TAT indicator are developed. Finally the ergonomics comprehensive quantitative is put forward by dimensionless dimension method. Aiming at the twin oil filters replacement and maintenance task in gear box module, the maintenance crew labour intensity and human injury risk are reduced by 46%, through using the simulation and evaluation method in this paper to optimize design scheme. So the method provides an effective solution for the crew operation simulation evaluation aiming at space interference and human injury risk. It has important significance in comprehensive optimization between crew operation, equipments layout and cabin space in ship design phase.

Keywords: Ship · Operation · Simulation · Evaluation · Accessibility · Ergonomics

1 Introduction

There are many types of equipments, complex pipeline and cable in the ship. So the overall space resource is poor. The crew is easy to encounter accessibility and ergonomics obstacles, when he is engaged in maintaining equipments in the limited space [1]. The accessibility comprehensively reflects the influence of working space, maintenance access, cabin layout on equipment use and maintenance [2]. It can be divided into contact accessibility and visual accessibility. The contact accessibility means that the crew is barely able to touch parts, unable or difficult to go through maintenance access. The visual accessibility means that the crew is unable to see operation because of insufficient visibility. Furthermore, from fatigue, strength, posture, energy

D. N. Cassenti (Ed.): AHFE 2019, AISC 958, pp. 187–199, 2020.
https://doi.org/10.1007/978-3-030-20148-7_18

consumption and other aspects, the human injury risk for work reasons is estimated by ergonomics analysis, including comfort and physical disturbance [3]. The comfort means that the crew can easily complete operation, or isn't so easy to feel tired for long time working. The physical disturbance means that the crew needs to adjust posture immediately, or meets with human body damage for prolonged operation, due to unreasonable environment or working condition.

At present the digital ship is only geometric ship by three dimensions computer aided design. It's focus on geometric representation for layout plan of equipment, structures, piping, cables, etc. It's mainly used to solve ship layout problems, with lack of consideration for crew use and maintenance operations. So it is unable to effectively detect and solve the design defects, such as the interference of crew operation space and the risk of human body injure. Siemens Jack provides functionality of human interference and visual inspection, Task Analysis Toolkit (TAT for short). TAT is the analytical package released in 1981, which is developed to improve work efficiency by National Institute for Occupational Safety and Health (NIOSH for short). But the above functions cannot accurately express the coordinate information of collision interference position. The vision algorithm has a few errors. TAT indicators are scattered. All of these cannot be adequate to the current demand of quantitative and refined design, conformity analysis evaluation in ship engineering. So this paper gives out an operation simulation and evaluation method based on accessibility and ergonomics through Jack secondary development. The method is applied to simulation evaluation and design improvement of the twin oil filters replacement and maintenance task in gear box module. The results of case application show that it has important significance in comprehensive optimization between crew operation, equipment layout and cabin space in ship design phase.

2   Contact Accessibility Analysis Method

According to Fig. 1, the arm joints model of virtual human is mainly controlled by $3°$ of freedom, take right hand, for instance, including θ_4, θ_5, θ_7. Among of them, θ_4 refers to the shoulder stretching of vertical direction, θ_5 refers to the shoulder stretching of horizontal direction, θ_7 refers to the elbow stretching. As shown in Table 1, the maximum activity range and comfort range of θ_4, θ_5, θ_7 are defined respectively, according to GB/T 15759-1995, "design and requirements for the templates of human body".

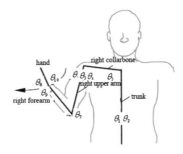

Fig. 1. Human arm control joint model

Table 1. Main activity area and comfortable posture adjustment range of the Chinese adults limb joints

Joint diagram	Joint name of body part	Joint maximum angle range	Joint comfort adjustment range
	Shoulder joint of upper arm to torso (θ_4)	-45°~+180°	+15°~+35°
	Shoulder joint of upper arm to torso (θ_5)	-40°~+140°	+40°~+90°
	Elbow joint of lower arm to upper arm (θ_7)	0°~+145°	+10°~+85°

The virtual human arm biggest and comfort reachable domain are proposed by using envelope generation method based on point-by-point scanning algorithm [4]. Then they are used to check the touch accessibility obstacle, through detecting if the reachable domains contain grab object or not [5]. The implementation process is as follows:

(1) The shoulder joint searches according to step size in the horizontal direction incrementally, and in the vertical direction diminishingly. Each time the joint angle changes, the 3D coordinate of arm scanning point is recorded immediately.
(2) When the horizontal search of shoulder joint reaches its limit, the shoulder joint is locked in the horizontal direction. Then the descending search of elbow joint is conducted, along with the descending search of shoulder joint in the vertical direction. Likewise, each time the joint angle changes, the 3D coordinate of scanning point is recorded immediately.
(3) According to Fig. 2, the recorded point data is carried out trough grid processing after scanning, and the triangles are used to represent the geometric surface of reachable domain. When the envelope surfaces of active areas of the above two activity areas are combined, a human arm range is generated (see Fig. 3).

As shown in Fig. 4, the virtual human touch accessibility analysis tool is designed and developed, based on the reachable domain envelope function of shoulder and elbow multi-degree-of-freedom parameters in Jack.

Fig. 2. Triangular face model

Fig. 3. Human arm touch reachable domain

Fig. 4. Analysis tool interface of touch reachable domain

3   Visual Accessibility Analysis Method

Through reviewing the correlation literature [6], the human best, good, effective field of view are confirmed respectively. When an object is within 3° range in vertical and horizontal direction of eye vision, its imaging falls on the fovea of retina exactly. It's the best field of vision to recognize objects. When an object is within 30° below range relative to the horizontal line of sight in vertical plane and 15° left and right range relative to the zero line in horizontal plane, its imaging is legible. It's the good field of view. The effective field of view is within 25° above and 35° below range relative to the horizontal line of sight in vertical plane. Meanwhile, it's within 35° left and right range relative to the zero line in vertical plane.

The crew usually operates with a wrench, screwdriver, etc., or disassembles and assembles by hand. So the stadia range is from 500 mm to 1500 mm. A ball is created in the center of virtual human eyes. Then it is attached to the left eye joint and moved by 1 stadia along the direction of sight line. The ball is visual focus which has the function of synchronous movement with eye joint. The motion range of eye joint is scanned point-by-point according to the step length along the horizontal and vertical direction respectively based on point-by-point scanning algorithm. The ball 3D coordinates are recorded simultaneously. According to Fig. 5, all the coordinates are graphed to generate the best, good and effective visual range of virtual human. Aiming at head rotation, or head and eye rotation, it's achieved by adjusting up, down, left, right angle parameters of the eyeball. They are used to check the visual accessibility obstacle, by detecting whether the visual field envelope contains observed object.

As shown in Fig. 6, the virtual human visual accessibility analysis tool is designed and developed, based on the cone envelope function of view parameters in Jack.

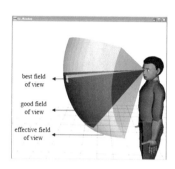

Fig. 5. Human visual reachable domain

Fig. 6. Analysis tool interface of visual reachable domain

4 Ergonomics Obstacle Analysis Method

4.1 TAT Indicators Integration

In this study, the ergonomics obstacle analysis method of use and maintenance task has been developed based on TAT indicators integration in Jack. According to Fig. 7, the ergonomics indicators system framework is established from the following aspects: fatigue, strength, posture, energy consumption. NIOSH, lower back analysis, static strength prediction and forcesolver are from the angles of force, torque and muscle force to evaluation. Manual handling limits, fatigue and recovery are from the ways of maximum muscle strength, fatigue time of muscle strength, recovery time of muscle fatigue to evaluation. Ovako working posture analysis, rapid upper limb assessment are from the difficulty level of various operating posture of human body to evaluation. Furthermore, metabolic energy expenditure is evaluated according to whether the calculated operating energy consumption exceeds the normal capacity of human body. It's mainly suitable for long time task [3].

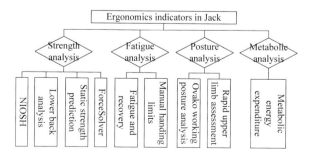

Fig. 7. Ergonomics indicators system framework of TAT

All indicators are integrated by the previous definitions of comfort and physical disturbance. The results are as follows:

Comfort evaluation $=$ a $\times$ NIOSH $+$ b $\times$ lower back analysis $+$ c $\times$ static strength prediction $+$ d $\times$ forcesolver $+$ e $\times$ ovako working posture analysis $+$ f $\times$ rapid upper limb assessment $+$ g $\times$ metabolic energy expenditure

$$(1)$$

physical disturbance evaluation $=$ a $\times$ NIOSH $+$ b $\times$ lower back analysis $+$ c $\times$ static strength prediction $+$ d $\times$ forcesolver $+$ e $\times$ ovako working posture analysis $+$ f $\times$ rapid upper limb assessment $+$ g $\times$ metabolic energy expenditure $+$ h $\times$ manual handing limits $+$ i $\times$ fatigue and recovery

$$(2)$$

In the formula (1) and (2), a, b, ···, h and i are the indicator weights respectively.

4.2   TAT Indicator Weights Allocation Based on the Therbligs

The action is divided into movement and operation class based on its characteristics of crew use and maintenance. The movement class action refers to position movement, posture change and adjustment. The operation class action refers to freehand and tool operation. These actions are divided into 10 therbligs on the whole. They are used for parameterized description of crew motion characteristics (see Table 2). In order to achieve action simulation, the above therbligs can be combined and invoked according to task demand [7–9]. As shown in Table 3, TAT indicator weights based on the therbligs are confirmed by expert investigation.

Table 2. Type, name and definition of the therbligs

Action type	Therblig name	Therblig definition
Movement class	Human_move	Azimuth change, for example normal walk, side walk, bend to walk, etc
	Carry	Azimuth change with the object
	Pose	Adjust to prepared posture before operation, for example upright standing, bend standing, sit, squat, kneel, lie prostrate, lie on the side, lie on the back, etc
Operation class	Get	Grasp the target object. Place the foot or eyes on the object
	Position	Control the tool or object before operation to align with point, line and face of the target object

(continued)

Table 2. (*continued*)

Action type	Therblig name	Therblig definition
	Hold	Maintain the existing posture or control the object so that it doesn't change orientation
	Place	Put away the tool or object after finishing an operation
	Release_resume	Release or remove the tool from the object, and return to the original or customary posture after finishing an operation
	Use_tool	Manipulate the object by using tool, for example knock, punch, twist, clip and pry, etc
	Operate	Disassemble the object, for example rotate, press, push and pull, etc

Table 3. TAT indicator weights allocation based on the therbligs

Therblig	Fatigue and recovery	Force-solver	Lower back analysis	Manual handing limits	Metabolic energy expenditure	NIO-SH	Ovako working posture analysis	Rapid upper limb assessment	Static strength prediction
Human_move	0.22	–	0.07	–	0.03	–	0.35	0.2	0.13
Carry	0.22	0.21	0.03	0.03	0.03	–	0.32	0.16	–
Pose	0.22	0.19	0.03	–	–	–	0.42	0.11	0.03
Get	0.22	0.16	0.03	–	–	–	0.35	0.11	0.13
Position	0.22	0.22	0.05	0.03	–	0.02	0.35	0.05	0.06
Hold	0.24	0.24	0.02	–	–	–	0.34	0.09	0.07
Place	0.24	0.16	0.04	–	–	–	0.32	0.13	0.11
Release_resume	0.2	0.28	0.05	–	–	–	0.41	0.03	0.03
Use_tool	0.19	0.27	0.13	0.03	0.05	–	0.21	0.08	0.04
Operate	0.16	0.24	0.02	0.04	0.05	0.05	0.28	0.12	0.04

4.3   TAT Indicators Comprehensive Quantitative Solution

The dimensionless dimension method is adopted to convert TAT indicators to rating scale ones. The calculation formula and rating rule of each TAT indicator are developed in Table 4 [10, 11]. Therefore, the ergonomics comprehensive score can be obtained by weighted summation.

Table 4. TAT indicators synthesis

Indicator	Scoring rule	Scoring formula
NIOSH	$0 \le SCORE \le 25$ is comfort range. It shows that the posture can be maintained for a long time and is still acceptable. $25 < SCORE \le 50$ is discomfort range. It shows that the posture is acceptable in a short time. $50 < SCORE \le 75$ shows that the posture needs to be improved as soon as possible. $75 < SCORE \le 100$ shows that the posture should be adjusted immediately or it will result in human injury risk.	$\begin{cases} SCORE = \left[100 \times LI\right], & \text{当} LI \le 1 \\ SCORE = 100, & \text{当} LI > 1 \end{cases}$ LI is the lifting index. LI>1 means excessive load and easy in low back pain risk. LI≤1 means that the current load is acceptable.
Lower back analysis		$\begin{cases} SCORE = \left[75 \times \dfrac{L4/L5}{3400}\right], & \text{当} L4/L5 \le 3400N \\ SCORE = \left[75 + 25 \times \dfrac{L4/L5 - 3400}{6400 - 3400}\right], & \text{当} 3400N < L4/L5 \le 6400N \\ SCORE = 100, & \text{当} L4/L5 > 6400N \end{cases}$ L4 / L5 is the spinal L4 / L5 stress.
Static strength prediction		$SCORE = \left[100 \times \left(1 - \prod_{i=1}^{7} A_i\right)\right]$ A_i is the percentage of people who can maintain the normal posture of Wrist, Elbow, ⋯, Knee, Ankle joint respectively.
Forcesolver		$SCORE = \left[100 \times \left(1 - \prod_{i=1}^{23} B_i\right)\right]$ B_i is the percentage of people who can maintain the normal posture of R Wrist Flx, L Wrist Flx, ⋯, R Angle, L Ankle joint respectively.
Ovako working posture analysis		$SCORE = 25 \times OWAS, \quad OWAS \in \{1,2,3,4\}$ OWAS is the working posture assessment grade.
Rapid upper limb assessment		$\begin{cases} SCORE = \left[25 \times \dfrac{RULA}{2}\right], & \text{当} RULA \in \{1,2\} \\ SCORE = 25 + \left[(50 - 25) \times \dfrac{RULA - 2}{2}\right], & \text{当} RULA \in \{3,4\} \\ SCORE = 50 + \left[(75 - 50) \times \dfrac{RULA - 4}{2}\right], & \text{当} RULA \in \{5,6\} \\ SCORE = 75 + \left[(100 - 75) \times \dfrac{RULA - 6}{3}\right], & \text{当} RULA \in \{7,8,9\} \end{cases}$ RULA is the rapid upper limb assessment grade.
Metabolic energy expenditure		$\begin{cases} SCORE = \left[25 \times \dfrac{E_{job}}{1.82}\right], & \text{当} E_{job} \le 1.82 \\ SCORE = 25 + \left[(75 - 25) \times \dfrac{E_{job} - 1.82}{5.25 - 1.82}\right], & \text{当} 1.82 < E_{job} \le 5.25 \\ SCORE = 75 + \left[(100 - 75) \times \dfrac{E_{job} - 5.25}{6.77 - 5.25}\right], & \text{当} 5.25 < E_{job} \le 6.77 \\ SCORE = 100, & \text{当} E_{job} > 6.77 \end{cases}$ E_{job} is the average total energy consumption during operation (unit: KJ / min). It's applied to unit conversion and classification with reference to the metabolic rate classification table for various activities in GB/T 18048-2008, "ergonomics of the thermal environment-determination of metabolic rate".

Manual handing limits	$\begin{cases} SCORE = 25, & \text{if the weight required for task} \\ \leq & \text{maximum weight for } 90\% \text{ people} \\ SCORE = 75, & \text{if else} \end{cases}$
Fatigue and recovery	$\begin{cases} SCORE = \left[25 \times \dfrac{SSCORE}{2}\right], & \text{当}SSCORE \in \{0,1,2\} \\ SCORE = 25 + \left[(75-25) \times \dfrac{SSCORE-2}{6-2}\right], & \text{当}SSCORE \in \{3,4,5,6\} \\ SCORE = 75 + \left[(100-75) \times \dfrac{SSCORE-6}{10-6}\right], & \text{当}SSCORE \in \{7,8,9,10\} \\ SCORE = 100, & \text{当}SSCORE>10 \end{cases}$ $SSCORE = \left[6.004 \times \left(\dfrac{T}{R_A}\right)^{-0.33}\right]$. T is the total activities time (unit: s). R_A is the fatigue recovery required time (unit: s).

5   Case Application Analysis

This paper chooses the twin oil filters replacement and maintenance task in gear box module to make case analysis. The task includes the following 7 operating procedures: switching the wrench, loosening the cover bolts, lifting the cover plate, taking out the filter element, installing the filter element, installing the cover plate, tightening the cover bolts. The total maintenance task simulation is proposed by the use_tool and operate therbligs. The fatigue and metabolism are almost negligible because of short task. Therefore, the weights allocation of other 7 TAT indicators are according to the values of use_tool and operate therbligs (see Table 3). There are two schemes analyzed in this study. Compared to scheme 1, the base height of twin oil filters equipment is lowered by 250 mm in scheme 2.

As shown in Figs. 8 and 9, the grab object is in the contact reachable domain, the observant is in the viewing range simultaneously during equipment maintenance. So there're no accessibility obstacles for the crew in either schemes.

Scheme 1   Scheme 2

Fig. 8. Contact accessibility analysis

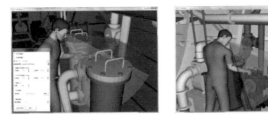

Scheme 1   Scheme 2

Fig. 9. Visual accessibility analysis

The TAT indicators of lifting the cover plate are discussed in Table 5 through case study (see Fig. 10). Then each indicator is put forward to grade value conversion based on the scoring formulas in Table 4. The results are shown in Table 6. Lifting the cover belongs to the operate therblig. So the comprehensive score of scheme 1 is 53 by computation according to the TAT indicator weights in Table 3. In the same way, scheme 2 is 30. It shows that the crew feels harder and has to stand on tiptoe during maintenance operations because the base of twin oil filters is too high in scheme 1. When the base height is adjusted, scheme 2 shows significant improvement.

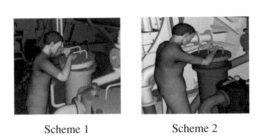

Scheme 1   Scheme 2

Fig. 10. The maintenance lifts the cover plate

Then the above method is applied to the other operational processes successively. Aiming at scheme 1 and scheme 2, the curves of comparative analysis results are drawn in Fig. 11. The average ergonomics comprehensive score of scheme 1 is 41. The score of lifting the cover plate and screwing the bolts are close to 50. It shows that the crew is discomfort and will be in injuries risk if the state lasts longer. The average ergonomics comprehensive score of scheme 2 is 22. At the same time the scores of all operating procedure are almost below 25. It means that the crew feels comfortable. Compared with scheme 1, scheme 2 reduces labour intensity and human injury risk by 46%。So it can better meet design requirements.

Table 5. Ergonomics value of lifting the cover plate

Indicator	Lifting the cover palte	
	Scheme 1	Scheme 2
Manual handing limits	The lifting upper limit of 90% people is 6 kg. It meets the requirements of lifting the cover	It's consistent with the left column
NIOSH	The lifting index LI = 1.120	The lifting index LI = 0.630
Ovako working posture analysis	The working posture assessment grade OWAS = 2	The working posture assessment grade OWAS = 1
Rapid upper limb assessment	When just lifting the cover, the rapid upper limb assessment grade RULA = 6 When lifting the cover to highest point, the rapid upper limb assessment grade RULA = 7	When just lifting the cover, the rapid upper limb assessment grade RULA = 4 When lifting the cover to highest point, the rapid upper limb assessment grade RULA = 5
Forcesolver	The joints percentage are as follows: R Wrist Flx = 100%, Wrist Flx = 100%, Wrist Dev = 79%, Wrist Dev = 100%, R Wr Supr = 100%, L Wr Supr = 100%, R Elbow = 100%, L Elbow = 100%, R Sh AbAd = 93%, L Sh AbAd = 100%, R Sh FwBk = 100%, L Sh FwBk = 100%, R Sh Hmrl = 18%, L Sh Hmrl = 100%, Trunk Flx = 93%, Trunk Bend = 100%, Trunk Twst = 100%, R Hip = 99%, L Hip = 98%, R Knee = 99%, L Knee = 100%, R Ankle = 94%, L Ankle = 98%	The joints percentage are as follows: R Wrist Flx = 75%, L Wrist Flx = 99%, R Wrist Dev = 100%, L Wrist Dev = 100%, R Wr Supr = 100%, L Wr Supr = 100%, R Elbow = 99%, L Elbow = 100%, R Sh AbAd = 96%, L Sh AbAd = 100%, R Sh FwBk = 100%, L Sh FwBk = 100%, R Sh Hmrl = 93%, L Sh Hmrl = 100%, Trunk Flx = 97%, Trunk Bend = 100%, Trunk Twst = 100%, R Hip = 99%, L Hip = 98%, R Knee = 100%, L Knee = 99%, R Ankle = 98%, L Ankle = 97%
Lower back analysis	The spinal stress L4/L5 force = 3316 N	The spinal stress L4/L5 force = 1878 N
Static strength prediction	The joints percentage are as follows: Wrist = 98.3%, Elbow = 98.9%, Shoulder = 87.8%, Torso = 96%, Hip = 98.4%, Knee = 99.1%, Ankle = 90.5%	The joints percentage are as follows: Wrist = 87%, Elbow = 99.6%, Shoulder = 99.5%, Torso = 99.3%, Hip = 99.3%, Knee = 99.8%, Ankle = 95.5%

Table 6. Ergonomics evaluation grade score of lifting the cover plate

Grade score	Lifting the cover plate						
	Manual handing limits	NIOSH	Ovako working posture analysis	Rapid upper limb assessment	Forcesolver	Lower back analysis	Static strength prediction
Scheme 1	25	100	50	79	89	73	28
Scheme 2	25	63	25	57	42	41	19

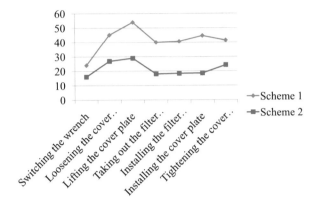

Fig. 11. Comparative analysis of the ergonomics comprehensive score for Scheme 1 and Scheme 2

6 Conclusion

The paper presents an operation simulation and evaluation method based on accessibility and ergonomics in ship limited space. Through Jack secondary development, the arm reachable domain and view frustum of virtual human are proposed by using envelope generation method based on point-by-point scanning algorithm firstly. They're applied to check the contact and visual accessibility obstacles. Then TAT indicator weights based on the therbligs are confirmed by expert investigation. Finally the dimensionless dimension method is adopted to realize the ergonomics comprehensive evaluation. The theoretical calculation and case application show that the method proposed in this paper is efficient and applicable in comprehensive optimization between crew operation, equipments layout and cabin space at the ship design stage.

References

1. Zhang, Y.M., Wei, Q.Q., Zeng, J.: Evaluation of overall warship operability and maintainability based on the technology of virtual reality. Chin. J. Ship Res. **8**(2), 6–12 (2013)
2. Fang, X.B., Tian, Z.D., Lin, R., et al.: A method for accessible domains computation and visulization in the case of using maintenance tools. Chin. J. Ship Res. **11**(11), 14–18, 41 (2016)
3. Chen, X., Niu, J.W., Jiang, Y.: Modeling, Simulation and Ergonomics Evaluation of Individual Soldier Equipment: Fundamentals. Science press, Beijing (2015)
4. Kallmann, M., Thalmann, D.: Modeling behaviors of interactive objects for virtual reality applications. J. Vis. Lang. Comput. **13**, 177–195 (2002)
5. Wang, S.S.: Specification of human action in virtual maintenance simulation. J. Syst. Simul. **2**(17), 507–509+512 (2005)
6. Zhu, X.Z.: Ergonomics. Xidian University Press, Xian (1999)

7. Fang, Q., Wang, S.S., Zhu, H., et al.: Ship maintainability design analysis technology based on digital prototyping [J]. Chin. J. Ship Res. **1**(11), 114–120 (2016)
8. Hao, J.P.: Virtual Maintenance Simulation Theory and Technology. National Defense Industry Press, Beijing (2008)
9. Li, J.H., Hao, J.P., Wang, S.S., et al.: Design and implementation of virtual human walking in Jack. Comput. Simul. **9**(26), 207–210 (2009)
10. Fan, W., Wang, J.F.: Layout optimization method for narrow working cabins. Chin. J. Ship Res. **11**(5), 19–27 (2016)
11. Niu, J.W., Zhang, L.: Human Factors Engineering Foundation and Application Examples in Jack. Publishing House of Electronics Industry, Beijing (2012)

Physical, Mental and Social Effects in Simulation

Surviving in Frankenstein's Lab- How to Build and Use a Laboratory that Applies Multiple Physiological Sensors

Sylwia Kaduk$^{(\boxtimes)}$, Aaron Roberts, and Neville Stanton

Human Factors Engineering, Transportation Research Group,
Faculty of Engineering and Physical Sciences, University of Southampton,
Building 176, Boldrewood Innovation Campus, Burgess Road,
Southampton SO16 7QF, UK
{s.i.kaduk, aprlcl3, n.stanton}@soton.ac.uk

Abstract. A growing area of research addresses how physiological data can be used to understand the state of an operator across a variety of operational environments, including semi-autonomous driving. Use of multiple physiological sensors is highly practical; however, it introduces a number of challenges related to signal noise reduction. During the design and construction of the laboratory in HI:DAVe it became clear that there was not much documentation concerning how to undertake such a task. Paper is assimilating the literature related to this to provide a user's guide for how someone can get to the stage of a fully functional Faraday Cage mid-fidelity driving simulator with an extensive physiological recording suite. It describes a process of psychophysiological laboratory construction and introduces a decision-tree to help with the choice of the most optimal noise-reduction strategies. The psychophysiological measures included are electroencephalography, electrooculography, electromyography, eye-tracking, voice analysis, electrocardiography, respiration, electrodermal response and oximetry.

Keywords: Psychophysiology · Operator state monitoring · Human factors ·
Signal noise reduction · Artefacts Reduction · Electroencephalography ·
Electromyography · Electrooculography · Eye-tracking · Electrocardiography ·
Respiration · Electrodermal response · Galvanic skin response ·
Acoustic voice analysis · Oximetry · Faraday Cage · Decision-tree

1 Introduction

1.1 Monitoring with Multiple Sensors

The development of automated technologies makes monitoring of the operator's state more important to ensure safety. One of the examples is semi-autonomous driving where the function of the driver is shifted into a more supervisory role. Unfortunately, people tend to perform poorly in tasks that require sustained attention. Driver state monitoring could ensure safety by surveilling if a driver is in the proper state and engaged in driving-related tasks [1]. The combination of multiple physiological

© Springer Nature Switzerland AG 2020
D. N. Cassenti (Ed.): AHFE 2019, AISC 958, pp. 203–214, 2020.
https://doi.org/10.1007/978-3-030-20148-7_19

recordings often gives higher accuracy than using just one. Moreover, a comparison of effectivity and accuracy of different measures enables the most optimal choice of the monitoring system. It could allow choice of the method most suitable for requirements of a particular situation in a manner of cost, efficacy, accuracy, speed and others. However, conducting a study using multiple sensors creates certain challenges, and is often avoided by researchers due to potential difficulties. Each type of physiological signal might be confounded by the different types of noise that should be reduced as much as possible [2]. A combination of different measures typically increases the amount of potential sources of noise that can effect such measurements. The control of all potential signal issues across methods requires a variety of strategies, which can increase the complexity of the experimental environment. The processes undertaken are often required to become more complex and cumbersome. A combination of different measures might also be affected by a combination of different noise sources; therefore, might require more cumbersome noise control. Additionally, a set-up of multiple sensors can be more complicated, especially if they interfere with each other. There is a gap in the literature regarding the set-up and the laboratory space preparation that would ensure the best data quality. The aim of this paper is to address this gap by providing a description of the optimal experimental set-up and laboratory construction for a multisensory recording on the example of multisensory, low-low fidelity driving simulator. It provides a decision-tree with a number of recommendations. The unique suite of recommendations is generated based on a specific set of methods chosen by a user. It is meant to be a practical tool enhancing a high-quality data collection.

1.2   HI:DAVe

Human Interaction: Designing Autonomy in Vehicles (HI:DAVe) is a research project that looks for the most optimal human-system interactions in semi-autonomous cars. One of the subprojects studies driver state monitoring using twelve different psychophysiological measures, electroencephalography (EEG), camera-based eye-tracking, electrooculography (EOG), electromyography (EMG), electrocardiography (ECG), respiration, electrodermal response (EDA), saliva-based cortisol and alpha-amylase analysis, acoustic voice analysis, oximetry and questionnaires. These measures were chosen after the completion of an extensive literature review revealing what measures have the maturity for use in applied environments in the short, medium and long term. Such a variety of the recorded signals required a cautious approach towards noise reduction. So-called signal noise, which is a recording of artefacts, can be partially avoided with a proper laboratory construction and experimental set-up. Unavoidable noise can be rejected from the data with different strategies of data pre-processing. However, algorithms that reject artefacts always cause a certain level of a signal loss; therefore, experimental set-up should be designed in a manner allowing the lowest amount of the artefacts possible (Fig. 1).

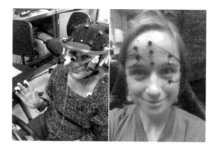

Fig. 1. On the left: A participant with multiple recording devices during pre-study equipment testing and calibration. She is wearing Enobio-20 EEG, Dekablis eye-tracker, head-mounted microphone, BioPac EMG and EOG electrodes on her face, BioPac ECG electrodes on her chest, BioPac respiration belt on her abdomen and BioPac EDA electrodes on her fingers. EMG electrodes are covered with a gauze; however, testing revealed interference between EEG cap, eye-tracker and electrodes. Pressing on the electrodes caused instability of the signal and increased impedance. It also created a high discomfort in the participant. Because of this, the interfering methods were divided into 3 separate experimental series: EEG, eye-tracker and EMG together with EOG. On the right: A researcher wearing only EOG and EMG electrodes without a gauze stabilization. Such montage allowed a better signal quality.

1.3 Types of Sensors and Sources of Noise

EEG, EMG, EOG, and ECG estimate an electric activity of the brain, muscles, retina and heart respectively. They use electrodes placed on the surface of the skin [3–6]. The EDA is a measure of electric potential on the skin surface assessed between two electrodes [7]. Eye-tracking is a wide group of methods of eye movements monitoring [8]. In case of HI:DAVe eye-tracker was a helmet with two cameras recording the eyes and one camera recording the visual field; however, there is also an option of eyes monitoring with desktop mounted camera [8]. There is a variety of devices measuring breathing amplitude and frequency, but in the case of HI:DAVe it was measured with a belt placed on the upper abdomen [2]. Acoustic voice analysis uses a voice recording to analyze acoustic characteristics of the speech [9]. Oximetry is the measure of blood oxygenation level and a pulse with use of near-infrared light generated by a device placed on a finger or an ear [2, 10].

Each of the methods has a specific set of potential artefact sources that need to be controlled in order to achieve as clean data as possible. In a case of acoustic speech analysis artefacts in the signal might be either caused by the noises other than speech or by changes in signal property caused by microphone, pre-amplifier, recorder or data conversion [9]. Electrophysiology is sensitive to any type of movements, muscle activity, electrodes displacement, sweat, and influence of surrounding electromagnetic signal, even the signal of the recording devices [2, 4, 11, 12]. Likewise, the measurement of the electrodermal activity can be disturbed by electromagnetic impulses, body movements, temperature changes or the electrodes displacement [7]. Respiratory signals can be changed with body movements [2]. The eye-tracking signal can be disturbed by blinking (in the case of gaze monitoring), eye-lashes and light instability [8]. Oximetry is susceptible to movement, other body signals, electromagnetic influences [13] and changes in the light [2]. Many of the noise sources are quite common in the environment, for example, electromagnetic impulses are generated by the surrounding electrical devices or electrical wires. Similarly, acoustic noise can be generated by objects like fans, crackling noise from the cables cables, elevators and others.

2 Methods

The paper is based on the laboratory construction experience and a thorough calibration process conducted with the HI:DAVe equipment. The main aim of the repetitive comparisons of the multiple montage strategies was to establish the most optimal set-up for the data quality. The literature was evaluated in search for the information about the set-up of the used measures, technical aspects of a Faraday Cage construction and noise insulation. The diligent testing process and literature search was conducted to obtain the most noise-free data collection in the HI:DAVe configuration.

3 Results and Discussion

3.1 Experimental Set-up- a Decision Tree

The decision tree is a result of a literature evaluation and hands-on comparison of different set-up strategies. It is a guide tool that allows choosing the most optimal experimental set-up for the measures selected for the experiment. This tool is an answer to the gap in the literature that would combine recommendations for set-up strategies while combining different physiological measuring devices. It includes EEG, EMG, EOG, ECG, EDA, camera-based eye-tracking, respiration belt, oximetry and the voice recording. The user should follow descending decision nodes and answer the questions asked in the nodes. The arcs represent answers 'yes' and 'no'. They lead to additional questions and finally to the unique set of the recommendations for the particular experiment. Recommendations are expressed as a list of numbers. Numbers are explained in the description to the Fig. 2. The explication of the recommendations is included in the parts 3.2 and 3.3. Due to the complexity of the decision-tree and the visibility issues the graph was divided into two parts A and B. A user following the arcs from the graph A to the graph B should continue following the arc with the same numerical description.

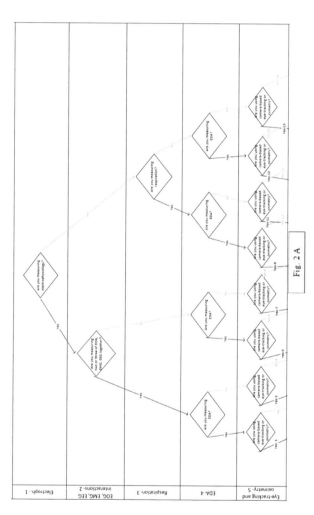

Fig. 2 A

Fig. 2. The flow chart presents a decision-tree leading to a number of recommendations for noise reduction in an experiment. Each level represents one decision, for example, if electrophysiology is used in the experiment. A black connector on the left represents a 'yes' choice and a grey connector on the right represents a 'no' choice. The bottom level consists of a unique set of recommendations for the chosen experimental design. Numbers represent the following recommendations: 1. Reduce acoustic noise in the lab, insulate walls from noise, use silent equipment. 2. Reduce electromagnetic influences, preferably use a Faraday Cage. 3. Prepare the skin for the electrophysiological electrodes. 4. Use the electrolyte for the electrophysiological electrodes. 5. Ask participants not to use soap, detergents, alcohol or hand-cream in the area where you apply EDA electrodes. 6. Reduce the movements of the participant. 7. Keep the stable temperature in the lab. 8. Keep the light in the lab at the stable level. 9. Ask participants not to use mascara. 10 Avoid the interactions between facial EMG, EOG and EEG electrodes, choose an alternative montage of EOG [14], lose electrodes montage of EEG instead of a cap, or divide the measures.

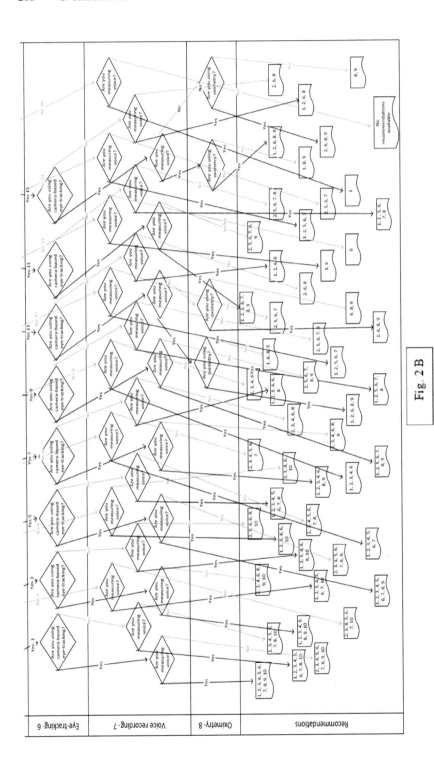

Fig. 2 B

Fig. 2. (*continued*)

3.2   Artefacts Reduction- Experimental Set-up

A number of techniques (e.g. artefact detection algorithms) exist to reduce noise and artefacts contained in physiological data. However, the accuracy of such tools is not always clear and the process typically leads to a loss/replacement of data, which can negatively impact the analysis process once such pre-processing has been completed. The best way to ensure high quality physiological recording is via a diligent experimental set-up and data collection process, reducing the level of noise contained in the raw data collected. [9]. It is for such reason the scope of this paper will only include set-up and a laboratory construction as the methods of noise reduction.

Electroencephalography. To avoid movement artefacts, participants are often required to fixate their gaze [15] and reduce movements [16]; however, it is hardly feasible in many experimental scenarios. For instance, driving tasks might require hand movements, whole body movements and vibration. Due to such difficulty, EEG devices are often used in a combination with additional devices to measure potential sources of signal disturbances, like EOG, EMG, gyroscope [17], eye-tracker [18] or a camera recording head movements [19]. In such instances, it is important to avoid putting electrodes too close to each other or pressing EOG/EMG electrodes with an EEG cap, because it might lead to a detachment of the electrodes and increased skin impedance. Any detachment of the electrodes or deficiency in the electrolyte can decrease a signal quality [20]. Skin impedance might be also increased by other factors, for example, sweat or not enough electrolyte used. To decrease the impedance of the skin there should be a thorough skin abrasion applied to remove a dead epidermis [21]. Impedance and signal quality might also depend on the equipment used. Even though there is a constant development of EEG technologies, devices recording signal with dry electrodes provide significantly worse data quality than gel or tap water based. Commercially distributed, cheap EEG devices with a small number of dry electrodes, also have significantly worse quality than scientific models [22]. In terms of the electrodes, montage caps give higher signal quality that headsets supporting electrodes [19].

Eye-Tracking. Most of the video-based eye-trackers require calibration process before and repeated during the experiment, which can be time consuming and break immersion during the experimental process. Some authors recommend planning a short experiment to avoid repetitive calibrations [8]. Aside from this, a number of simple steps can reduce the likelihood of common signal disturbances. For example, eyelashes are often a problem, participants should not wear mascara for the experiment. A further consideration is that, the laboratory should have a stable light intensity [8].

Electrooculography. EOG is often measured together with EEG or EMG. Many EEG devices include additional EOG electrodes that use the same grounding and reference as EEG. However, if a separate EOG device is used, electrodes for vertical eye-movements might interfere with forehead electrodes or a cap in EEG. An additional challenge is a placement of ground electrodes that is mostly recommended on the bony area in the middle of the forehead [5]. These electrodes can interfere with EEG electrodes, EEG cap and the electrodes of the forehead EMG. Therefore, it is recommended to use those devices separately. If they need to be combined, it is acceptable to use alternative EOG electrodes placement with two electrodes placed in the external eye edges and only one ground electrode in middle of the forehead [14]. Previous work describes experimentally validated, acceptable set-ups of the EOG electrodes [14]. Same as with other electrophysiological methods an electrode site should be prepared with a thorough skin abrasion process [21].

Electromyography. EMG electrodes can be placed anywhere on the skin surface. It is recommended not to combine forehead EMG and frontal cortex EEG or an EEG cap. HI:DAVe observed a decrease in signal quality and a more rapid impedance increase over time when forehead EMG electrodes were additionally pressed by the EEG cap. In a case, when EEG and forehead EMG need to be combined, the solution might be using a loose electrodes EEG montage. Same as with other electrophysiological methods an electrode site should be prepared with a thorough skin abrasion process [21].

Electrocardiography. The skin impedance highly influences the quality of the signal received by electrodes. Therefore, it is recommended to use wet electrodes with highly the conductive electrolyte to maximally reduce the impedance [11]. Same as with other electrophysiological methods an electrode site should be prepared with a thorough skin abrasion process [21].

Respiration. Respiration measured with an abdominal belt is quite an artefact-resilient method. However, like many other physiological signals, it can be disturbed by the noise created by the body movements. Hence, it is recommended that the participant remains still during the experiment [2].

Electrodermal Response. As EDA is a measurement related to the sweat glands activity, laboratory that uses this method should put special attention into keeping a stable temperature in the room [23]. Different guides recommend placing EDA electrodes on the non-dominant hand if the measurement is taken from the fingers, palm, or the wrist. Unlike with electrophysiology, the skin should not be scraped before the recording. Use of soaps, alcoholic substances and other detergents can disturb recording and because of that participants should be asked to clean their hands only with water before the experiment and not to use the hand-cream [24].

Acoustic Speech Analysis. External noise can be reduced with a noise-insulation of the lab space [9]. This will be described in details in part 3.3. Other research devices, that sometimes need to stay inside the insulated room, might produce low-frequency noise, for example, with fans or cables buzzing. If possible, choice of the other devices should take into account the level of the low-frequency noise that they generate [9]. Especially the recording equipment should be as noise-free as possible. The pre-amplifier should have a balanced XLR to minimalize artefacts caused by cables. It should, also, have a high gain, broad dynamic range, high SNR (signal to noise ratio), and phantom power [9]. The microphone should be either head-mounted or kept in the stable, very close distance from the mouth. The distance of the four centimeters is preferable and allows to reduce lots of external noise [9]. It is important to choose a microphone that has a wide and flat frequency curve to avoid different responses to the different frequencies of the speech [9]. Omnidirectional microphones mostly have a more even response to different frequencies; however, they should be used only in very quiet laboratories [25].

Oximetry. Oximetry is an optical method; hence, it is susceptible to the noise related to sources of the light. It is recommended to keep the light at the constant level. Movements of the participant can also disturb the measurement, so they should be reduced to the minimum [2, 10]. If the sensor is placed on the finger, it is beneficial to put it on the less active hand. It is not recommended to use finger sensors on the ears and vice versa, due to a decrease in signal quality [26].

3.3   Artefacts Reduction- Lab Construction

The laboratory environment for a voice recording should be well noise-insulated. Some researchers use anechoic chambers [25]; however, they are expensive and predominantly used solely for voice recording purposes. In a case of multisensory recording, the laboratory should meet multiple requirements to reduce different noise types. Therefore, it is optimal to build an isolation booth with different types of insulation depending upon the sensory recording(s) being focused upon. If recording voice for example, the booth, should be constructed using materials that provide noise insulation, ideally with double walls or walls insulated with acoustic material, and a floating floor [25]. It is optimal to remove loud devices from of the booth, such as air-conditioning fans, loud lights and PCs with loud fans [25].

In an experiment using electrophysiological methods, electrodermal activity, or oximetry it is beneficial to reduce the surrounding electromagnetic signals [7, 11, 13] by constructing a booth into a Faraday Cage [27]. Additionally, in the case of EDA measurements, the temperature inside the booth should be kept on the stable level [23].

The optical measuring methods, such as oximetry and camera-based eye-trackers, are susceptible to sources of light and because of this, it is recommended to keep the light inside the laboratory at the constant level [2, 8].

The laboratory constructed for HI:DAVe is a noise-insulated Faraday Cage currently configured with a mid-fidelity driving simulator. However, it is possible to re-configure the simulation suite to facilitate studies using physiological recordings across many domains (e.g. flight simulator), as the fundamental issues, concerning the quality of data collection would remain the same. Figure 3 presents photos of the outside noise insulation and an inside electromagnetic insulation. The walls inside the booth were later covered with fire-retardant plastic to reduce light effects caused by the aluminum foil. The booth was constructed from plywood and a fire-retardant cortex-like plastic. PCs, BioPac signal receiver and an oximeter were placed outside of the booth to reduce their electromagnetic influence on the signal. The holes were drilled in the walls to put receiving antennas and power cables through them. The receiving antennas from the BioPac receiver were extended with RP-SMA cables and Enobio-20 USB receiver was extended with a USB extension cable. Power cables and a voice recorder that had to stay inside were wrapped in aluminium foil.

Fig. 3. A: Outside of the laboratory booth insulated with fire-retardant acoustic foam to reduce acoustic noise in the voice recording. B: Inside of the laboratory booth insulated from electromagnetic noise with several layers of a heavy-duty aluminium foil. C: The way to connect aluminium foil with a cable. Cable has only a ground wire inside. The cable is led outside of the booth through the hole drilled in the wall and switched to the socket.

To ensure a reduction in electromagnetic interference the Faraday Cage had to be evenly covered with a conductive material. It does not have to be a sealed unit, it can be also covered with mesh or wire with the size of holes respective to the level of undesired frequencies. Materials recommended for electrophysiology are copper or several layers of heavy-duty aluminium foil [27, 28]. Even though small cages can properly shield without a grounding, much larger Faraday Cages need to be grounded to maintain shielding properties [28]. In the case of HI:DAVe's Faraday Cage aluminium foil from two walls were formed into 'pony-tails', clipped with a metal clip was used to connect these to the cables. The cables were plugged to the electrical sockets outside of the laboratory booth for grounding purposes. The cables were led outside through the holes drilled in the walls. Touching the aluminium walls of the Faraday Cage might cause an a tiny shock due to the static potential being transported from the body or clothes to the ground. Therefore, it is recommended to put non-conductive shielding on the walls. This is, also, beneficial as it reduces light reflexes created by the aluminium that might disturb eye-tracking recording. It is important to remember that all of the materials (e.g. staples) used to attach things to the walls should be highly conductive, so they do not disturb the Faraday Cage effect. In the case of HI: DAVe plastic sheets were attached to the aluminium walls with construction staples made of metal.

The current work presented a decision tree which can be utilized when undertaking the construction of a space to be used for an applied experimentation using physiological recoding techniques. The construction of a noise insulated faraday caged laboratory was detailed alongside recommendation for signal noise reductions.

Acknowledgements. We would like to thank Sean Tennant for the technical support provided during the design and construction of the Faraday cage. We would also like to thank Jodie Woodgate for her help during testing and calibrating psychophysiological equipment. This work was supported by Jaguar Land Rover and the UK-EPSRC grant EP/N011899/1 as part of the jointly funded Towards Autonomy: Smart and Connected Control (TASCC) Programme.

References

1. Kyriakidis, M., et al.: A human factors perspective on automated driving. Theor. Issues Ergon. Sci. 1–27 (2017)
2. Sweeney, K.T., Ward, T.E., McLoone, S.F.: Artifact removal in physiological signals—Practices and possibilities. IEEE Trans. Inf Technol. Biomed. **16**(3), 488–500 (2012)
3. Chang, K.-M.: Arrhythmia ECG noise reduction by ensemble empirical mode decomposition. Sensors **10**(6), 6063–6080 (2010)
4. Phinyomark, A., Phukpattaranont, P., Limsakul, C.: The usefulness of wavelet transform to reduce noise in the SEMG signal. In: EMG Methods for Evaluating Muscle and Nerve Function. InTech (2012)
5. Reddy, M.S., Chander, K.P., Rao, K.S.: Complex wavelet transform driven de-noising method for an EOG signals. In: 2011 Annual IEEE India Conference (INDICON). IEEE (2011)
6. Urigüen, J.A., Garcia-Zapirain, B.: EEG artifact removal—state-of-the-art and guidelines. J. Neural Eng. **12**(3), 031001 (2015)
7. Taylor, S., et al.: Automatic identification of artifacts in electrodermal activity data. In: 2015 37th Annual International Conference of the IEEE Engineering in Medicine and Biology Society (EMBC). IEEE (2015)
8. Duchowski, A.T.: Eye tracking methodology. Theory Pract. **328**, 614 (2007)
9. Plichta, B.: Best practices in the acquisition, processing, and analysis of acoustic speech signals. University of Pennsylvania Working Papers in Linguistics, **8**(3), 16 (2002)
10. Ram, M.R., et al.: A novel approach for motion artifact reduction in PPG signals based on AS-LMS adaptive filter. IEEE Trans. Instrum. Meas. **61**(5), 1445–1457 (2012)
11. Kirst, M., Glauner, B., Ottenbacher, J.: Using DWT for ECG motion artifact reduction with noise-correlating signals. In: 2011 Annual International Conference of the IEEE Engineering in Medicine and Biology Society, EMBC. IEEE (2011)
12. Rahman, M.Z.U., Shaik, R.A., Reddy, D.R.K.: Efficient sign based normalized adaptive filtering techniques for cancelation of artifacts in ECG signals: application to wireless biotelemetry. Signal Process. **91**(2), 225–239 (2011)
13. Chong, J.W., et al.: Photoplethysmograph signal reconstruction based on a novel hybrid motion artifact detection–reduction approach. Part I: motion and noise artifact detection. Ann. Biomed. Eng. **42**(11), 2238–2250 (2014)
14. Lopez, A., et al.: A study on electrode placement in EOG systems for medical applications. In: 2016 IEEE International Symposium on Medical Measurements and Applications (MeMeA). IEEE (2016)
15. Plöchl, M., Ossandón, J.P., König, P.: Combining EEG and eye tracking: identification, characterization, and correction of eye movement artifacts in electroencephalographic data. Front. Hum. Neurosci. **6**, 278 (2012)

16. Islam, M.K., Rastegarnia, A., Yang, Z.: Methods for artifact detection and removal from scalp EEG: a review. Neurophysiol. Clin./Clin. Neurophysiol. **46**(4–5), 287–305 (2016)
17. O'Regan, S., Faul, S., Marnane, W.: Automatic detection of EEG artefacts arising from head movements using EEG and gyroscope signals. Med. Eng. Phys. **35**(7), 867–874 (2013)
18. Noureddin, B., Lawrence, P.D., Birch, G.E.: Online removal of eye movement and blink EEG artifacts using a high-speed eye tracker. IEEE Trans. Biomed. Eng. **59**(8), 2103–2110 (2012)
19. Bang, J.W., Choi, J.-S., Park, K.R.: Noise reduction in brainwaves by using both EEG signals and frontal viewing camera images. Sensors **13**(5), 6272–6294 (2013)
20. Nolan, H., Whelan, R., Reilly, R.: FASTER: fully automated statistical thresholding for EEG artifact rejection. J. Neurosci. Methods **192**(1), 152–162 (2010)
21. Burbank, D.P., Webster, J.G.: Reducing skin potential motion artefact by skin abrasion. Med. Biol. Eng. Comput. **16**(1), 31–38 (1978)
22. Pinegger, A., et al.: Evaluation of different EEG acquisition systems concerning their suitability for building a brain–computer interface: case studies. Front. Neurosci. **10**, 441 (2016)
23. Society for Psychophysiological Research Ad Hoc Committee on Electrodermal Measures, et al.: Publication recommendations for electrodermal measurements. Psychophysiology, **49** (8), 1017–1034 (2012)
24. Cacioppo, J.T., Tassinary, L.G., Berntson, G.: Handbook of Psychophysiology. Cambridge University Press, Cambridge (2007)
25. Hunter, E.J., et al.: Acoustic voice recording, "I am seeking recommendations for voice recording hardware…". Language, **6**, 66–70 (2007)
26. Haynes, J.M.: The ear as an alternative site for a pulse oximeter finger clip sensor. Respir. Care **52**(6), 727–729 (2007)
27. Fathima, N., Umarani, K.: Reduction of noise in EEG signal using Faraday's cage and wavelets transform: a comparative study. IJESC, **6**(7) (2016)
28. Cutmore, T.R., James, D.A.: Identifying and reducing noise in psychophysiological recordings. Int. J. Psychophysiol. **32**(2), 129–150 (1999)

Modeling of Airflow and Dispersion of Aerosols in a Thanatopraxy Room

Jihen Chebil[1(✉)], Geneviève Marchand[2], Maximilien Debia[3], Loïc Wingert[2], and Stéphane Hallé[1]

[1] École de technologie supérieure, Montreal, Canada
{jihen.chebil.1,stephane.halle}@etsmtl.ca
[2] Institut de recherche Robert-Sauvé en Santé et en Sécurité du Travail, Montreal, Canada
{Genevieve.Marchand,Loic.Wingert}@irsst.qc.ca
[3] Université de Montréal, Montreal, Canada
maximilien.debia@umontreal.ca

Abstract. Workers performing post-mortem examination and embalming can be exposed to a risk of infection. Based on the hierarchy of preventive measures, ventilation is the most effective solution for reducing worker exposure since elimination and substitution cannot be implemented. The objective of this paper is to evaluate by numerical simulation (CFD) the effect of the ventilation rate on the behaviour of an aerosol in a thanatopraxy room. Airflow pattern in the room were validated with the tracer gas step-down technique. Airflow rate was demonstrated to have an influence on the mass concentration in the breathing zone. However deposition fraction on the floor is almost independent of the ventilation rate. The ventilation strategy has a significant impact. The mass fraction in breathing zone is reduced by 30% by adjusting the airflow rate from the supply grids compared to the ceiling diffusers.

Keywords: Thanatopraxy · Ventilation · Simulation · Aerosol

1 Introduction

Thanatopraxy is performed to delay the degradation of a dead body and preserve it up to the time of burial or cremation. The procedures used to preserve the body consist of several steps including the injection of disinfectant and preservative products, and the drainage of fluids and body gases in order to remove the visible effects of thanata-morphosis. It is complemented by aesthetic treatments, which can range from very light makeup to reconstruction [1–3]. The cadaver represents a potential risk of infection for the workers assigned to these tasks [4]. Potentially dangerous infections include, for example, Mycobacterium tuberculosis, hepatitis A, B and C, viruses, and human immunodeficiency virus [5–7]. When the body is being manipulated, these microorganisms can become airborne and be inhaled by the worker. Thanatopraxy procedures represent the same level of biological risk as autopsy procedures since both expose the worker to the same contamination source, that is, the dead body [5, 8].

© Springer Nature Switzerland AG 2020
D. N. Cassenti (Ed.): AHFE 2019, AISC 958, pp. 215–223, 2020.
https://doi.org/10.1007/978-3-030-20148-7_20

Many studies on the airborne transmission of infectious respiratory diseases demonstrate the relationship between ventilation and infection through inhalation [9, 10]. These studies show that an efficient ventilation system can minimize the propagation of airborne infectious pathogens in hospital rooms [11–13]. The performance of a ventilation system is affected by several parameters, one of which is the air change per hour (ACH), which is defined as the ratio between the volumetric airflow (m^3/h) and the volume of the room (m^3) [14]. Increasing airflow reduces the concentration of a pollutant in the ambient air and reduces the risk of airborne infection [15]. From a practical point of view, the ventilation system must provide good air quality while consuming minimal energy. ASHRAE standard 170 (2008) recommends 12 ACH for a morgue and autopsy room [16]. Standards in France recommend 4 ACH for a thanathopraxy room [5].

The objective of this paper is to evaluate by numerical simulation the effect of ventilation rate on the aerodynamic behaviour of an aerosol in a thanatopraxy room. The numerical model known as computational fluid dynamics (CFD) is based on the Navier-Stokes equations coupled with a mass transport equation for the bioaerosol. These equations are resolved using the Fire Dynamic Simulator software.

2   Methodology

2.1   The Thanatopraxy Room

The thanatopraxy room under investigation is located in Montreal (Québec, Canada). This room, presented schematically in Fig. 1, has two post-mortem tables, two access doors, cabinets, a counter and a closet. The room is divided into two sections: an entrance zone or transition zone of 23 m^3 and the post-mortem zone of 89 m^3. The total volume of this room is 112 m^3. A manikin located inside the chamber represents the deceased. Air is supplied to the room through four wall supply grilles (S1 to S4) of 0.28 m × 0.12 m and a square ceiling diffuser (S5) of 0.6 m × 0.6 m located in the transition zone. Air is extracted by two 1 m × 0.08 m grilles located near the floor and two 0.44 m × 0.16 m grilles located between the counter and cabinets at 1.08 m from the floor.

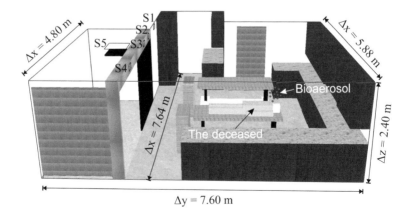

Fig. 1. Configuration of the thanatopraxy room

2.2   CFD Modelling

Modelling the air movements and aerosol advection/diffusion was performed using Fire Dynamics Simulator software (FDS, version 6.7). The software developed by the National Institute of Standard and Technology [17] is based on the Large Eddy Simulation method. The governing equations for the gas phase are the Navier-Stokes equations that are solved on a rectilinear numerical grid. Isothermal conditions are assumed for all cases.

The particle diameter is set to 3.47 μm. This value corresponds to the average diameter measured with an ultraviolet aerodynamic particle sizer spectrometer (TSI 3314) during the closing of an incision. At this size, particles can remain airborne with a potential for transmission by inhalation. Thus, it is reasonable to assume they will follow the behaviour of a passive scalar and can therefore be represented using a scalar transport equation, which incorporates the effects of Brownian motion, turbulent diffusion, and gravitational settling. This model called the drift-flux model was developed by Zhao et al. [18].

$$\frac{\partial}{\partial t}(\rho Z) + \nabla \cdot \left(\rho Z \left[\mathbf{U} + \mathbf{U}_{dep}\right]\right) = \nabla \cdot (\rho D \nabla Z). \tag{1}$$

Here, Z represents the mass fraction of the amount of contaminant in the air, expressed in (kg/kg of air). $\mathbf{U}$ is the velocity vector of the airflow and the air density is represented by ρ. The variable D is the diffusion coefficient. The aerosol deposition velocity to surface, ($\mathbf{U}_{dep}$), is the sum of gravitational settling velocity and diffusion-turbulence velocity. The spatial domain is discretized in simple orthogonal grids with grid spacings of 0.04 m to ensure feasibility of computation.

3   Results and Discussions

3.1   Numerical Model Validation

The tracer gas decay method was used to determine the number of air change per hour in the thanatopraxy room and to validate the CFD code. Complete pre-mixing of the SF_6 tracer gas was performed prior to the measurements using mixing fans. The SF_6 concentrations in parts per million (ppm) were measured at three locations in the room using an AutoTrac portable electron capture chromatograph (AutoTrac model 101, Lagus Applied Technology Inc., California, U.S.) after the pre-mixing period and for a duration of 24 min. The concentration measurements as a function of time were compared to the averaged concentration in the room as determined by the CFD (Fig. 2). The numerical result of the SF_6 concentration as a function of time is very similar to the experimental data. Theses results validate the capacity of our numerical model to represent the ventilation conditions of a real environment.

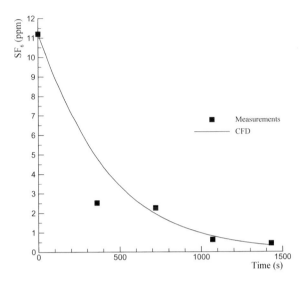

Fig. 2. Comparison between SF_6 concentrations measured experimentally and determined by CFD.

3.2 Mass Fraction in the Breathing Zone

The source of infectious material considered in our simulation occurs when the worker closes an incision made at the neck of the deceased to prevent leakage of body fluids. The duration of this operation and the particles diameter emitted were set respectively to 3 min and 3.47 μm. The behaviour of the particles emitted is determined for five ACH (4.64, 5.81, 6.97, 8.36 and 10.03). The first parameter chosen to evaluate the effect of ACH is the aerosol's mass fraction in the breathing zone. The breathable zone (V_{BZ}) is defined as an internal volume delimited by a minimum height of 0.08 m, a maximum height of 1.8 m and located at 0.6 m from the walls. Figure 3 presents the volume-weighted mean of the aerosol mass fraction in the breathing zone, $\bar{Z}_{BZ}$, normalized by the total mass of particles emitted during the exposure scenario ($m_{ejected}= 2.88 \times 10^{-3}$ mg).

$$MF_{BZ} = \frac{\rho_{air} \times V_{BZ} \times \bar{Z}_{BZ}(t)}{m_{ejected}}. \tag{2}$$

In this figure, the grey area corresponds to the emission period of the aerosol, which is between the first and fourth minute of simulation. There is a rapid increase in the mass fraction as soon as the aerosol source is activated. The mass fractions reach their maximum values between 3 min 45 s and 4 min 30 s depending on the ACH. Whatever the ventilation rate, the mass fraction decreases exponentially from the fifth minute until the end of the simulation.

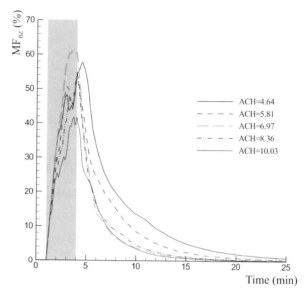

Fig. 3. Normalized mass fraction in the breathing zone for five ACH

The time required for aerosol extraction depends on the ventilation rate. Table 1 compares the average mass fraction in the breathing zone between minute 4 (end of particle emission) and minute 25. The mass fractions are normalized by the average mass fraction obtained at the lowest ventilation rate (ACH = 4.64). By comparing the non-dimensional mass fraction in the thanatopraxy room, air change rates were not found to be proportionately effective in reducing aerosol concentration. Increasing ventilation rates from 4.64 to 5.81 reduced the mass fraction by 31%, but the reduction is only 15% from 6.97 to 10.03 ACH.

Table 1. Non-dimensional mass fraction in the breathing zone

ACH	4.64	5.81	6.97	8.36	10.03
$\bar{Z}_{BZ}/\bar{Z}_{BZ,ACH=4.64}$	1.0	0.69	0.46	0.44	0.39

3.3   Aerosol Deposition

A worker can be infected by inhaling airborne particles from the air, but direct or indirect contact with a contaminated surface is also a potential route of entry leading to infection. Figure 4 presents the total mass of deposited particles on the horizontal surfaces normalized by the total mass of emitted particles during the exposure scenario. We noted that the deposition fraction (DF) is almost independent of the ventilation rate for ACH = 4.64, 5.81 and 6.97. DF drops from 1.8% to 1.5% at ACH = 8.36 and then to 1.4% at ACH = 10.03.

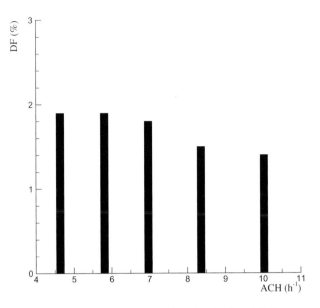

Fig. 4. Aerosol deposition fraction for five ACH

3.4   Ventilation Strategy

The thanatopraxy room is ventilated by five openings. Each opening does not have the same impact on the aerosol concentrations and deposition. Figure 5 shows the airflow streamlines associated with each opening. Openings S1 to S4 are located in the post-mortem zone and opening S5 provides ventilation mainly in the transition zone. Wall grille S4 generated airflow where the fresh air moves along the ceiling to hit the cabinets and then goes towards the deceased. This airflow pattern disperses the aerosol from the deceased throughout the room including the transition zone. The airflow from the ceiling diffuser (S5) moves from the transition zone to the post-mortem zone and then to the extraction grilles. In the post-mortem zone, the flow generated by S5 is weakly turbulent and almost unidimensional. This airflow pattern is similar to the airflow generated by a laminar flow ventilation system.

The air leaving the supply grilles S1 to S4 comes from the same ventilation duct. A pressure change in the duct will affect the amount of air at these outlets. However, the ceiling square diffuser (S5) is connected to an independent duct. In the actual room, the total volumetric airflow at the wall grilles is approximately the same as the volumetric airflow from the ceiling diffuser. Two additional ventilation strategies were simulated by CFD to determine if it could be possible to facilitate the extraction of aerosol by keeping the ACH constant. Only the air fractions supplied by the grilles and the diffuser are modified. Table 2 shows two ventilation strategies added to the original configuration in which the proportions of air to wall grilles and ceiling diffuser have been modified. The ACH is 5.81 which is equivalent to 0.181 m^3/s (Q_t).

The second ventilation strategy promotes dispersion of the aerosol throughout the room (Fig. 6). In the breathing zone, the aerosol non-dimensional mass fraction is 20%

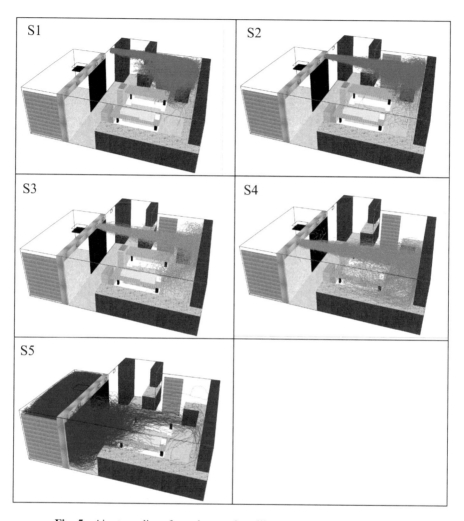

Fig. 5. Air streamlines from the supply grilles and diffuser (ACH = 5.81)

Table 2. Ventilation strategies

Strategy	$\frac{Q_{grilles}}{Q_t}$	$\frac{Q_{diffuser}}{Q_t}$	$\frac{Q_{infiltration}}{Q_t}$
1 (actual)	46%	48%	6%
2	64%	30%	6%
3	30%	64%	6%

higher than the current configuration and 30% higher than the value obtained for the third ventilation strategy. The intensity of the turbulence and the mixing conditions in the post-mortem zone are favoured by the increase of airflow through the wall grilles. This scenario delays aerosol extraction, which explains why the average concentration is higher, compared to strategies 1 and 3. The increase in turbulence intensity also explains why particles tend to settle down more effectively under this ventilation strategy.

In the third ventilation strategy, the flow is highly turbulent in the transition zone and the average air velocity is higher than in strategies 1 and 2. However, turbulence is mainly limited to this zone. As the flow passes from the transition zone to the post-mortem zone, it has stabilized and the intensity of turbulence is lower than in strategies 1 and 2.

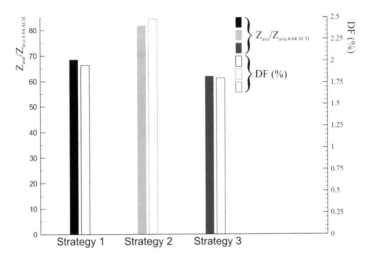

Fig. 6. Non-dimensional mass fraction and deposition fraction for three ventilation strategies

4   Conclusions

Aerosol dispersion in a thanatopraxy room was studied numerically. Particles were injected in the room from a surface of 0.04×0.04 m at the neck of a deceased. The mass fraction in the breathing zone and the deposition fraction were compared for five air changes per hour ranging from 4.64 to 10.03. Two additional ventilation strategies were simulated at ACH = 5.81. The aims was to determine if the air fractions supplied by the grilles and the diffuser could affect the aerosol concentration and deposition.

The results show that:

- The mass fraction in the breathing zone is reduced by 31% from ACH = 4.64 to ACH = 5.81. This reduction is only 15% from ACH = 6.97 to ACH = 10.03.
- The ventilation rate has little influence on the settling of particles. DF is between 1.4 and 1.9% regardless of the ventilation rate.
- The ventilation strategy has a significant impact. The mass fraction in breathing zone is reduced by 30% by adjusting the airflow rate from the supply grids compared to the ceiling diffusers.

Acknowledgments. The authors would like to thank Yves Beaudet for his participation in the study. This project was financially supported by the *Institut de Recherche Robert Sauvé en Santé et en Sécurité de Travail* (IRSST).

References

1. Anger, J.-P., et al.: Revue générale La thanatopraxie: une technique utile pour conserver les corps, mais qui peut gêner l'expertise toxicologique médico-légale. Ann. Toxicol. Anal. **20** (1), 1–10 (2008)
2. Lemonnier, M.: Thanatopraxie et thanatopracteurs: étude ethno-historique des pratiques d'embaumement. Montpellier, 3 (2006)
3. Lemonnier, M., Pesquera, K.: Thanatopraxie: technique, histoire et pratique au quotidien. Sauramps médical (2007)
4. Le Bacle, C., Duclovel-Pame, N., Durand, E.: Influenza aviaire, grippe aviaire et menace de pandémie: un nouvel enjeu en santé au travail. Doc. Med. Trav. **106**, 139–169 (2006)
5. Guez-Chailloux, M., Puymérail, P., Le Bâcle, C.: La thanatopraxie: état des pratiques et risques professionnels. In: Dossier médico-technique 104TC105, INRS, pp. 449–469 (2005)
6. Emritloll, Y., Payre, C., Vansaten, S.: Risques sanitaires liés à l'exhumation des corps humains et des carcasses d'animaux pour les travailleurs en France métropolitaine. Ecole des Hautes études en santé publique (2008)
7. Balty, I., Caron, V.: Biological and chemical hazards encouraged by Gravediggers. Occup. Health **130**, 25 (2012)
8. Colleter, R., et al.: Study of a 70-year-old French artificial mummy: autopsy, native and contrast CT scans. Int. J. Forensic Med. **132**(5), 1405–1413 (2018)
9. Chow, T.-T., Yang, X.-Y.: Performance of ventilation system in a non-standard operating room. Build. Environ. **38**(12), 1401–1411 (2003)
10. Yagoub, S.O., El Agbash, A.: Hospital-Delivery and Nursing Rooms (2010)
11. Streifel, A.: Hospital Epidemiology and Infection Control, Chapter 8, 2nd edn. Lippincott Williams & Wilkins, Philadelphia (1999)
12. Kaushal, V., Saini, P., Gupta, A.: Environmental control including ventilation in hospitals. Nursing **5**, 12 (2004)
13. Beggs, C.B., et al.: The ventilation of multiple-bed hospital wards: review and analysis. Am. J. Infect. Control **36**(4), 250–259 (2008)
14. Rim, D., Novoselac, A.: Ventilation effectiveness as an indicator of occupant exposure to particles from indoor sources. Build. Environ. **45**(5), 1214–1224 (2010)
15. Memarzadeh, F., Xu, W.: Role of air changes per hour (ACH) in possible transmission of airborne infections. In: Building Simulation. Springer (2012)
16. Ashrae, A.: ASHRAE/ASHE Standard 170-2008 Ventilation of health care facilities, Atlanta (2008)
17. McGrattan, K., et al.: Fire dynamics simulator technical reference guide, verification. Technical report, NISTIR special publication, 1018-2, 6th edn., vol. 2 (2017)
18. Zhao, B., et al.: Particle dispersion and deposition in ventilated rooms: testing and evaluation of different Eulerian and Lagrangian models. Build. Environ. **43**(4), 388–397 (2008)

Assessment of Students' Cognitive Conditions in Medical Simulation Training: A Review Study

Martina Scafà[✉], Eleonora Brandoni Serrani, Alessandra Papetti,
Agnese Brunzini, and Michele Germani

Università Politecnica delle Marche, Via Brecce Bianche 12,
60131 Ancona, Italy
{m.scafà,a.brunzini}@pm.univpm.it,
{e.brandoni,a.papetti,m.germani}@staff.univpm.it

Abstract. Although performance measures are strongly used in the field of medical education to evaluate skills of trainees and medical students, the assessment of their cognitive state is relatively "uncommon". This fact is disadvantageous if we consider the introduction of technologies as physical medical simulators and augmented/virtual reality devices, which may represent an improvement in the students' immersion in the simulated scenario or, conversely, a potential risk of a serious information overload. Therefore, a precise assessment of the cognitive conditions is an essential element of the design process of a medical training session. This study aims to provide the current state in literature on the assessment of cognitive state during medical simulation training sessions. It provides critical insights on the validity and reliability of current metrics and helps in the selection of measurements tools when applied in simulation-based training contexts.

Keywords: Medical simulation · Training · Cognitive conditions ·
Human factors – stress and workload

1 Introduction

The increase in professional activities that have a "mental dimension" has encouraged the development of cognitive ergonomics. While the traditional "physical" ergonomic focuses on our bodies, cognitive ergonomics is interested in what happens in our brain: how information is perceived, understood and interpreted, and therefore what determines decision-making processes [1]. Cognitive ergonomics also integrates theories and principles of ergonomics, neuroscience and human factors to provide information on the function and behavior of the brain in training phase [2]. Cognitive ergonomics focuses on the analysis of the cognitive process and is used to support the skills and limitations of human beings in their interaction with a system, taking into consideration: attention, perception errors, strategies, workloads, information visualization, decision support, human-machine interaction, situation awareness and training [3]. For the medical students the training phase is a decisive moment to learn better and to practice their studies. However, this phase can generate excessive levels of acute stress,

© Springer Nature Switzerland AG 2020
D. N. Cassenti (Ed.): AHFE 2019, AISC 958, pp. 224–233, 2020.
https://doi.org/10.1007/978-3-030-20148-7_21

thus compromising their performance status [4]. With the use of simulation-based training, medical students can be provided with direct assessment of their skills. Training on high-fidelity simulators has emerged as an effective way of complementing the clinical training of medical students and residents. In theory, simulation training appears to offer training conditions that are optimal for the acquiring of clinical skills [5]. In this context, mannequin-based simulators are often employed as they offer the possibility to create a wide range of standardized clinical scenarios and this fact puts the trainees into the position to be exposed to adverse events, improving their diagnostic and decision-making skills to respond with minimal time and errors [6]. Both technical and "non-technical" skills, however, can be adversely affected by the high demand on the cognitive resources of the learners caused by these complex learning environments [7]. Additionally, the introduction of new technologies in these context (as physical medical simulators and augmented/virtual reality devices) may represent both an improvement in the user's immersion and involvement in the simulated scenario, but also a potential risk of a serious information overload during training [8]. We should consider whether elements such as high fidelity and the increased emotions of students can influence the learning of some trainees [9]. During these simulation sessions, trainees could feel discomfort, fatigue and stress. Commonly recognized stressors include technical complications, time pressure, distractions, interruptions, and increased workload [10]. From this prospective, the assessment of cognitive workload becomes a key component in the medical students' performance and the proper management of their cognitive conditions can be an essential aspect of medical education. In this review it is decided to consider a specific aspect of cognitive stress: mental workload [11]. The analysis aims to cover the missing in the literature of the following information:

- classification of all the existing methods and tools for the analysis of mental load;
- an analysis of mental workload during training.

The aim of this paper is to propose a new literature review focused on the assessment of mental workload in medical education with a focus on simulation-based training models.

2 Materials and Methods

The review has been conducted by using the ScienceDirect, Scopus and ReaseachGate databases as sources of scientific papers, as well as considering articles found in the bibliography of the analyzed papers. The search strategy employed the following terms with the appropriate combinations: "mental workload", "students", "simulation", "cognitive load", and "medical education". In order to be included in the review, studies had to fulfill the following criteria: (a) the population studied consisted of trainees, novices or medical students; (b) the study design was experimental; (c) the study included at least one measure of cognitive workload; and (d) the language of publication was English.

The full set of references have been collected, analyzed and categorized according to the following criteria:

– general information: title, authors, years;
– paper typology: proposal of theoretical methods, practical approaches, review paper, case study;
– specific information: objective, main findings, conclusions, limitations of the study.

Among all the papers found, 40 articles were selected, as they were considered more significant and more interesting for this review. Figure 1 reports the temporal distribution of references used in this study.

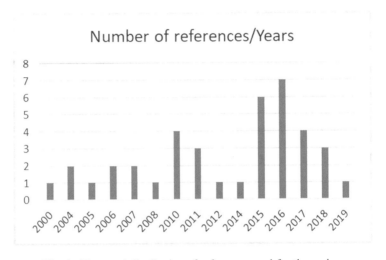

Fig. 1. Temporal distribution of references used for the review

3 Mental Workload

3.1 What Is Mental Workload?

Mental workload (MWL) is probably one of the most researched constructs in ergonomics and human factors with several applications in aviation, automobile industry, rail industry and increasingly considered in Human-Computer Interaction as well as clinical medicine [12]. Despite over 40 years of research, there is still no clear and universally accepted definition of human mental workload [13].

Mental workload is often described by terms such as 'cognitive load', 'cognitive effort', 'mental strain', and 'mental effort' and there are too many models adopted by practitioners, with an evident limitation both in their application and in their comparison making. The principal reason for measuring mental workload is to quantify the mental cost of performing a task in order to predict operator and system performance, and it is a fundamental construct for exploring the interaction of people with technological devices [14].

The concept of mental workload is based on the assumption that each person has a relatively fixed cognitive capacity. This capacity is likened to a pool from which resources can be drawn to meet the demands of all ongoing tasks [15]. Given this assumption, the mental workload for a given task is the ratio of mental resources required to the total resources available, on a moment-to-moment basis. Put another way, mental workload is inversely related to spare capacity when performing the task of interest [16]. However, this is a simplistic view because mental workload "emerges from the interaction between the requirements of a task, the circumstances under which it is performed and the skills, behaviors and perceptions of the operator" [17].

Several reviews attempted to organize the significant amount of knowledge behind measurement procedures [18]. Generally, researchers agree in classifying MWL into three main broad categories: self-assessment, task performance and physiological measures.

The category of self-assessment measures, often defined as subjective or self-assessment, is based on the subjective perceived experience of the interaction operator-system and is obtained from the direct estimation of task difficulty. The self-assessment measures have always attracted many researchers because there is the belief that only the individual concerned in the task can provide an accurate judgment with respect to the MWL experienced [19].

The class of task performance measures assumes that the mental workload of an operator, when interacting with a system, acquires importance only if it influences system performance. These measures are usually divided in primary and secondary task measures. In primary task methods, the performance of a clinician is monitored and analyzed according to changes in the demands of the primary task under execution. Examples of common measurement parameters are response, reaction time, accuracy and error rate, estimation time, objective speed and signal detection [20].

The category of physiological measures considers physiological responses of the operator's body that are believed to be correlated to MWL. They are aimed at interpreting psychological processes by analyzing their influence on the state of the body and not by measuring perceptual subjective ratings or task performance [21].

3.2 Assessment of Methods and Tools for Mental Workload Analysis

In the literature there are different tools for the assessment of mental load [22]. These tools can be used in three different methods of mental load analysis: *(i)* subjective measures method, *(ii)* performance measurement method and *(iii)* psychophysiological parameters measures method [23]. The following paragraphs present the most complete classification of all tool used in different models.

Subjective Measures Method. The *subjective measurement methods* provide information on how students subjectively evaluate various aspects of mental load in the training phase [24]. The subjective measures are an attempt to systematically interrogate the subjects on their awareness of the difficulty of the task. These methods use psychometric scales or questionnaires to analyze how medical students feel in the training phase [25].

The tool more used for subjective measurement is *NASA Task Load Index (NASA-TLX)*. It is a multidimensional assessment tool that rates perceived workload in order to assess a task, system, or team's effectiveness or training aspects of performance [26]. Muresan et al. used NASA-TLX for testing mental workload of forty-one medical students after introduced virtual reality drills for intracorporeal suturing training [27]. The *Surgery Task Load Index (SURG-TLX)* tool is used to provide diagnostic information about the impact of various sources of stress on the perceived demands of trained surgical operators [28]. Wucherer et al. for example tested mental workload of their innovative simulator-based methodology for medical training using SURG-TLX tool. After these tests they evidenced the necessity to develop realistic simulation environments that prepare young medical to respond to emergent events in the operating room [29]. While methods such as NASA-TLX are more suitable for the analysis of the mental load of a single task, the *Multiple Resources Questionnaire (MRQ)* tool is used to measure the cognitive load in more complex training processes [30]. Carswell et al. used an innovative vocal version of MRQ to evaluate sixty-four undergraduates that performed fifteen of laparoscopic training tasks [31]. Helton tested *Short Stress State Questionnaire (SSSQ)* through two studies providing initial psychometric and validation evidence of a short multidimensional self-report measure of stress state. His aim was to prove validity of information on the SSSQ regarding its sensitivity to task-stressors [32]. The *Borg score of the Perceived Exertion scale (BORG)* is a tool for measuring perceived mental fatigue during physical work. William explain that during the test asks to participants to rate their exertion during the activity, combining all sensations and feelings of stress and fatigue. They are told to disregard any one factor such as leg pain or shortness of breath but to try to focus on the whole feeling of exertion [33]. Haji et al. presented an experiment that through *Subjective rating of mental effort (SRME)* tool investigate the effects of variations in task complexity on novices' cognitive load and learning during simulation-based procedural skills training [34].

Performance Measurement Method. The *performance measurement method* identifies which characteristics of the task performed by the student can be used as an index of the imposed mental load. This method assumes that the MWL of an operator interacting with a system is relevant only if it affects system performance. The behavior of the operator is used to determine the workload. For example, lowered and/or irregular performance may indicate that the worker is or is reaching unacceptable levels of MWL. The *primary task method* tries to deduct the mental load from the performance on the activity of interest. It uses techniques to directly quantify the ability of the operator to perform the primary task at an acceptable level. However, it is not obvious that primary task measures have much direct association with the workload: performance errors do not necessarily indicate a high load imposed by the main activity. For this reason, the *secondary task method* is used the most [35]. In this type of analysis, the operator is required to perform a secondary activity concurrently with the main activity of interest. In particular, the student is asked to perform two tasks at the same time, one is used for the mental load measured (primary task), the other (secondary task) is used to calculate the mental load associated with the primary task [36]. Haji et al. presented a report a two-stage study that studies the sensitivity of subjective assessments of mental effort and the performance of the secondary task of medical students. The training involved the following tasks: single task (monitoring of the virtual patient's heart rate) and double task (ligation of surgical nodes on a benchtop simulator during monitoring the virtual patient's heart rate) [37].

Psychophysiological Parameters Measures Method. The *psychophysiological parameters* can be indirect indicators of mental workload.

In fact, changes associated with different levels of mental load have been reported in cardiac, ocular, respiratory and cerebral systems. In particular, the following parameters can be monitored:

- Heart Rate (HR)
- Heart Rate Variability (HRV)
- Respiratory Frequency
- Gaze Entropy
- Gaze Velocity

These parameters are monitored using wearable devices such as: smart heart sensor for monitoring HR, HRV and Respiratory Frequency and smart glasses for monitoring gaze entropy and velocity. For this reason, procedures for the detection of psychophysiological indices are not always accepted by students who may consider them invasive. In fact, the measurement monitors the physical reactions to which the student is subjected [38]. However, there are applications in the literature. Di Stati et al. gaze entropy and velocity of surgical trainees and attending surgeons during two surgical procedures thought wearable eye tracker device use [39]. Theodoraki et al. instead presented a case in which students were connected to a biofeedback device that measures heart rate, heart rate variability, respiratory rate and EMG masticator. Stress situations have been identified by an increase in heart rate and a decrease in heart rate variability [40].

4 Discussion

The cognitive load measurement methods used in the reviewed studies are shown in Table 1. Most studies (78%) applied only one method to measure the cognitive load, and the remaining (22%) used two methods. Similarly, most studies applied only one tool (67%), while the leftover authors (33%) used a total of two tools as shown in Table 2.

Table 1. Number of methods used in each paper.

Number of methods applied	Percentage of use in papers
One methods	78%
Two methods	22%
Three methods	0

Table 2. Number of tools used in each paper.

Number of tools applied	Percentage of use in papers
One tool	67%
Two tools	33%
Three tools	0
Four tools	0

All the papers have been classified to analyze which methods and tools are most used in medical training. All the percentages shown in Table 3 have been calculated out of the totality of 40 papers. Among the 3 different methods that are recognized to be used, the self-assessment method is the most applied (72%); the 56% of paper chose to apply the NASA Task Load Index (NASA-TLX); the 8% used a modified version of NASA-TLX, the Surgery Task Load Index, developed and validated specifically to capture the surgical context. Furthermore, to control for potential fluctuations of fatigue across the experiment, in 6% of studies an adapted version of the Borg Rating of Perceived Exertion Scale (BORG) was used. The BORG ranges between 6 and 20, higher scores indicating higher perceived mental fatigue. Multiple Resources Questionnaire (MRQ) and Short Stress State Questionnaire (SSSQ) were used in the 8% of paper. To measure cognitive load, 15% of cases recorded subjective ratings of mental effort (SRME) after each trial by asking participants to indicate the amount of mental effort invested in achieving the LP on a 9-point Likert scale.

The method of task performance measures was applied in 39% of paper and the tools used by all of them is secondary task measure.

The method of Psychophysiological measures was applied only in 11% of studies. The 6% of paper examined measured three cardiovascular parameters during the FESS surgery: the heart rate (HR), the respiratory frequency and the heart rate variability (HRV). The other 6% considered gaze entropy and velocity.

The most widely used tool for mental workload measurement is NASA-TLX. However, this tool has proved its validity in many research studies, remains a subjective evaluation tool. And more generally, the subjective measurement method is used in a very high percentage. The main limit remains therefore that the evaluation could be susceptible to the operator's injury. In contrast, methods used to measure the physiological parameters are used in lower percentages. However, in the clinical setting, physiological measures represent natural indicators of mental workload because human work requires physiological activity. However, only a small amount of studies has taken physiological measures since they require tools and equipment that are often physically hostile. Furthermore, the analysis of the data collected by these tools is a complex process that requires data analysis experts. The measure of performance, on the other hand, shows difficulties in assessing human psychic and psychometric performance in certain working conditions, such as, for example, carrying out several work tasks. This method requires considerable experience and basic knowledge and it may involve further development of software and hardware.

Table 3. Percentage of different methods' use.

Subjective measures (72%)		Performance measurement (39%)		Psychophysiological parameters measures (11%)	
NASA Task Load Index (NASA-TLX)	56%	Primary task measure	0	HR	6%
Surgery Task Load Index (SURG-TLX)	6%	Secondary task measure	39%	HRV	6%
Multiple Resources Questionnaire (MRQ)	6%			Respiratory frequency	6%
Short Stress State Questionnaire (SSSQ)	6%			Gaze entropy	6%
Borg Rating of Perceived Exertion Scale (BORG)	6%			Gaze velocity	6%
Subjective rating of mental effort (SRME)	11%				

5 Conclusion

The paper presents a review of the methods and tools most used for mental workload assessment. In particular, the research was focused on the mental load of medical students during the training phase. The following critical issues have emerged:

- the method used most is the subjective measurements,
- there is no method that calculates the mental load through the calculation of objective parameters;
- most of the studies are conducted in medical field.

Thus, future developments should have two main objectives. Firstly, mental load studies should have a more engineering approach. Secondly, the main goal should be to develop an objective method that allows the monitoring of measurable meters and the development of algorithms that will analyze the data and make it easy to read.

References

1. Studer, R.K., Danuser, B., Gomez, P.: Physicians' psychophysiological stress reaction in medical communication of bad news: a critical literature review. Int. J. Psychophysiol. **120**, 14–22 (2017)
2. Parasuraman, R.: Neuroergonomics: brain, cognition, and performance at work. Curr. Dir. Psychol. Sci. **20**, 181–186 (2011)
3. Parasuraman, R., Sheridan, T.B., Wickens, C.D.: Situation awareness, mental workload, and trust in automation: viable, empirically supported cognitive engineering constructs. J. Cogn. Eng. Decis. Mak. **2**, 140–160 (2008)

4. Dias, R.D., Ngo-Howard, M.C., Boskovski, M.T., Zenati, M.A., Yule, S.J.: Systematic review of measurement tools to assess surgeons' intraoperative cognitive workload. Glob. Surg. **105**(5), 491–501 (2018)

5. Fraser, K., Ma, I., Teteris, E., Baxter, H., Wright, B., McLaughlin, K.: Emotion, cognitive load and learning outcomes during simulation training. Med. Educ. **46**(11), 1055–1062 (2012)

6. Van Merriënboer, J.J.G., Sweller, J.: Cognitive load theory in health professional education: design principles and strategies. Med. Educ. **44**(1), 85–93 (2010)

7. Charles, R.L., Nixon, J.: Measuring mental workload using physiological measures: a systematic review. Appl. Ergon. **74**, 221–232 (2019)

8. Atalay, K.D., Can, G.F., Erdem, S.R., Müderrisoglu, I.H.: Assessment of mental workload and academic motivation in medical students. J. Pak. Med. Assoc. **66**, 574 (2016)

9. Du, W., Kim, J.H.: Performance-based eye-tracking analysis in a dynamic monitoring task. In: International Conference on Augmented Cognition, pp. 168–177 (2016)

10. Arora, S., Sevdalis, N., Nestel, D., Woloshynowych, M., Darzi, A., Kneebone, R.: The impact of stress on surgical performance: a systematic review of the literature. Surgery **147**, 318–330 (2009)

11. Bosse, H.M., Mohr, J., Buss, B., Krautter, M., Weyrich, P., Herzog, W., Junger, J., Nikendei, C.: The benefit of repetitive skills training and frequency of expert feedback in the early acquisition of procedural skills. BMC Med. Educ. **15**, 22 (2015)

12. Longo, L.: Mental workload in medicine: foundations, applications, open problems, challenges and future perspectives. In: 2016 IEEE 29th International Symposium on Computer-Based Medical Systems (CBMS), Belfast and Dublin, Ireland, pp. 106–111 (2016)

13. Cain, B.: A Review of the Mental Workload Literature. Defense technical information center, Toronto, Canada (2007)

14. Longo, L.: Formalising human mental workload as a defeasible computational concept (2014)

15. Wickens, C.D., Gordon, S.E., Liu, Y.: An introduction to human factors engineering, pp. 64–90, Upper Saddle River (2004)

16. Carswell, C.M., Clarke, D., Seales, W.B.: Assessing mental workload during laparoscopic surgery. Surg. Innov. **12**, 80–90 (2005)

17. Hart, S.G.: Nasa-task load index (NASA-TLX); 20 years later. In: Proceedings of the Human Factors and Ergonomics Society Annual Meeting, vol. 50, no. 9, pp. 904–908 (2006)

18. Xie, B., Salvendy, G.: Review and reappraisal of modelling and predicting mental workload in single- and multi-task environments. Work Stress **14**(1), 74–99 (2000)

19. Woods, B., Byrne, A., Bodger, O.: The effect of multitasking on the communication skill and clinical skills of medical students. BMC Med. Educ. **18**, 76 (2018)

20. International Encyclopaedia of Ergonomics and Human Factors. https://www.taylorfrancis.com/books/e/9780849375477

21. Lee, G.I., Lee, M.R.: Can a virtual reality surgical simulation training provide a self-driven and mentor-free skills learning? Investigation of the practical influence of the performance metrics from the virtual reality robotic surgery simulator on the skill learning and associated cognitive workloads. Surg. Endosc. **32**(1), 62–72 (2017)

22. Boet, S., Sharma, B., Pigford, A., Hladkowicz, E., Rittenhouse, N., Grantcharov, T.: Debriefing decreases mental workload in surgical crisis: a randomized controlled trial. Surgery **161**(5), 1215–1220 (2017)

23. Scerbo, M.W., Britt, R.C., Montano, M., Kennedy, R.A., Prytz, E., Stefanidis, D.: Effects of a retention interval and refresher session on intracorporeal suturing and knot tying skill and mental workload. Surgery **161**(5), 1209–1214 (2016)

24. Gardner, A.K., Clanton, J., Jabbour, I.I., Scott, L., Scott, D.J., Russo, M.A.: Impact of seductive details on the acquisition and transfer of laparoscopic suturing skills: emotionally interesting or cognitively taxing? Surgery **160**(3), 580–585 (2016)
25. Britt, R.C., Scerbo, M.W., Montano, M., Kennedy, R.A., Prytz, E., Stefanidis, D.: Intracorporeal suturing: transfer from fundamentals of laparoscopic surgery to cadavers results in substantial increase in mental workload. Surgery **158**(5), 1428–1433 (2015)
26. Colligan, L., Potts, H.W.W., Finn, C.T., Sinkin, R.A.: Cognitive workload changes for nurses transitioning from a legacy system with paper documentation to a commercial electronic health record. Int. J. Med. Inform. **84**(7), 469–476 (2015)
27. Muresan, C., Lee, T.H., Seagu, J., Park, A.E.: Transfer of training in the development of intracorporeal suturing skill in medical student novices: a prospective randomized trial. In: 11th World Congress of Endoscopic Surgery, Japan, pp. 537–541 (2008)
28. Wilson, M.R., Poolton, J.M., Malhotra, N., Ngo, K., Bright, E., Masters, R.S.W.: Development and validation of a surgical workload measure: the surgery task load index (SURG-TLX). World J. Surg. **35**(9), 1961–1969 (2011)
29. Wucherer, P., Stefan, P., Abhari, K., Fallavollita, P., Weigl, M., Lazarovici, M., Winkler, A., Weidert, S., Peters, T., De Ribaupierre, S., Eagleson, R., Navab, N.: Vertebroplasty performance on simulator for 19 surgeons using hierarchical task analysis. IEEE Trans. Med. Imaging **34**(8), 1730–1737 (2015)
30. Boles, D.B., Bursk, J.H., Phillips, J.B., Perdelwitz, J.R.: Predicting dual-task performance with the multiple resources questionnaire (MRQ). Hum. Factors: J. Hum. Factors Ergon. Soc. **49**, 32–45 (2007)
31. Carswell, C.M., Lio, C.H., Grant, R., Klein, M.I., Clarke, D., Seales, W.B., Strup, S.: Hands-free administration of subjective workload scales: acceptability in a surgical training environment. Appl. Ergon. **42**(1), 138–145 (2010)
32. Helton, W.S.: Validation of a short stress state questionnaire. Hum. Factors: J. Hum. Factors Ergon. Soc. **48**(11), 1238–1242 (2004)
33. Williams, N.: The borg rating of perceived exertion (RPE) scale. Occup. Med. **67**(5), 404–405 (2017)
34. Haji, F.A., Cheung, J.J.H., Woods, N., Regehr, G., De Ribaupierre, S., Dubrowski, A.: Thrive or overload? The effect of task complexity on novices' simulation-based learning. Med. Educ. **50**(9), 955–968 (2016)
35. Sørensen, S.M.D., Mahmood, O., Konge, L., Thinggaard, E., Bjerrum, F.: Laser visual guidance versus two-dimensional vision in laparoscopy: a randomized trial. Surg. Endosc. **31**(1), 112–118 (2017)
36. Blanco, M., Biever, W.J., Gallagher, J.P., Dingus, T.A.: The impact of secondary task cognitive processing demand on driving performance. Accid. Anal. Prev. **38**(5), 895–906 (2006)
37. Haji, F.A., Khan, R., Regehr, G., Drake, J., De Ribaupierre, S., Dubrowski, A.: Measuring cognitive load during simulation-based psychomotor skills training: sensitivity of secondary-task performance and subjective ratings. Adv. Health Sci. Educ. **20**(5), 1237–1253 (2015)
38. Hu, J.S.L., Lu, J., Tan, W.B., Lomanto, D.: Training improves laparoscopic tasks performance and decreases operator workload. Surg. Endosc. **30**(5), 1742–1746 (2016)
39. Di Stasi, L.L., Díaz-Piedra, C., Ruiz-Rabelo, J.F., Rieiro, H., Sanchez Carrion, J.M., Catena, A.: Quantifying the cognitive cost of laparo-endoscopic single-site surgeries: gaze-based indices. Appl. Ergon. **65**, 168–174 (2017)
40. Theodoraki, M.N., Ledderose, G.J., Becker, S., Leunig, A., Arpe, S., Luz, M., Stelter, K.: Mental distress and effort to engage an image-guided navigation system in the surgical training of endoscopic sinus surgery: a prospective, randomised clinical trial. Eur. Arch. Otorhinolaryngol. **272**(4), 905–913 (2015)

Investigating Human Visual Behavior by Hidden Markov Models in the Design of Marketing Information

Jerzy Grobelny and Rafał Michalski[✉]

Faculty of Computer Science and Management, Wrocław University of Science and Technology, 27 Wybrzeże Wyspiańskiego, 50-370 Wrocław, Poland
{jerzy.grobelny, rafal.michalski}@pwr.edu.pl
http://www.JerzyGrobelny.com,
http://www.RafalMichalski.com

Abstract. The research demonstrates the use of hidden Markov models (HMMs) in analyzing fixation data recorded by an eye-tracker. The visual activity was registered while performing pairwise comparisons of simple marketing messages. The marketing information was presented in a form of digital leaflets appearing on a computer screen and differed in the components' arrangement and graphical layout. Better variants were selected by clicking on them with a mouse. A simulation experiment was performed to determine best HMMs in terms of information criteria. Seven selected models were presented in detail, four of them graphically illustrated and thoroughly analyzed. The identified hidden states along with predicted transition and emission probabilities allowed for the description of possible subjects' visual behavior. Hypotheses about relations between these strategies and marketing message design factors were also put forward and discussed.

Keywords: Eye tracking · Cognitive modeling · Visual presentation · Digital signage · Advertisement · Human factors · Ergonomics

1 Introduction

Understanding the way people process visual information plays an extremely important role in providing ergonomic recommendations. The effectiveness, efficiency, and satisfaction concerned with graphical messages of various kinds should be taken into account during the design process.

Eye-tracking has recently become a popular technique that facilitate the understanding of human attentional behavior in a variety of contexts. For instance, Grebitus and Roosen [1] examined the influence of the number of attributes and alternatives on the information processing in discrete choice experiments. Muñoz Leiva et al. [2] focused on the roles of engagement and the banner type in advertising effectiveness. Recently, Michalski and Grobelny [3] examined package design graphical factors. Ergonomic analysis of different variants of digital control panels were complemented by an eye-tracking study in [4]. In the work [5], in turn, eye-movement data were used

© Springer Nature Switzerland AG 2020
D. N. Cassenti (Ed.): AHFE 2019, AISC 958, pp. 234–245, 2020.
https://doi.org/10.1007/978-3-030-20148-7_22

for investigating medicine information flyers. More applications can be found, for example, in the literature review presented by Huddleston et al. [6].

The main advantage of applying eye-trackers is the possibility of measuring peoples' visual activities in an objective way. Unfortunately, the registered line of sight provides information only on the so called overt attention [7]. Researchers constantly pursue methods that would extend the analyses also to covert attention, which cannot be directly observed by eye-tracking. One of the possible approaches involves the use of Markov models. Already in 1986, Ellis and Stark [8] proposed using stochastic first-order Markov process for this purpose. Some further studies also followed this idea, extending it to hidden Markov models (HMMs) where hidden states are interpreted as manifestations of covert attention (e.g., [9–11]). Lately, Chuk et al. [12] used HMM analysis in face recognition tasks. Grobelny and Michalski [13] applied a series of HMMs for visual activity data recorded while performing monitoring tasks on digital control panels. Comparable methodology was involved in the analysis of pairwise comparisons of virtual packages [14].

The present study is a continuation of these studies and employs similar approaches. The general objective is to extend our knowledge about human visual behavior while assessing different versions of advertisement leaflets. HMMs parameters are estimated based on observation fixation sequences derived from the eye tracking data, assumed number of states, and a vocabulary corresponding to the defined areas of interests (AOIs).

2 Overview of the Experiment

The eye-movement activity used in the present research come from the work of Michalski and Ogar [15]. The main goal of that study was to verify the influence of graphic elements of the flyer presenting details about internet packages on preferences of potential customers. Four variants of the leaflet were designed differing in the location of information components and graphical grouping of the leaflet content. The first factor was examined on two levels: package price was situated either directly under the package title in the upper area of the flyer, or at its very bottom. As a consequence of the two possible locations of the price information, the text *Order* changed its place accordingly (compare Fig. 1). The second effect was also specified on two levels. The three types of internet packages were separated either by white, empty spaces or spaces in the color of the flyer background (grey). In the rest of this paper, the former level is called *divided* whereas the second one – *solid*.

Graphical layouts were subjectively assessed by binary pairwise comparisons. Each participant evaluated all possible pairs of investigated stimuli, thus, in one experiment there were six comparisons.

The experiment was conducted in an isolated room equipped with a desk, typical office chair, keyboard, optical computer mouse, and 21" monitor. Subjects' behavior was monitored through a one-way mirror and registered by video cameras. Their visual activity was recorded by a modern SMI RED500 stationary infrared eye-tracker system. The device records eyeball movements at 500 Hz sample rate with 0.4° accuracy.

As a result of a quality analysis of the gathered data, results from 49 out of total 71 experiment participants were explored. There were 16 female and 33 male students from Wrocław University of Science and Technology (Poland). Their age ranged between 19 and 31 with the mean equal 21.9 and standard deviation of 3.0.

Generally, the preferences outcomes revealed higher rates for variants with divided graphical structures and those, with price information placed in the upper part of the leaflet. The visual activity analysis involved heat maps, descriptive statistics of processing and searching measures, along with analyses of variance for fixation durations, number of fixations, and pupil diameters.

3   Modeling Visual Activity by HMMs

Markov models were introduced by a Russian mathematician Markov in 1913 [16]. They deal with states and transitions between them specified by likelihoods of their occurrences. A hidden version of Markov models involves unobserved states that can be identified indirectly. In the present research, a discrete, first-order HMM was employed. It usually consists of a set of hidden states, a group of observations for every state, states' transition probability matrix A with probabilities of switching from one state to another one, emission probabilities matrix B, and the starting likelihoods π. A detailed introduction to HMM is provided by Rabiner [17].

For the experiment briefly described in Sect. 2, a set of AOIs were defined for experimental conditions in a way, graphically demonstrated in Fig. 1. Definitions of all AOIs used in further simulations, their abbreviations, and appearance in comparisons are put together in Table 1. The sequences of fixations registered in those AOIs for all examined subjects were next applied to train HMMs parameters (A, B, π). The eye tracking data were first elaborated in the SMI BeGaze 3.6 software, next processed in TIBCO Statistica 13.3 package [18]. The HMMs were derived according to the Baum-Welch algorithm [19] taking advantage of procedures and functions developed by Murphy [20]. The maximum number of iterations equaled 1000, and the convergence threshold: 0.0001. HMMs calculations were conducted in Matlab R2018b [21].

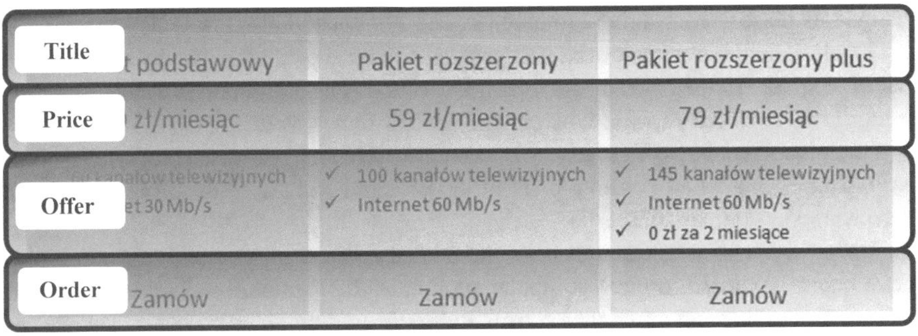

Title	t podstawowy	Pakiet rozszerzony	Pakiet rozszerzony plus
Price	zł/miesiąc	59 zł/miesiąc	79 zł/miesiąc
Offer	ałów telewizyjnych et 30 Mb/s	✓ 100 kanałów telewizyjnych ✓ Internet 60 Mb/s	✓ 145 kanałów telewizyjnych ✓ Internet 60 Mb/s ✓ 0 zł za 2 miesiące
Order	Zamów	Zamów	Zamów

Fig. 1. Defined AOI for a sample stimulus (*divided* and *price up*)

Table 1. Definitions of AOIs, their abbreviations, and appearance in comparisons

	Top leaflet				Bottom leaflet			
	AOI abbreviation	Grouping	Arrangement	AOI type	AOI abbreviation	Grouping	Arrangement	AOI type
Comparison 1	T_S_PU_TI	Solid	Price up	Title	B_D_PU_TI	Divided	Price up	Title
	T_S_PU_PR	Solid	Price up	Price	B_D_PU_PR	Divided	Price up	Price
	T_S_PU_OF	Solid	Price up	Offer	B_D_PU_OF	Divided	Price up	Offer
	T_S_PU_OR	Solid	Price up	Order	B_D_PU_OR	Divided	Price up	Order
Comparison 2	T_S_PD_TI	Solid	Price down	Title	B_D_PD_TI	Divided	Price down	Title
	T_S_PD_OR	Solid	Price down	Order	B_D_PD_OR	Divided	Price down	Order
	T_S_PD_OF	Solid	Price down	Offer	B_D_PD_OF	Divided	Price down	Offer
	T_S_PD_PR	Solid	Price down	Price	B_D_PD_PR	Divided	Price down	Price
Comparison 3	T_S_PU_TI	Solid	Price up	Title	B_S_PD_TI	Solid	Price down	Title
	T_S_PU_PR	Solid	Price up	Price	B_S_PD_OR	Solid	Price down	Order
	T_S_PU_OF	Solid	Price up	Offer	B_S_PD_OF	Solid	Price down	Offer
	T_S_PU_OR	Solid	Price up	Order	B_S_PD_PR	Solid	Price down	Price
Comparison 4	T_D_PU_TI	Divided	Price up	Title	B_D_PD_TI	Divided	Price down	Title
	T_D_PU_PR	Divided	Price up	Price	B_D_PD_OR	Divided	Price down	Order
	T_D_PU_OF	Divided	Price up	Offer	B_D_PD_OF	Divided	Price down	Offer
	T_D_PU_OR	Divided	Price up	Order	B_D_PD_PR	Divided	Price down	Price
Comparison 5	T_S_PU_TI	Solid	Price up	Title	B_D_PD_TI	Divided	Price down	Title
	T_S_PU_PR	Solid	Price up	Price	B_D_PD_OR	Divided	Price down	Order
	T_S_PU_OF	Solid	Price up	Offer	B_D_PD_OF	Divided	Price down	Offer
	T_S_PU_OR	Solid	Price up	Order	B_D_PD_PR	Divided	Price down	Price
Comparison 6	T_D_PU_TI	Divided	Price up	Title	B_S_PD_TI	Solid	Price down	Title
	T_D_PU_PR	Divided	Price up	Price	B_S_PD_OR	Solid	Price down	Order
	T_D_PU_OF	Divided	Price up	Offer	B_S_PD_OF	Solid	Price down	Offer
	T_D_PU_OR	Divided	Price up	Order	B_S_PD_PR	Solid	Price down	Price

3.1 Simulation Experiment

A simulation experiment was designed and conducted to determine the most appropriate number of hidden states, necessary to model the visual behavior registered while performing pairwise comparisons. Overall, 30 conditions were examined. They were differentiated by five possible hidden states (from 2 to 6) and six comparisons (5 states × 6 comparisons).

The HMM estimations depend on starting likelihoods, therefore, 100 single simulations were carried out for each experimental condition with random starting values. All the models were assessed according to Akaike's Information Criterion (AIC) [22], Bayesian Information Criterion (BIC) [23], and log-likelihood values. Obtained minimum values for these criteria suggested the most relevant number of hidden states given the occurrences of fixations in specified areas of interests. The summary of outcomes obtained in these simulations for all 30 conditions are put together in Table 2.

Table 2. The HMM simulation results for all comparisons. Values in brackets denote standard deviations

No	Comparison	No of states	AIC Mean	AIC Min	BIC Mean	BIC Min	Log-likelihood Mean	Log-likelihood Max
1.	**1.**	2	3770 (97)	3737	3878 (97)	3845	-1863 (48)	-1847
2.		3	3671 (31)	3648	**3849 (31)**	3825	-1800 (15)	-1788
3.	**S_PU <-> D_PU**	4	3605 (44)	3564	**3860 (44)**	3820	-1750 (22)	-1730
4.		5	3554 (50)	3497	3899 (50)	3841	-1707 (25)	-1678
5.		6	**3528 (41)**	**3487**	3970 (41)	3930	-1674 (20)	**-1653**
6.	**2.**	2	4485 (106)	4415	4597 (106)	4526	-2221 (53)	-2185
7.		3	4351 (70)	4315	4533 (70)	4498	-2140 (35)	-2122
8.	**S_PD <-> D_PD**	4	4262 (64)	4206	**4525 (64)**	4469	-2079 (32)	-2051
9.		5	4223 (49)	4172	4577 (49)	4527	-2041 (24)	-2016
10.		6	**4200 (44)**	**4153**	4656 (44)	4609	-2010 (22)	**-1987**
11.	**3.**	2	3892 (79)	3858	4000 (79)	3966	-1924 (39)	-1907
12.		3	3727 (45)	3706	**3904 (45)**	**3883**	-1828 (22)	-1817
13.	**S_PU <-> S_PD**	4	3684 (28)	3657	3940 (28)	3913	-1790 (14)	-1777
14.		5	3656 (27)	3628	4001 (27)	3972	-1758 (14)	-1744
15.		6	**3648 (25)**	**3616**	4091 (25)	4058	-1734 (12)	**-1718**
16.	**4.**	2	4294 (105)	4233	4405 (105)	4344	-2125 (53)	-2094
17.		3	4150 (57)	4113	**4331 (57)**	**4294**	-2039 (28)	-2020
18.	**D_PU <-> D_PD**	4	4070 (39)	4041	**4332 (39)**	4304	-1983 (19)	-1969
19.		5	4028 (36)	3989	4381 (36)	4342	-1944 (18)	-1924
20		6	**4019 (38)**	**3970**	4473 (38)	4423	-1920 (19)	**-1895**
21	**5.**	2	3983 (113)	3888	4092 (113)	3997	-1970 (57)	-1922
22		3	3745 (67)	3709	**3924 (67)**	**3887**	-1837 (33)	-1818
23	**S_PU <-> D_PD**	4	3671 (65)	3631	3929 (65)	**3889**	-1783 (33)	-1763
24		5	3622 (44)	3574	3970 (44)	3922	-1741 (22)	-1717
25		6	**3607 (37)**	**3563**	4054 (37)	4010	-1714 (18)	**-1691**
26	**6.**	2	3777 (86)	3736	3885 (86)	3844	-1866 (43)	-1846
27		3	3654 (61)	3607	**3831 (61)**	3784	-1791 (30)	-1767
28	**D_PU <-> S_PD**	4	3583 (29)	3559	3838 (29)	3814	-1739 (15)	-1728
29		5	3548 (41)	3509	3891 (41)	3853	-1704 (21)	-1685
30		6	**3536 (31)**	**3496**	3978 (31)	3938	-1678 (16)	**-1658**

Bearing in mind that smaller values of AIC and BIC indicate more suitable models, the presented results suggest that in comparisons 1 and 2, four-states HMMs are necessary. For all remaining conditions, three-states HMMs are sufficient enough. In the fifth comparison, there is a very small difference in BIC values for 3- and 4-states models. These conclusions are drawn based on the BIC, which is known for recommending more parsimonious solutions with smaller number of states and parameters.

Both AICs and log-likelihood criteria outcomes, suggest 6-states HMMs. The best solutions according to BICs are presented in detail and comprehensively examined in the next section. They are also compared with the preference analysis results.

3.2 Analysis and Discussion of the Best HMMs

In all tables presenting HMM estimated parameters, the second rows include initial states probabilities (π), next four (or three) rows consist of between-states transition probabilities (A), and the remaining rows contain emission probabilities (B). The best four- and three-states HMMs proposals for all six comparison types are given in Tables 3, 4, 5, 6, 7, 8 and 9. Overall, we illustrated the models as simplified graphs for three, four-states models that differed considerably in their structure and probability distributions (Fig. 2, 3, 5), and one of the remaining three three-states models, which are very similar to each other.

Table 3. Four states HMM for the first comparison

	S1	S2	S3	S4
π	0.86	0.02	0.12	0.00
S1	0.66	0.19	0.15	0.09
S2	0.12	0.65	0.02	0.16
S3	0.15	0.00	0.62	0.05
S4	0.07	0.16	0.22	0.70
[1]:T_S_PU_TI	0.00	0.00	0.39	0.00
[2]:T_S_PU_PR	0.11	0.00	0.55	0.00
[3]:T_S_PU_OF	0.52	0.02	0.03	0.01
[4]:T_S_PU_OR	0.31	0.00	0.00	0.02
[5]:B_D_PU_TI	0.06	0.00	0.00	0.37
[6]:B_D_PU_PR	0.00	0.06	0.03	0.56
[7]:B_D_PU_OF	0.00	0.73	0.00	0.02
[8]:B_D_PU_OR	0.00	0.18	0.00	0.02

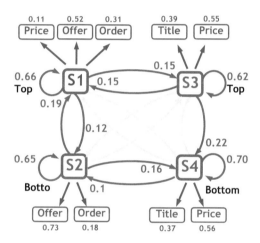

Fig. 2. Four states HMM for the first comparison

The common feature of the all the obtained models is the occurrence of hidden states that can be related to compared leaflets, located either on top or at the bottom of the screen. In four-states models, two hidden states are associated with processing the upper flyer, the two remaining – with the lower leaflet. In the three-states models, two states are related with fixations on the upper variant, whereas one state corresponds to the observation of the lower leaflet.

Analyzing 4-states models for comparisons 1, 2, and 5, it is easy to notice that the states are related to visually examining top and bottom parts of leaflets appearing on top or at the bottom of the screen. One of the three-states models is presented in Fig. 4 and concerns comparison 3. It can be observed that the hidden state associated with the lower flyer is a kind of a juxtaposition of corresponding two states from four-states models. That is, it covers all AOIs from the leaflet located in the lower area of the monitor.

Table 4. Four states HMM for the second comparison

	S1	S2	S3	S4
π	0.83	0.00	0.15	0.03
S1	0.59	0.19	0.12	0.15
S2	0.21	0.67	0.11	0.04
S3	0.03	0.11	0.71	0.16
S4	0.17	0.03	0.07	0.65
[1]:T_S_PD_TI	0.00	0.00	0.00	0.33
[2]:T_S_PD_OR	0.06	0.00	0.00	0.50
[3]:T_S_PD_OF	0.57	0.00	0.00	0.17
[4]:T_S_PD_PR	0.30	0.00	0.06	0.00
[5]:B_D_PD_TI	0.08	0.00	0.58	0.00
[6]:B_D_PD_OR	0.00	0.29	0.36	0.00
[7]:B_D_PD_OF	0.00	0.47	0.00	0.00
[8]:B_D_PD_PR	0.00	0.24	0.00	0.00

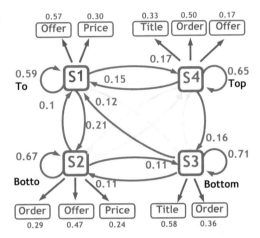

Fig. 3. Four states HMM for the second comparison

The arrays of transition probabilities between hidden states provide some insights on the possible patterns or strategies employed by subjects while performing experimental tasks. For instance, in the model from Fig. 2, taking into account the most probable transitions, one can note that subjects' visual strategy involved the analysis of the key elements of the upper flyer first (S1 → S3), and then move to the lower variant (S3 → S4, and S1 → S2). In the model derived for comparison 2, in turn, participants explored predominantly upper parts for leaflets located both at the top and bottom of the screen (S1 → S2). One should also pay attention to relatively small differences between estimated probabilities that suggest application of various patterns by individual persons or the existence of hybrid visual strategies.

Comparisons 3, 4, and 6 are represented by three-states models. All of them are qualitatively and quantitatively very close to the graphical illustration of the HMM for comparison 3, illustrated in Fig. 4. This model indicate the employment of partial comparison strategy. Offer and Order AOIs in state S1 for the upper leaflet are most often compared with their equivalents situated in the lower area of the screen. In next fixations, subjects examined the remaining, supplementary AOIs. Such a strategy could follow the sequences S1 → (Offer, Order), S1 → S2, S2 → (Order, Offer), S2 → S1, S1 → S3, S3 → (Title, Price), S3 → S2, S2 → (Title, Price).

However, examining Fig. 4, one can come up with a more or less similarly probable strategy of building a full picture of the upper leaflet first, and then move to the lower one: S1 → (Offer, Order), S1 → S3, S3 → (Title, Price), S3 → S2, S2 → (Order, Offer), S2 → (Title, Price).

Table 5. Three states HMM for the third comparison

	S1	S2	S3
π	0.89	0.00	0.11
S1	0.66	0.22	0.16
S2	0.19	0.75	0.16
S3	0.15	0.03	0.68
[1]:T_S_PU_TI	0.00	0.00	0.41
[2]:T_S_PU_PR	0.06	0.00	0.56
[3]:T_S_PU_OF	0.51	0.00	0.02
[4]:T_S_PU_OR	0.31	0.00	0.00
[5]:B_S_PD_TI	0.10	0.15	0.00
[6]:B_S_PD_OR	0.03	0.34	0.01
[7]:B_S_PD_OF	0.00	0.34	0.00
[8]:B_S_PD_PR	0.00	0.17	0.00

Table 6. Three states HMM for the fourth comparison

	S1	S2	S3
π	0.93	0.00	0.07
S1	0.57	0.18	0.19
S2	0.20	0.75	0.14
S3	0.24	0.07	0.67
[1]:T_D_PU_TI	0.00	0.00	0.27
[2]:T_D_PU_PR	0.01	0.01	0.62
[3]:T_D_PU_OF	0.60	0.00	0.10
[4]:T_D_PU_OR	0.32	0.00	0.00
[5]:B_D_PD_TI	0.06	0.21	0.00
[6]:B_D_PD_OR	0.01	0.33	0.00
[7]:B_D_PD_OF	0.00	0.31	0.00
[8]:B_D_PD_PR	0.00	0.14	0.00

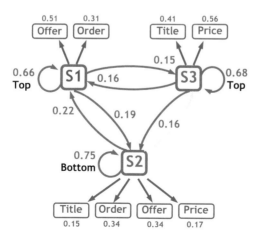

Fig. 4. Three states HMM for the third comparison.

Table 7. Three states HMM for the fifth comparison

	S1	S2	S3
π	0.98	0.00	0.02
S1	0.65	0.15	0.14
S2	0.18	0.81	0.13
S3	0.16	0.04	0.73
[1]:T_S_PU_TI	0.00	0.00	0.40
[2]:T_S_PU_PR	0.06	0.00	0.55
[3]:T_S_PU_OF	0.57	0.00	0.04
[4]:T_S_PU_OR	0.30	0.00	0.00
[5]:B_D_PD_TI	0.07	0.24	0.00
[6]:B_D_PD_OR	0.00	0.36	0.00
[7]:B_D_PD_OF	0.00	0.29	0.00
[8]:B_D_PD_PR	0.01	0.11	0.00

Table 8. Three states HMM for the sixth comparison

	S1	S2	S3
π	0.90	0.03	0.07
S1	0.67	0.19	0.18
S2	0.15	0.76	0.15
S3	0.18	0.05	0.67
[1]:T_D_PU_TI	0.00	0.00	0.45
[2]:T_D_PU_PR	0.09	0.00	0.54
[3]:T_D_PU_OF	0.55	0.00	0.00
[4]:T_D_PU_OR	0.27	0.03	0.01
[5]:B_S_PD_TI	0.09	0.22	0.00
[6]:B_S_PD_OR	0.00	0.33	0.00
[7]:B_S_PD_OF	0.00	0.31	0.00
[8]:B_S_PD_PR	0.00	0.12	0.00

Table 9. Four states HMM for the fifth comparison

	S1	S2	S3	S4
π	0.97	0.03	0.00	0.00
S1	0.65	0.13	0.04	0.19
S2	0.18	0.74	0.06	0.04
S3	0.04	0.05	0.72	0.05
S4	0.14	0.08	0.18	0.72
[1]:T_S_PU_TI	0.00	0.38	0.00	0.00
[2]:T_S_PU_PR	0.04	0.55	0.00	0.00
[3]:T_S_PU_OF	0.58	0.06	0.00	0.00
[4]:T_S_PU_OR	0.30	0.00	0.02	0.00
[5]:B_D_PD_TI	0.07	0.00	0.73	0.01
[6]:B_D_PD_OR	0.00	0.00	0.24	0.41
[7]:B_D_PD_OF	0.00	0.00	0.00	0.42
[8]:B_D_PD_PR	0.00	0.00	0.00	0.16

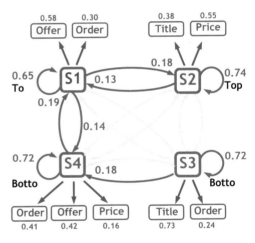

Fig. 5. Four states HMM for the fifth comparison

The detection of two states related with the upper leaflet in every comparison seems to be caused by a general principles of processing visual information by humans [7]. One of the phenomenon observed by researchers is the tendency of starting the analysis from the top part of the visual stimulus [24–26]. Another, theoretical model of attention called zoom-lens [27] may be used here to explain the subjects' visual behavior. First fixations registered in the present experiment correspond to the first step in this model which concerns the general, usually preattentive, processing of the stimulus. In this phase, people probably take advantage of the peripheral vision intensively to organize the whole picture. Such an explanation especially fits to high values of predicted emission probabilities for the Offer AOI in all comparisons. This AOI is located directly at the center of each leaflet. The emission probabilities show that this *general* state may concern either the upper (e.g., in comparison 1) or the lower flyer (in comparisons 2 and 3).

The identification of one or two states for processing the lower leaflet is somewhat surprising. If we assume that bigger number of hidden states is associated with more complex visual tasks [9], the explanation may be twofold.

First, in comparisons 1, 2, and 5, two states were devoted to examining divided leaflets placed at the bottom of the screen. The flyers clear divisions were more expressive in comparison with their solid counterparts. This probably attracted attention in the peripheral vision and required more processing even while focusing on the upper variant.

Secondly, in these three four-states models, the lower versions were considered by subjects as more convincing than the upper ones [19]. In the experiment, participants were to select the more preferred option by mouse clicking on it. Thus, one may presume that one of the states associated with the leaflet at the bottom should represent this acceptance and confirmation stage, since clicking with a mouse requires visual control. It is possible that this task complicates processing of the better leaflet by generating an additional, hidden state.

Irrespective of the presented hypotheses, the obtained and discussed models suggest the existence of a contextual diversification of visual strategies applied during pairwise comparisons of simple stimuli. The differences between HMMs can be coupled with leaflet designs (examined factors), their relative locations (top, bottom), as well as with subjects' preferences expressed towards presented stimuli. The demonstrated results certainly encourage to further explore human visual behavior in various other marketing contexts.

4   Conclusions

The paper shows how modern eye-tracking equipment may be used in examining human visual behavior while processing various aspects of marketing message designs. The collected data can be analyzed in a standard way. However, applying a more sophisticated approach like HMMs, may lead to a discovery of not trivial visual activity patterns. Generally, classic scan path analysis approaches provide information only about overt attention manifestations. HMMs, in turn, allow for making analyses of hidden states that can be linked with covert attention [7]). Therefore, modelling by HMM facilitates deeper and broader search for visual activity patterns concerned with decision and attentional processes.

Acknowledgments. The research was partially financially supported by Polish National Science Centre Grant No. 2017/27/B/HS4/01876.

References

1. Grebitus, C., Roosen, J.: Influence of non-attendance on choices with varying complexity. Eur. J. Mark. **52**(9/10), 2151–2172 (2018). https://doi.org/10.1108/EJM-02-2017-0143
2. Muñoz Leiva, F., Liébana-Cabanillas, F., Hernández-Méndez, J.: Etourism advertising effectiveness: banner type and engagement as moderators. J. Serv. Mark. **32**(4), 462–475 (2018). https://doi.org/10.1108/JSM-01-2017-0039
3. Michalski, R., Grobelny, J.: An eye tracking based examination of visual attention during pairwise comparisons of a digital product's package. In: UAHCI 2016. Part I. LNCS, vol. 9737, pp. 430–441 (2016). https://doi.org/10.1007/978-3-319-40250-5_41
4. Michalski, R.: Information presentation compatibility in the simple digital control panel design – eye-tracking study. Int. J. Occup. Saf. Ergon. (2017). https://doi.org/10.1080/10803548.2017.1317469
5. Ozkan, F., Ulutas, H.B.: Using eye-tracking data to evaluate medicine information leaflets on-screen. J. Math. Stat. Sci. **3**(12), 364–376 (2017)
6. Huddleston, P.T., Behe, B.K., Driesener, C., Minahan, S.: Inside-outside: using eye-tracking to investigate search-choice processes in the retail environment. J. Retail. Consum. Serv. **43**, 85–93 (2018). https://doi.org/10.1016/j.jretconser.2018.03.006
7. Findlay, J.M., Gilchrist, I.D.: Active vision. The psychology of looking and seeing. Oxford University Press, New York (2003)
8. Ellis, S.R., Stark, L.: Statistical dependency in visual scanning. Hum. Factors: J. Hum. Factors Ergon. Soc. **28**(4), 421–438 (1986). https://doi.org/10.1177/001872088602800405
9. Hayashi, M.: Hidden Markov models to identify pilot instrument scanning and attention patterns. In: 2003 IEEE International Conference on Systems, Man and Cybernetics, vol. 3, pp. 2889–2896 (2003). https://doi.org/10.1109/ICSMC.2003.1244330
10. Liechty, J., Pieters, R., Wedel, M.: Global and local covert visual attention: evidence from a Bayesian hidden Markov model. Psychometrika **68**(4), 519–541 (2003). https://doi.org/10.1007/BF02295608
11. Simola, J., Salojärvi, J., Kojo, I.: Using hidden Markov model to uncover processing states from eye movements in information search tasks. Cogn. Syst. Res. **9**(4), 237–251 (2008). https://doi.org/10.1016/j.cogsys.2008.01.002
12. Chuk, T., Chan, A.B., Hsiao, J.H.: Understanding eye movements in face recognition using hidden Markov models. J. Vis. **14**(11), 1–14 (2014). https://doi.org/10.1167/14.11.8
13. Grobelny, J., Michalski, R.: Applying hidden Markov models to visual activity analysis for simple digital control panel operations. In: Proceedings of 37th International Conference on Information Systems Architecture and Technology. ISAT 2016, Part III, Advances in Intelligent Systems on Computing, vol. 523 (2017). https://doi.org/10.1007/978-3-319-46589-0_1
14. Grobelny J., Michalski R.: Zastosowanie modeli Markowa z ukrytymi stanami do analizy aktywności wzrokowej w procesie oceny wirtualnych opakowań techniką porównywania parami. Zeszyty Naukowe Politechniki Poznańskiej. Organizacja i Zarządzanie, vol. 73, pp. 111–125 (2017). https://doi.org/10.21008/j.0239-9415.2017.073.08
15. Michalski, R., Ogar, A.: Wpływ struktury graficznej prostych ulotek reklamowych na preferencje potencjalnych klientów - badanie okulograficzne. Zeszyty Naukowe Politechniki Poznańskiej. Organizacja i Zarządzanie (2019). http://www.zeszyty.fem.put.poznan.pl/
16. Markov, A.A.: An example of statistical investigation of the text Eugene Onegin concerning the connection of samples in chains. Bull. Imperial Acad. Sci. St. Petersburg **7**(3), 153–162 (1913). (in Russian). Unpublished English translation by Morris Halle (1955). Nitussov, A. Y., Voro-pai, L., Custance, G., Link, D. (trans.): Science in Context **19**(4), 591–600 (2006)

17. Rabiner, L.R.: A tutorial on hidden Markov models and selected applications in speech recognition. Proc. IEEE **77**(2), 257–286 (1989). https://doi.org/10.1109/5.18626
18. TIBCO Software Inc.: Statistica (data analysis software system), version 13 (2017). http://statistica.io
19. Baum, L.E.: An inequality and associated maximization technique in statistical estimation for probabilistic functions of Markov processes. In: Shisha, O. (ed.). Proceedings of the 3rd Symposium on Inequalities, pp. 1–8. University of California, Los Angeles (1972)
20. Murphy, K.: Hidden Markov Model (HMM) Toolbox for Matlab (1998, 2005). www.cs.ubc.ca/~murphyk/Software/HMM/hmm.html
21. Mathworks: Matlab (R2018b) (2018). http://www.mathworks.com
22. Akaike, H.: Information theory as an extension of the maximum likelihood theory. In: Petrov, B.N., Csaki, F. (eds.) Second International Symposium on Information Theory, pp. 267–281. Akademiai Kiado, Budapest (1973)
23. Schwarz, G.: Estimating the dimension of a model. Ann. Stat. **6**(2), 461–464 (1978). https://doi.org/10.1214/aos/1176344136
24. Jeannerod, M., Gerin, P., Pernier, J.: Deplacements et fixations du regard dans l'exploration libre d'une scene visuelle. Vis. Res. **8**, 81–97 (1968)
25. Chédru, F., Leblanc, M., Lhermitte, F.: Visual searching in normal and brain-damaged subjects (contribution to the study of unilateral inattention). Cortex **9**(1), 94–111 (1973)
26. Simion, F., Valenza, E., Cassia, V.M., Turati, C., Umiltà, C.: Newborns' preference for up–down asymmetrical configurations. Dev. Sci. **5**(4), 427–434 (2002). https://doi.org/10.1111/1467-7687.00237
27. Eriksen, C.W., James, J.D.S.: Visual attention within and around the field of focal attention: a zoom lens model. Percept. Psychophys. **40**(4), 225–240 (1986). https://doi.org/10.3758/BF03211502

Dynamics of Interactions

Zbigniew Wisniewski$^{(\boxtimes)}$, Aleksandra Polak-Sopinska,
Malgorzata Wisniewska, and Marta Jedraszek-Wisniewska

Faculty of Management and Production Engineering,
Lodz University of Technology, Piotrkowska 266, 90-924 Lodz, Poland
{zbigniew.wisniewski,aleksandra.polak-sopinska,
malgorzata.wisniewska}@p.lodz.pl, martl9@wp.pl

Abstract. The article summarizes premises for the construction of simulation models of dynamics of task performing teams. It presents theoretical foundations and discusses models which in further research will be used to analyze group dynamics. The modelling assumes that it is possible to describe human behavior in man – man and man – machine interactions by means of control theory tools. Here, man has been considered a reactive object as characterized within behavioral framework and what follows are descriptions of basic types of dynamical blocks which can be used to form objects: people, working environment. The analysis was performed in the research lab SYDYN.

Keywords: Interactions · Human-environment interaction ·
Human-human interaction

1 Introduction

A direct correlation between a response (action) of a human to the corresponding stimulus is postulated by the behavioral concept of personality. An identical model is applied to the analysis of responses of man viewed as a reactive being. Skorny [1, p. 46] argues that this may become a basis of behavior control by adequately manipulating stimuli correlated with the knowledge of relationships holding between stimuli and anticipated responses. The behavioral model is especially useful in analyzing animal behavior as well as simple acts performed by man. The behavioral approach constitutes a cybernetic model of man according to Wiener [2], Mazur [3, 4] and other models of behavior derived from cybernetics [5–9].

Performing acts and tasks emergent from the assumed role is a consequence of changes occurring in the environment. Those changes result from the undertaken action. Acts in turn are conditioned by contexts in which they are performed [10]. The conditioning of acts and their effectiveness in contexts may be explained by the phenomenon of feedback, where an action leads to a change of context in which it is taking place, and the context determines the course of the action itself [1, p. 79].

Man does not function in the environment as an involuntary tool but displays conscious activity manifesting itself in the ability to transform the nature of the environment – be it directly or through intermediate actions performed as part of deliberate plans. The purpose of innovative actions is to effect changes in the context of

© Springer Nature Switzerland AG 2020
D. N. Cassenti (Ed.): AHFE 2019, AISC 958, pp. 246–258, 2020.
https://doi.org/10.1007/978-3-030-20148-7_23

acts performed by an individual. Hence, it should be noted that such a formulation of imperative to act runs counter [11] to the behavioral concept of man regarded as a reactive (controllable) being.

This seeming contradiction is illusory as is evident in the behavior of a worker tasked with a goal to achieve (a task to do) and equipped with an instruction of how to achieve it. One of the most important features of an act (and thus an action) is a goal. A man who is performing an act is aware of its goal, the current level of its attainment as well as the effectiveness of the performance over a given period of time. Consequently, acts are processes that are directed and performed in a conscious manner. The existence of those goals prompts changes in the intensity of stimuli affecting the man in the process of achieving the goals. Stimuli frequently function not only as triggers to initiate actions but also as motivators. The actual performance of actions aimed at achieving specific goals may hence lead to a change in intensity of motivators. The case in point would be a situation in which a worker has been given a goal to attain, is working towards reaching the goal and is aware of the shortening distance to achieving the goal. The primary goal which triggered the activity (the distance to goal) is modified automatically in the course of the task completion and constitutes a motivator of variable intensity. Even if its intensity decreases over time, a drop in effectiveness of actions does not necessarily follow.

An analysis of all forms of human activity demands investigating their dependence on the process of stimulation towards acting. This dependence of acting on motives appears to prove the hypothesis that man is not controlled merely by external stimuli and contexts. Man consciously modifies interactions with environment and effects changes in the environment.

It could thus be concluded that a worker who has been given instructions on how to execute a task illustrates behavioral concepts. Within these theories, the worker's conduct may be described as a response to certain requests. Conversely, due to volatility of context in which the acts are to be performed and which is affected by the worker's activity, initiative and intelligence as manifested in selecting various action strategies, it could be argued that the worker does not act as an unwitting reactive subject.

A useful approximation at a model of behavior would need to combine basic features of both the discussed concepts: man undertakes acts in response to stimuli but the manner of acting as well as the dynamics of response to the stimuli depend not only on personality traits (behaviorism) but also on the amassed emotions, aspirations, and needs, as well as the external context emergent from the performed activities and acts.

2 Foundations of Behavior Dynamics

In order to describe behavior within behavioral framework, which constitutes a simplification of reality, a decomposition of complex systems into elementary constituents must be performed. There is no need to create concepts of such elementary components, as ready-made tools are available within control theory: a set of basic dynamical elements constitute building blocks of complex systems as well as offer transmittances describing their mutual correlations. The following basic dynamical elements can be used in the description of interaction dynamics [12]. It is, however, necessary to treat them as models of fragments of reality, therefore their references to real aspects will be presented. This is due to the possibility of finding a phenomeno-logical imitation. If, for example, during the analysis of a group's reaction to a request in the form of an imposed new quality standard, it is possible to treat individuals as individual elements, it is easier to simulate the group's behavior. Of course, in such a case it is necessary to have proper knowledge of the dynamics of individual persons (elements). Occasionally, however, it is not possible to easily relate the phenomena determining the dynamical features of the system with their equivalents in the physical sense [13]. This is not a special hindrance to the control process itself, although the intuitive pairing of the dynamical features of the model with their physical equivalents facilitates the understanding of the entire control system.

2.1   Proportional Block

In organizational practice it is rarely possible to attribute the dynamics of this element to an object or a group of factors. Especially in the objects where the human factor plays a decisive role, it is difficult to expect directly proportional responses to a given input signal. Due to the *quasi*-static character of this element, its features are attributed to systems observed in the long term.

The characteristics of the proportional block can be applied to physical objects, where there is no accumulation of energy and the inertia is negligible. Examples of objects and situations in which there may be a proportionality relationship between the level of output and input signals are presented below.

– Productivity of the manufacturing unit: dependence of the output on the number of employed staff.
– An employee and his or her ability to perform work depending on the level of remuneration.
– Organizational unit (team, department, plant) and its revenues depending on the value of the production output.

These examples represent certain simplifications of reality, however, in specific situations it is possible to treat objects as proportional elements, provided that the aforementioned limiting conditions are satisfied.

2.2   First Order Inertia Block

This block occupies a special position in the analysis of interaction dynamics. It is the simplest form of the dynamical block which accurately represents the nature of relationships occurring in many systems with the dominant role of man.

Examples of situations and objects identified as first order inertial elements in observation are as follows.

- Adaptation of an employee to new requirements in familiar work; the work is carried out in routine fashion, the employee has mastered the procedures for its performance, but there comes a change in the requirements concerning efficiency, quality, etc.
- Behavior of a small group (team) in response to changes in requirements regarding efficiency, quality, etc. without changing the object of work.
- Acquisition of new skills by an employee, a team.
- Level of process capability indicators, fractions of deficiencies during the implementation of a new method of process supervision.

2.3   Second Order Inertia Block

It is a kind of block which is similar to the first order inertia block, but with an additional degree of inertia. This is manifested by the fact that in the initial period of the response the characteristics have a zone of insensitivity. This does not mean a lack of response but rather "lazy initialization". There occurs a response in the form of an increase in effect but initially it is much slower than in the case of the first order inertial block.

There occurs a certain class of situations which due to their specificity demonstrate exceptional affinity to the characteristics of the second order inertial block. These are the situations in which there is a stimulus of a previously unknown character and an employee (group) proceeds with the implementation with a low initial efficiency. Examples of such situations include:

- delegating a new task to an employee - the situation occurs only when the employee is confronted with a new problem and must either come up with a way to complete the task or learn to perform it according to the instructions received;
- delegating a task without specifying a provisional timeframe – if the employee is not given any deadline upon accepting the task, he or she may fail to undertake adequate action at the initial stages of the task execution; the deadline is established only later (based on the received request or an analysis of previous tasks and the accumulated experience); further course is coincident with the dynamics of the first order inertial block [14, 15].
- acquiring a new skill by an employee or a team – depending on the specificity of context there may occur an instant, enthusiastic commencement of learning and training (inertial first-order) or there is an apparent degree of "lag" typically caused by the necessity of familiarizing oneself with a more demanding task (inertial second-order).

2.4   Oscillating Block

In some situations oscillations may occur in response to the stimulus. Experience shows that a person, for example in car control systems, technological equipment, a machine tool, an airplane, achieves the set goal in exactly this manner. It is possible to generalize such cases in the following way: there is a pressure to correctly perform the task and achieve the goal, simultaneously, there is lack of experience in using a given technical means and knowledge of its dynamic characteristics, still there is a possibility to use the full range of settings of such a device. In order to achieve the goal, in case of incomplete knowledge of the device dynamics, the operator tries to maximize the performance of the effector. The goal is not achieved at the desired speed, due to the inertia of the system [16]. When reaching a state close to the target, the operator changes the effector's operation to the opposite, but because of inertia, the system continues for some time to maximize the effect, exceeding the target value. After this sequence, the controlled parameter begins to decrease in value, the operator tries to change the direction yet again, but the inertia causes the main parameter to oscillate around the target for a certain period of time. An example of such an action may be an attempt to drive a vehicle along a specific trajectory. At the beginning of learning, the driver (operator) will always create a trajectory oscillating around the target, especially after a change of direction, which serves as a stimulus here. Another example is the manual control of the temperature of a furnace.

2.5   Integrator

Integrating elements occur where accumulation happens (of mass, energy, information). Given there exist objects acting as a buffer, they can be assumed to have the characteristics of the integrator. An example can be a mass storage unit, a tank. There is a simple dependence of the filling level on the supply flow and time.

2.6   Delay Block

In practice, delay is perfectly implemented as a part of transport processes. Transport of mass, energy or information (without accumulation, loss) is modelled with the use of delay blocks.

3   Systemic Interactions of Man

Applying the concept of the dynamic nature of the system, it is possible to specify the basic functional models of man - environment interactions. Out of 36 possible system configurations [17, p. 45], those which are of the greatest importance in management and constitute more than 60% of the systems existing in everyday human activity [18, p. 69] will be presented. Models of human systems in interaction with the environment will be based on studies by Staniszewski, Ziemba, Jarominek [18–21], which largely derive from the original concepts of man as an autonomous system created by Mazur [3, 4, 22].

3.1 Man – Man System

Simplified interactions among people are illustrated in Fig. 1. The influence of individuals on one another is effected by output signals y^1, and y^2. These signals are converted by each person's logical potential $H_w(s)$. Upper indices: "^{1}" and "^{2}" denote the first and second person respectively.

The transmittances shown in Fig. 1 have the following interpretation [18]:

$$H_a(s) = H_1(s) H_2(s) \tag{1}$$

$$H_b(s) = H_3(s) H_2(s) \tag{2}$$

$$H_c(s) = H_5(s) H_6(s) H_7(s) \tag{3}$$

$$H_d(s) = H_4(s) e^{-s\tau_4} \tag{4}$$

where:

$H1(s)$ –   perception dynamics,
$H2(s)$ –   perception enhancement,
$H3(s)$ –   decision signal formation dynamics,
$H4(s)$ –   central nervous system feedback dynamics with a delay of τ_4,
$H5(s)$ –   executive action dynamics when effected on an individual,
$H6(s)$ –   information processing dynamics in man to man communication,
$H7(s)$ –   control system dynamics; in this case a person subject to the influence of another

The output signals y^1 and y^2 correspond to human articulation. They may manifest themselves in the form of speech, behavior, publication, etc.

The presented structure is a simplification of the real autonomous human system. There exists analogy of this type of structure to control structures of large-scale systems [23]. The actual complexity does not allow for deterministic, accurate description of working processes and control processes occurring in man. It is only by way of observation that the character of the structure rather than its precise form can be conjectured. Thus, several blocks have been singled out, whose dynamical characteristics can be described through psychological experiments. For the purpose of the analysis of human interaction in an organization, the macro-scale view of man's mentality appears to be sufficient.

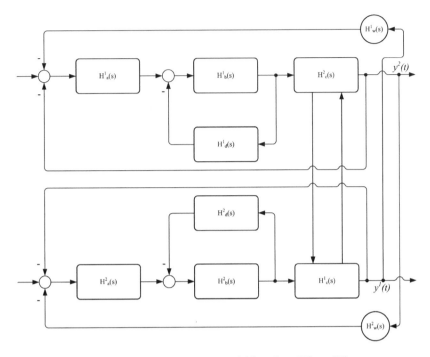

Fig. 1. Man – man system model based on [18, p. 70]

This also implies that since accurate determination of the features of individual systems that form such a structure is not possible (and if it were successful, there is no guarantee that the structure would be stationary), it is necessary to treat man as any other observed system. The consequence of this is the adoption of the principle of identifying traits on the basis of observations of reactions to stimuli. Knowledge of the approximate structure of human dependence on another person is valuable, but due to descriptive difficulties hardly applicable in management practice. In order to develop a model, appropriate experiments may be designed with a view to revealing the structure and parameters of particular elements of this system in more detail. For the purpose of identifying the role of a person in a team, environment or surroundings, it suffices to know the response to stimuli and to have an approximate idea of the structure of such a relationship (as in Fig. 1).

3.2   Man – Group System

The model of such a system is applicable to, for example, the arrangement of a superior - a group of subordinates. The basic dependencies in such a system are determined by the use and control of such parameters as [18, p. 73]:

- a set of aptitude parameters of individual members of the group that interacts with the manager,
- a set of parameters describing the aptitude and mental state of the manager,

- a set of impacts of the external environment on the manager
- a set of possible behaviors (responses) of the decision-maker,
- a set of constituent impacts on individual members of the group,
- a set of group members' responses to stimuli,
- a set of limitations on the external characteristics of the individual members of the group.

Man - group of subordinates systems are characterized by the following features:

- high adaptability to external factors and stimuli affecting the system,
- ability to perpetuate acquired experience, dependencies,
- ability to optimize internal communication paths,
- appreciation of the role of the decision-maker in the case of beneficial effects of the current system's operation.

The system of man (decision-maker) - group is the most commonly found relationship in management practice. Due to the hierarchical structure of most organizations and the role of power centers in them, majority of management activities are performed in the presented arrangement. The implementation of changes and innovations occurs most frequently along the decision-maker - subordinate path. The reaction to the stimulus in such a system manifests itself as a response to a command given by the superior. Analysis of the dynamics of relationships occurring within the man-group systems may contribute to improving the effectiveness of enforcement of tasks delegated by the decision-maker. Much like in the man - man system, it is not possible to recognize the subtle relationships between the members of the group or between them and their superior. However, it is important to understand the characteristics that can be identified on a macroscopic scale. Such identification is achieved through experiments involving the observation of reactions to stimuli appearing in the system.

3.3   Man – Environment System

The role of man in the man - environment system is to interpret and impact the environment. Recognition of the environment is based on direct or indirect perception via technical means. The environment defines the framework for human behavior. It is the part of the surroundings that determines the rules of conduct: law, physics, norms, customs, schemes, etc.

The flow of information related to control in the man - environment system is based on the rule of four phases [18, p. 76]: measurement, evaluation, development of control signals, execution of the work process. The model of the system whose aim it is to recognize and identify the environment is presented in Fig. 2. There are two information loops in the system: the control process loop and the identification and control loop which allows a person to control the correctness of the system's operation.

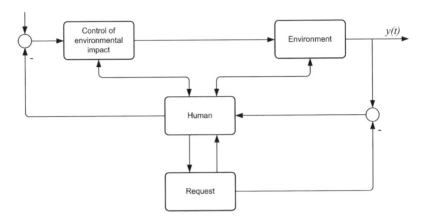

Fig. 2. Man-environment system model to satisfy the goal of identifying the environment. [18, p. 75]

Approximate representation of the dynamical characteristics of this system is expressed by the transmittance:

$$H(s) = \frac{H_3(s)\,e^{-s\tau_1}\,H_2(s)}{1 \pm H_1(s)\,H_2(s)\,H_3(s)\,e^{-s\tau_1}} \tag{5}$$

where:

$H_1(s)$ – dynamic characteristics of man in the process of measurement and identification of environment,

$H_2(s)$ – characteristics of influence block over the measurement process,

$H_3(s)$ – characteristics of measurement process,

$e^{-s\tau_1}$ – delay in operation of the environment measurement system.

A person who participates in the process of identifying and measuring the environment is a part of a closed system. Together they find themselves in an environment that is different and natural for such a system. There are interactions between the person and the observed environment, because the observation cannot be performed without quantitative changes both in the environment itself and in the man. On the basis of perception (observation results) the person can perform a number of actions, initiating changes in the observed process. Such actions are always of a regulatory nature and affect the dynamic state of the process [21]. This is justified as each rational and purposeful human action can be classified as a regulatory action. After all, the pursuit of a specific goal consists in exerting such an impact on the surroundings (environment) via the use of energy and other means that the difference between the current state of the environment and the target state is minimized.

In the man-environment system there occurs a problem with collecting and processing information, namely, interference. In integrated structures, the impact of interference is eliminated in a number of ways. In the case of practical implementation in operational management systems, the impact of interference is factored in by entangling it in the values of relevant signals. Control in such a case is reduced to observing the signals present in the system with no interference assumed. Its actual influence is discernible in the dynamical characteristics of individual elements. This is a valid solution, albeit exclusively for interferences satisfying the conditions of non-variability in time (at least in terms of amplitude, frequency or number of interfering signals). In such cases it is possible to treat the system as undisturbed as their influence is accounted for by the dynamical characteristics of objects. However, in the case of man-environment system, the role of interference is greater and cannot be reduced to a change in its dynamics. This is due to the fact that the essence of impact in the discussed system is the interaction of man with the environment. In this case, the influence of the measurement process on the state of the system is of considerable importance.

Two approaches are applied. The first one consists in the operator preselecting signals affected by interference and isolating signals which are useful for control. The other approach is to use filters in the system to automatically separate the signal from interference.

3.4   Man – Machine System

This configuration is one of the most important basic structures of operational and tactical management systems. This is due to the fact that most of the production processes are performed with the use of man-operated technical devices. The particular importance of this system stems from one more fact. By observing the human-machine relationship, it is possible to learn more about human dynamics. In the context of observation of such systems as: man - man, man - group, man - environment there is no possibility of objective presetting of the level of signals for testing and identification of dynamic properties of an individual human. He or she always interacts with the other element of each system, which is as non-deterministic and unpredictable as the observed person. The situation of coupling a human being with a machine affords the comfort of identifying and programming the device's dynamic features quite precisely. Human reactions that may affect the dynamic state of the device are negligible. Therefore, the man-machine system can be treated as a reference set for determining the dynamical features of a human being.

Naturally, the practical application of the collected data on the dynamic features of man is more important. The presented system has two advantages: in some circumstances it can be used to determine the dynamical attributes of a human being, while in other cases it constitutes implementation of normal operational and tactical management procedures in the execution of production processes.

Perceptual, motoric and psychological features of a person are not permanent, but depend on their preparation for work, knowledge, habits, health condition, age, motivation to act, role in a group, level of satisfaction of needs at a given moment and many others. They also depend on the relationship with the environment, working conditions, as well as the characteristics of the machines they use. It can therefore be argued that a person's action is characterized by instability due to the variability of his or her characteristics. Fortunately, the main features that determine its dynamic characteristics in relation to the machine and the environment (people) do not change in a shorter time than the shift control cycle.

There is no universal model of human dynamics and it is unlikely that in the near future such a model will emerge, which is why it is so important to identify the person's role and interactions in each specific case. The approximations used to describe the dynamics of machine operators (drivers, train drivers, construction machine operators and industrial processes operators) are based on the principle that man performs the role of a regular dynamic element tasked with a specific control program to be executed.

$$H(s) = \frac{b_m s^m + b_{m-1} s^{m-1} + \ldots + b_1 s + b_0}{a_n s^n + a_{n-1} s^{n-1} + \ldots + a_1 s + a_0} e^{-s\tau} \tag{6}$$

The numerator represents the impact on a person. It is a corrective element, determining the ability of a man to self-regulate and adapt his or her own characteristics to other dynamic characteristics of the system. If the operator is required to act in advance, such abilities are determined by the value of amplification factors and time constants in the numerator. They determine the behavior of the operator when confronted with changes in the speed of the excitation signal and specify the urgency of taking action in response to the changes in speed.

The denominator determines the ability of behavior and reaction to stimuli. It is an effector element and describes the delay between the signal already released in the operator and the actual action. The dynamics of the effector is described by a second order differential equation. At low frequencies it can be approximated by a first order equation. This is due to the fact that, predominantly, a person's reaction to the excitation is displacement rather than force [18, p. 90].

The term following the fraction represents a delay in response. It describes the time required for excitation, transit of the initiating signal and interpretation of information as well as the calculation necessary to initiate the action.

It is a generalized form of dynamic characterization of a human being. In order to determine the actual dynamic characterization of the operator, it is necessary to conduct an identification experiment. It consists in giving a specific stimulus and observing the reaction through time. When determining such a characteristic, it is advisable that the inducing signal be random. The presence of feedback should also be factored in. Failure to satisfy any of these conditions results in a misrepresentation of the results since a person will exercise his or her ability to improve control results in the control process. For example, if the stimulus is not randomized, the operator will predict its occurrence with increasingly higher success, leading to a significant improvement in the quality of control, which will be unrelated to his or her dynamical characteristics but will be the result of his or her anticipatory abilities.

4 Summary

The notion of motivation is connected with a person's inner determinations defining the direction of his or her actions. One of important stimulators of human activity is curiosity. It provokes an exploratory need for information. The motivational function of cognitive mechanisms appears to be confirmed by the inadequacy of the behavioristic model of conduct [24]. The postulate that the model of behavior depends on the internal image of the world - an action plan - may manifest itself as functions of needs and instincts. In the reality of organization (i.e. related to human professional activity) the role of action plans is, however, more commonly assumed (apart from impulsive and emotional actions) by mechanisms created on the basis of the acquired individual experiences and group cooperation, as well as accomplishment of subsequent goals.

The discrepancy between the goal and the current status of the object is the basic motivating mechanism of an action to eliminate such a difference. The elimination will always proceed in accordance with the dynamical characteristics of objects and the entire system. The conscious action of man apparently eludes deterministic description but interactions with surrounding elements situate man as any other dynamic object - as an element of a dynamic system with corresponding dynamical characteristics and as such it can be quantified. It is not always known, however, what equation to use for its quantification.

Man is naturally capable to restore stability to a system and, consequently, possesses, among other things, an astonishing ability to adapt to changing conditions. This is also manifested in the ability to perform processes with the use of machines according to an agreed program. Here, man attains positive results owing to inherent characteristics, but also through appropriate configuration of the control system. Its most important feature is the presence of a feedback loop, which is provided, for example, in the dynamic route through the man-environment system. Such a configuration enables control over the volatility of the system based on a program and elimination of deviations in case of the occurrence of interference or new initiating stimuli. Such a system, however, has its limitations resulting, as in any system, from the limitations of parameters that affect the process, time, cost, energy and many others. Finally, the limitations are rooted in natural barriers to the psychology and physiology of a human being as well as sociological behavior of the group in which he or she is, or which he or she manages.

References

1. Skorny, Z.: Mechanizmy regulacyjne ludzkiego dzialania. PWN, Warszawa (1989)
2. Wiener, N.: Cybernetyka, czyli sterowanie i komunikacja w zwierzeciu i maszynie. PWN, Warszawa (1971)
3. Mazur, M.: Cybernetyczna teoria ukladow samodzielnych. PWN, Warszawa (1966)
4. Mazur, M.: Cybernetyka i charakter. Panstwowy Instytut Wydawniczy, Warszawa (1976)
5. Cadwallader, M.L.: The cybernetic analysis of change in complex social organizations. Am. J. Soc. **65**(2), 154–157 (1959)

6. Carver, C.S.: A cybernetic model of self–attention processes. J. Pers. Soc. Psychol. **37**(8), 1251–1281 (1979)
7. Carver, C.S., Scheier, M.F.: Control theory. A useful conceptual framework for personality–social, clinical, and health psychology. In: The Self in Social Psychology, pp. 299–316 (1999)
8. Geyer, F., van der Zouwen, J. (eds.): Sociocybernetics: Complexity, Auto-poiesis, and Observation of Social Systems. Greenwood Press, Westport (2001)
9. Mindell, D.A.: Between human and machine: feedback, control, and computing before cybernetics. Johns Hopkins University Press, Baltimore (2002)
10. Festinger, L.: A theory of cognitive dissonanse. Stanford University Press, Evanston (1957)
11. Powers, W.T.: Feedback: beyond behaviorism. Science **179**, 351–356 (1973)
12. Wisniewski, Z., Polak-Sopinska, A., Wisniewska, M., Wrobel-Lachowska, M.: Dynamics of interactions – motivation. Adv. Intell. Sys. Comput. **605**, 155–163 (2018)
13. Staniszewski, M., Legutko, S., Krolczyk, J.B., Foltys, J.: The model of changes in the psychomotor performance of the production workers. Teh. Vjesn. **25**, 197–204 (2018)
14. Polak-Sopinska, A.: Workplace adjustments for people with disabilities. A case study of a research company. Part I – barriers for people with disabilities. Adv. Intell. Sys. Comput. **606**, 335–347 (2018)
15. Polak-Sopinska, A.: Workplace adjustments for people with disabilities. A case study of a research company. Part II - adjustment recommendations. Adv. Intell. Sys. Comput. **606**, 335–347 (2018)
16. Polak-Sopinska, A., Wisniewski, Z., Jedraszek-Wisniewska, M.: HR staff awareness of disability employment as input to the design of an assessment tool of disability management capacity in large enterprises in Poland. Procedia Manuf. **3**, 4836–4843 (2015)
17. Wisniewski, Z.: Wdrazanie zmian w organizacji. Ujecie dynamiczne. Wydawnictwo Politechniki Lodzkiej, Lodz (2010)
18. Staniszewski, R.: Teoria systemow. Ossolineum, Wroclaw (1988)
19. Staniszewski, R.: Cybernetyka systemow projektowania, Wroclaw (1980)
20. Staniszewski, R.: Cybernetyczna teoria projektowania. Ossolineum, Wroclaw (1986)
21. Ziemba, S., Jarominek, W., Staniszewski, R.: Problemy teorii systemow. Ossolineum, Wrocław (1980)
22. Mazur, M.: Jakosciowa teoria informacji. WNT, Warszawa (1970)
23. Gawronski, R.: Problemy Bioniki w systemach wielkich. Wydawnictwo Ministerstwa Obrony Narodowej, Warszawa (1975)
24. Miller, N.E., Galanter, E., Pribram, K.H.: Plans and the structure of behavior. J. Holt, New York (1960)

Application for Occupational Safety

Risk Perception with and Without Workers Present in Hazard Recognition Images

Jennica L. Bellanca[1(✉)], Brianna Eiter[1], Jonathan Hrica[1],
Robert Weston[2], and Terry Weston[2]

[1] Pittsburgh Mining Research Division, National Institute for Occupational
Safety and Health, Centers for Disease Control and Prevention,
Pittsburgh, PA, USA
{JBellanca, BEiter, JHrica}@cdc.gov
[2] Center for Business and Industry, South Central College, Mankato, MN, USA
{Robert.Weston, Terry.Weston}@southcentral.edu

Abstract. Effective training is a critical component in improving mineworkers' ability to identify and assess hazards and ultimately reduce accidents and injuries. Previous research has identified the importance of context and cultural relevance as significant factors in the effectiveness of training. Research from adult education, social marketing, and public health education also suggest that the inclusion of a personal connection to training may improve communication. However, there is limited empirical knowledge about the effects of training content composition. Specifically, it is unclear how the inclusion of workers in training materials may affect how mineworkers' perceive a hazard. The goal of this paper is to examine if the presence of workers in the context of hazards affects how mineworkers' assess the risks. Researchers collected risk assessments from participants in the laboratory and during mandatory, annual refresher training. The results have implications for training material development and workplace examination timing.

Keywords: Risk assessment · Training · Hazards · Mining

1 Introduction

With 38 fatal injuries occurring between October 1, 2013, and December 31, 2014, the number of fatalities in the metal/nonmetal mining industry nearly doubled from the previous 15 months [1]. This sharp increase in fatal injuries led to an increased focus on hazard recognition in order to improve hazard mitigation. This focus supports the idea that mineworkers must be able to first recognize that a condition or behavior can cause harm that it is a hazard and that mineworkers must be able to appropriately assess the risk of the hazard [2]. Previous research has identified both hazard knowledge and risk perception as critical competencies of hazard recognition [3]. Risk assessment has also been shown to influence mitigation behaviors [4], and an increased level of risk perception has been associated with a decrease in risky behaviors [4–7]. However, mineworkers still fail to recognize a significant number of hazards [8]. Research suggests that there is a great deal of variability in how workers perceive risk [7, 9].

D. N. Cassenti (Ed.): AHFE 2019, AISC 958, pp. 261–273, 2020.
https://doi.org/10.1007/978-3-030-20148-7_24

Therefore, effective training is a key component to improving mineworkers' ability to recognize and assess hazards and ultimately reduce accidents and injuries.

Effective training is engaging, authentic, and includes relevant and understandable content. Research has shown that training that is more engaging (e.g., behavioral modeling) leads to greater knowledge acquisition and reductions in accidents and injuries [10]. As a part of maintaining engagement, research has also shown that adult learners need to perceive the training content to be personally relevant to apply it [11]. Problems with relevant content may come in the form of context or understanding. For example, a study by Wilkins on construction workers' perceptions of the Occupational Safety and Health Administration's (OSHA's) 10-h construction safety course found a correlation between workers expressing lack of relevance and a lower score on the knowledge assessment. This difference exemplifies a disconnect between the training and the critical concepts [12]. Research has shown that when trainees perceived training to be useful and valuable they are more likely to apply it in the workplace [13, 14]. Previous studies also found that training that is more realistic improves training transfer [15, 16]. Ultimately, improving mineworkers' ability to recognize and assess hazards through training may lead to better safety outcomes.

The same principles that govern effective training suggest that the inclusion of a person in the context of a hazard may improve hazard knowledge training. For example, research from social marketing suggests that a personal connection and clear context may improve communication [17]. The addition of a person in hazard imagery may allow the trainees to more easily put themselves in the situation. Including a person in a hazard image may also improve the trainees' ability to predict the consequences of a hazard by clarifying the context. Isolated hazards often require discussion or text to describe possible outcomes, but language and literacy can be barriers in safety training [12, 17]. The addition of a person may also affect cultural relevance. This is important because previous work suggests that appropriate cultural relevance improves engagement and retention, and a mismatch may have a negative impact [18]. Additionally, hazard images including people may directly improve training strategies. Behavioral modeling an engaging form of training that includes demonstration and practice is most effective with the inclusion of both positive and negative examples [19]. Including a person in context with the hazard may serve as a negative example. However, it remains unclear how the presence of a person in the context of a hazard affects risk perception, as there is a lack of consensus on the effect of social and cultural framing on risk perception [20].

Therefore, the goal of this study was to determine the effect of the inclusion of a person in hazard images on risk assessment, including overall risk and its principal components, probability and severity. Because research suggests that training effectiveness [12] and risk assessment [9] may change based on experience, this study also explores the role mining experience plays on the effect of a person's presence in hazard images on risk assessment.

2 Methods

2.1 Data Collection

The data discussed in this paper comes from two different sources. The first source was a larger, more in-depth laboratory study that drove the development of a broader training study. Researchers selected hazards from the laboratory study to be used in the training study to gather a larger sample of participants and explore identified effects. As such, the analysis of the laboratory study is a secondary analysis. Only part of both data collection efforts will be examined here.

Each study is briefly described below. For a more complete description, see [9]. Table 1 presents the participant demographics from both studies. Age and mining experience are presented as categorized in the training study for comparison, where the categories were defined as [18–29, 30–39, 40–49, 50–59, 60+] and [0–2, 3–10, 11–20, 21–30, 30+] years respectively.

Laboratory Study. The initial data collection took place in a virtual reality laboratory as a part of a larger study examining hazard recognition and risk perception of mineworkers. As a part of an Institutional Review Board (IRB) and Office of Management and Budget approved protocol, NIOSH researchers recruited 51 participants with various types and levels of surface mining experience; this included safety professionals, experienced miners, inexperienced miners, and students. Safety professionals had at least two years of environmental, health, or safety experience with a mine operator or government agency. Experienced mineworkers had more than two years of mining experience. Inexperienced mineworkers had some but less than two years of mining experience. Lastly, students were participants enrolled in a mining-related program and not otherwise classified [8].

The laboratory study first required participants to search for hazards in life-size panoramic images, performing a simulated workplace exam [8], [21]. Next, during a debrief of their search performance, participants completed risk assessments of all 102 hazards included in panoramic images. This paper focuses on the risk assessments of a subset of hazards, matched sets of slip, trip, and fall (STF) hazards with and without mineworkers present in the images.

Table 1. Participant demographics from laboratory study and training study

Study	Group	N	Median age category	Median mining experience category
Lab	Safety professionals	13	40–49	11–20
	Experienced	11	40–49	11–20
	Inexperienced	13	18–29	0–2
	Students	14	18–29	0–2
Training		1,232	40–49	3–10

Training Study. In order to get a broader picture of risk perception across the surface mining industry, a second data collection was performed during MSHA-mandated annual health and safety refresher training (30 CFR § 46/48). The fourth and fifth authors of this paper facilitated all the data collection sessions during their normal training classes in their capacity as certified MSHA trainers. As such, data collection was a convenience sample, including mine companies, mine service companies, engineering firms, and open public classes. A total of 1,508 participants in 52 training sessions were recruited to participate in an IRB approved protocol. Sixty-six declined to participate and 210 participants were eliminated due to missing data, leaving a total of 1,232. Participants consented to participate and recorded their answers via an audience response system (clickers). Participants performed risk assessments on four of the hazards from the laboratory study. This paper focuses on one matched set of STF hazards with and without mineworkers present in the images.

2.2   Hazards

The original panoramic images from the laboratory study were of typical locations at a surface limestone mining operation, including the pit, plant, shop, and roadway. The images included hazards that matched the type, severity, and prevalence of hazards in MSHA's accident and injury database. Subject matter experts reviewed the images and hazards for accuracy and realism. To the extent possible, hazards were staged, but otherwise they were edited or enhanced (see [21]). Any people included in the images were dressed in typical mine site attire, except where omitted as a hazard.

Fig. 1. *Fall to walk way or working surface.* The left image displays a slip, trip, and fall (STF) hazard without a person, where the pipes and cables inside the orange rectangle are the hazard. The right image displays a STF hazard with a person, where a bucket being used as a step is the hazard. Both hazards were included in the laboratory and training data collections.

Researcher chose to focus on STF hazards, because they are one of the most prevalent types of hazards in metal/nonmetal mining [8] and can be clearly demonstrated with and without mineworkers. In the original laboratory study, the panoramic images included 102 hazards; 23 of the original hazards were STF. From the 23, researchers identified matched sets of hazards by selecting ones with the same accident type [22] and similar average risk assessment ratings by the safety professionals in the laboratory study. Researchers included safety professional rating in the selection criteria in order to minimize perceived differences in the hazards because exact matches were not available. Researchers specifically chose safety professionals as a reference because no objective or gold standard for risk assessment exists, but they are considered experts, as they are responsible for the health and safety of all the mineworkers at a mine site. This resulted in the identification of two matched sets of hazards in the laboratory study (Figs. 1 and 2), one of which was incorporated into the training study (Fig. 1).

Fig. 2. *Fall from equipment.* The left image displays a slip, trip, and fall (STF) hazard without a person, where the bottom rung of the ladder on the haul truck is missing. The right image displays a STF hazard with a person, where the mineworker cannot maintain three points of contact while climbing the ladder because he is carrying a lunchbox. Both hazards were included in the laboratory study only.

2.3   Risk Assessment

The risk assessment procedure was similar for both data collections. Researchers first reviewed the risk assessment scales with the participants (Table 2), [7]. Then, for each hazard, the participants received a brief description of what the hazard was (e.g., Fig. 1 - left: in this picture, you can see the pipes and cables in front of the main entrance to the shop). Next, researchers asked participants to rate: (1) how *severe* you think the likely resulting accident would be, (2) the *likelihood* that an accident will occur, and (3) your *overall risk* assessment on a 5-point scale. Researchers instructed participants to go with their first instincts and rate the hazards as quickly as possible.

Table 2. Risk assessment rating scales [7]

Scale	Numerical value				
	1	2	3	4	5
Accident severity	No injury	Minor injury No leave	Injury ≥ 3 days leave	Non-fatal Major injury	Fatal
Accident probability	Very unlikely	Fairly unlikely	Average likelihood	Fairly likely	Very likely
Overall risk	Very low	Low	Medium	High	Very high

2.4 Statistical Analysis

Multinomial population-averaged statistics were used to determine the effect of the inclusion of a person as a part of a STF hazard image on risk assessment controlling for experience across the mining population. For the performed statistical analysis, all risk assessment rating data was assumed to be ordinal. To do so, each dimension of the risk assessment was modeled using generalized estimating equations (GEE) with a multi-nomial distribution, cumulative logit link function, and an independent correlation structure. GEE were selected because they are deemed to be more robust to variable or unknown correlations [23]. The laboratory data was modeled as a full factorial inter-action between group (student, inexperienced, experienced, safety professional) and the inclusion of a person (person, no person). Interaction effects with hazard type (fall to walk way or working surface, fall from equipment) were initially included in the model, but were removed once they were found to be insignificant. Odds ratios were calculated for any significant effect with safety professional and person as the refer-ences. Similarly, the training data was modeled using a full factorial interaction between years of mining experience (0–2, 3–10, 11–20, 21–30, 30+) and the inclusion of a person (person, no person). All data was analyzed using IBM SPSS Statistics Software (Cary, NC). Alpha was 0.05.

3 Results

3.1 Laboratory Study

The GEE models of the laboratory data indicate that participants rated the accident probability of a hazard without a person significantly lower ($p = 0.030$) (Fig. 3). There was also a significant group effect for both risk ($p = 0.035$) and severity (0.009).

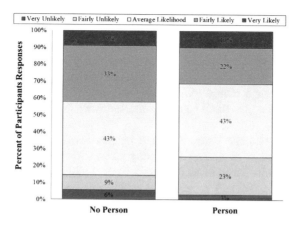

Fig. 3. Graph depicting the accident probability ratings breakdown from the laboratory study, where participants rated hazards without people significantly higher.

The parameter estimates from the GEE models reveal the following significant relationships. The odds of participants rating hazard images *without people* with a *higher probability* of occurring was 2.66 (95% CI 1.15 to 6.16; p = 0.023) times that of hazards with people. Figure 3 displays this significant difference through the breakdown of the accident probability ratings. The results also indicate that the odds of students rating STF hazards as *less risky* than safety professionals was 3.60 (95% CI 1.35 to 9.52; p = 0.010). Figure 4 displays the group effect and the general trend of risk increasing with experience. Lastly, the model estimated that the odds of students rating STF hazards as *less severe* than safety professionals was 5.59 (95% CI 2.00 to 15.90; p = 0.001). Severity similarly increases with experience level in the laboratory study.

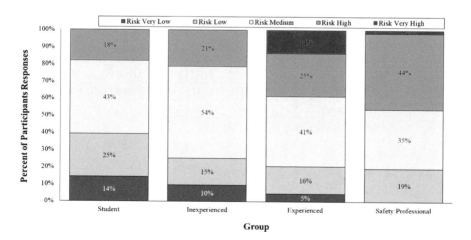

Fig. 4. Graph depicting the overall risk ratings breakdown across experience groups for the laboratory study, where the students rated risk significantly lower than safety professionals.

3.2   Training Study

The GEE models of the training data indicate that participants rated the accident probability of a hazard without a person significantly lower (p < 0.001) and the severity significantly higher (p < 0.001). However, there was no significant person effect on the overall risk ratings (p = 0.794). The results also showed that there was a significant experience (p = 0.001) and person-experience interaction effect (p = 0.021) for severity.

The parameter estimates from the GEE models reveal the following significant relationships. The odds of participants rating hazard images *without people* with a *higher probability* of occurring was 1.62 (95% CI 1.17 to 2.27; p = 0.004) times that of hazards with people. Table 3 presents the coefficients and odds ratio of probability model, where the significant person effect is just over 0.5. Figure 5 displays the probability response breakdown for hazards with and without people. The difference in ratings between the hazards appears larger than observed in the laboratory study (Fig. 3).

Table 3. Training study probability parameter estimates

	Coefficient			Odds ratio		
	Estimate	95% CI		Estimate	95% CI	
Experience						
30+	−0.16	−0.58	0.26	0.85	0.56	1.30
21–30	−0.34	−0.73	0.06	0.72	0.48	1.06
11–20	−0.23	−0.57	0.11	0.79	0.56	1.11
2–11	−0.16	−0.47	0.16	0.86	0.62	1.17
0–2	Reference					
Person						
Person	−0.54*	−0.83	−0.25	0.58*	0.44	0.78
No Person	Reference					
Interaction Effects						
30+ x Person	0.05	−0.39	0.50	1.06	0.68	1.64
21–30 x Person	0.42	−0.03	0.87	1.52	0.97	2.38
11–20 x Person	0.01	−0.38	0.41	1.01	0.68	1.50
2–11 x Person	0.06	−0.30	0.43	1.07	0.73	1.54

* Indicates significance at $\alpha = 0.05$

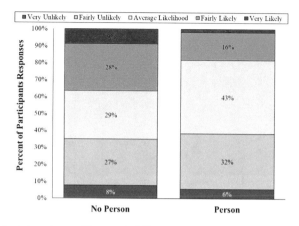

Fig. 5. Graph depicting the accident probability ratings breakdown from the training study, where participants rated hazards without people significantly higher

The results also estimate that the odds of participants rating hazard images *without people* as *less severe* was 2.54 (95% CI 1.90 to 3.40; p < 0.001). However, this effect appears to decreases as experience increases as demonstrated by the coefficients in Table 4. In fact, the odds of participants with 30+ years of mining experience rating hazard images *without people* as *more severe* than those with 0–2 years of experience was 2.24 (95% CI 1.35 to 3.75; p = 0.002), but there was no significant difference in the severity rating of participants with 30+ years of mining experience between hazards with and without people. Figure 6 displays this interaction effect for severity in the training study through the ratings breakdown across experience and between hazards with and without a person.

Table 4. Training study severity parameter estimates

	Coefficient		Odds ratio	
	Estimate	95% CI	Estimate	95% CI
Experience					
30+	1.09*	0.66	1.52	2.98*	1.94	4.58
21–30	0.61*	0.22	1.01	1.85*	1.24	2.76
11–20	0.43*	0.07	0.79	1.54*	1.08	2.20
2–11	0.26	−0.07	0.59	1.30	0.93	1.80
0–2	Reference				
Person					
Person	0.93*	0.64	1.225	2.54*	1.90	3.40
No Person	Reference				
Interaction Effects					
30+ x Person	−0.81*	−1.32	−0.30	0.44*	0.27	0.74
21–30 x Person	−0.40	−0.84	0.04	0.67	0.43	1.04
11–20 x Person	−0.26	−0.67	0.15	0.77	0.51	1.17
2–11 x Person	−0.12	−0.49	0.24	0.88	0.61	1.28

* Indicates significance at $\alpha = 0.05$

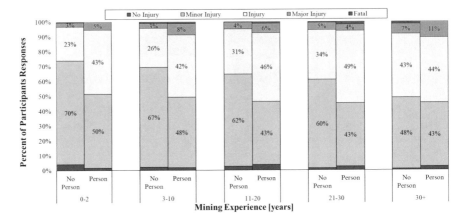

Fig. 6. Stacked plots depicting the severity ratings breakdown across experience from the training study for the hazards with and without a person. The significant interaction effect is apparent by the difference between no person and person for each experience level except 30+ years.

4 Discussion

Research from adult education, social marketing, and public health education suggest that the inclusion of a person in the context of hazards as a part of hazard recognition training materials may improve its knowledge transfer, but it is unclear how this difference affects risk assessment. Researchers are exploring data from two studies to characterize this effect. The results suggest that the inclusion of a person in the context of a hazard has roughly equal and opposite changes in probability and severity ratings but no change in overall risk. Additionally, the results indicate that there may be a confounding effect of specific types of experience.

Overall, participants in both studies rated the probability of hazards without people significantly higher than those with people. This result suggests that there may be a tradeoff between hazard situational specifics and an increase in incident possibilities. Having a person in the midst of a hazard may suggest that the person is closer to safety, leading to the lower accident probability ratings. Alternatively, mineworkers may be experiencing projection bias [24] or false consensus bias [25] in that they are assuming the person in the image has the same increased visibility or awareness of the hazard as they do as observers, again decreasing the probability of an accident. Mineworkers should be careful of attributional biases and avoid allowing them to affect their perceptions. Risk assessment is intended to be "objective estimations based upon well-defined criteria, relating to the probability of an event, its potential consequences, and levels of exposure to these consequences" as opposed to subjective risk perception [20].

The results of this study also suggest that for the training participants, less experienced mineworkers rated the severity of hazards without people lower. The decrease in severity may be due to the lack of clear definition of the hazardous situation without a person. This makes the hazard outcome less salient that is, less cognitively available and thus participants tended to rate the severity lower [26]. However, this effect is not present in more experienced participants. One possible reason for the difference is that experienced participants tend to be older. Older adults may be more sensitive to STF hazards because they tend to be more risk adverse [27]. Alternatively, the increased morbidity and mortality of falls for older adults [28] and the increased likelihood of a previous personal injury may have an amplification effect on their severity perception [26].

Another interesting finding was that the results from the laboratory and training studies differ on severity and risk. This difference suggests that there may be a difference between general experience and safety-specific experience as has been demonstrated previously [7–9]. Despite similarities in years of mining experience, safety professionals have a different focus and type of experiences than experienced mineworkers. Furthermore, the laboratory study included mining students that lack any practical experience that even an inexperienced mineworker would have. It is also important to note that the age and experience makeup of the laboratory study is lower than that of the training study.

Lastly, the results indicate that for the training participants, the opposite changes in probability and severity appeared to offset each other, resulting in no change in overall risk. However, it is unclear whether the magnitude of the changes in probability and severity are hazard specific.

As a secondary analysis, this paper explores the changes in participants risk perceptions based on the context of hazard images. Though the primary focus was the inclusion of a person, this study is limited in that exact matching hazards were not used. It is possible that the effect observed in this analysis were due to other differences between the images. Specifically, tripping on pipe is different that falling off a bucket. Additionally, the number and type of hazards included in this study was small. This study only included two sets of two types of STF hazards. Researchers should conduct an additional directed study using a tighter comparison design with additional hazard types and hazard classifications.

The generalizability of the laboratory and training data may be limited by the convenience sampling. Participants for the laboratory study self-selected to participate and the training recruitment was limited to the clientele of the fourth and fifth authors. Furthermore, the cultural background of the participants is unknown. As suggested in the literature, cultural relevance can influence perception in positive and negative ways [20]; additional studies should control for this.

5 Conclusions

Overall, the results of this work indicate that the context of a hazard can alter trainees' risk perceptions. These results begin to suggest that the inclusion of a person in hazard images may reduce probability ratings and increase severity ratings. More work is needed to directly attribute these findings. However, this does not mean that adding in workers no longer has the potential to improve training materials. Including context and variability in training material is an important aspect to building up trainees' exposure and understanding. The results also highlight the differences across hazards and experience levels. Together, this underscores the importance of standardizing and objectifying risk assessment.

Disclaimer. The findings and conclusions in this paper are those of the authors and do not necessarily represent the official position of the National Institute for Occupational Safety and Health, Centers for Disease Control and Prevention. Mention of company names or products does not constitute endorsement by NIOSH.

References

1. MSHA: Overview of Deaths in Metal and Nonmetal Mining from October 2013 to December 2016. Mine Safety and Health Administration (2017). http://MSHA.gov
2. Bahn, S.: Workplace hazard identification and management: the case of an underground mining operation. Saf. Sci. **57**, 129–137 (2013)
3. Eiter, B.M., Helfrich, W., Hrica, J., Bellanca, J.L.: From the laboratory to the field: developing a portable workplace examination simulation tool. AHFE 2018. In: Cassenti, D. (ed.) Advances in Human Factors in Simulation and Modeling, AHFE 2018. Advances in Intelligent Systems and Computing, vol. 780. Springer, Cham (2019)
4. Hunter, D.R.: Risk Perception and Risk Tolerance in Aircraft Pilots (No. DOT/FAA/AM-02/17). Federal Aviation Administration, Washington, DC, Office of Aviation Medicine (2002)
5. Brewer, N.T., Chapman, G.B., Gibbons, F.X., Gerrard, M., McCaul, K.D., Weinstein, N.D.: Meta-analysis of the relationship between risk perception and health behavior: the example of vaccination. Health Psychol. **26**(2), 136 (2007)
6. Brown, I.D., Groeger, J.A.: Risk perception and decision taking during the transition between novice and experienced driver status. Ergonomics **31**(4), 585–597 (1988)
7. Perlman, A., Sacks, R., Barak, R.: Hazard recognition and risk perception in construction. Saf. Sci. **64**, 22–31 (2014)
8. Eiter, B.M., Bellanca, J.L., Helfrich, W., Orr, T.J., Hrica, J., Macdonald, B., Navoyski, J.: Recognizing mine site hazards: identifying differences in hazard recognition ability for experienced and new mineworkers. In: Cassenti, D. (ed.) Advances in Human Factors in Simulation and Modeling, AHFE 2017. Advances in Intelligent Systems and Computing, vol. 591. Springer, Cham (2018)
9. Bellanca, J.L., Eiter, B.M., Hrica, J., Weston, T., Weston, R.: The effect of work experience on risk assessment skills. In: The Interservice/Industry Training, Simulation and Education Conference (I/ITSEC 2018), Orlando, FL, pp. 1–14 (2018)
10. Burke, M.J., Sarpy, S.A., Smith-Crowe, K., Chan-Serafin, S., Salvador, R.O., Islam, G.: Relative effectiveness of worker safety and health training methods. Am. J. Public Health **96**(2), 315–324 (2006)
11. Aik, C.T., Tway, D.C.: Elements and principles of training as a performance improvement solution. Perform. Improv. **45**(3), 28–32 (2006)
12. Wilkins, J.R.: Construction workers' perceptions of health and safety training programmes. Constr. Manag. Econ. **29**(10), 1017–1026 (2011)
13. Burke, L.A., Hutchins, H.M.: Training transfer: an integrative literature review. Hum. Resour. Dev. Rev. **6**, 263–296 (2007)

14. Velada, R., Caetano, A., Michel, J.W., Lyons, B.D., Kavanagh, M.: The effects of training design, individual characteristics and work environment on transfer of training. Int. J. Train. Dev. **11**, 282–294 (2007)
15. Grossman, R., Salas, E.: The transfer of training: what really matters. Int. J. Train. Dev. **15**, 103–120 (2011)
16. Kraiger, K.: Perspectives on training and development. In: Handbook of Psychology, Industrial and Organizational Psychology, vol. 12, pp. 171–192. Wiley, Hoboken (2003)
17. Brunette, M.J.: Development of educational and training materials on safety and health. Fam. Community Health **28**(3), 253–266 (2005)
18. Castro, F.G., Barrera Jr., M., Holleran Steiker, L.K.: Issues and challenges in the design of culturally adapted evidence-based interventions. Ann. Rev. Clin. Psychol. **6**, 213–239 (2010)
19. Taylor, P.J., Russ-Eft, D.F., Chan, D.W.: A meta-analytic review of behavior modeling training. J. Appl. Psychol. **90**(4), 692 (2005)
20. Weyman, A.K., Clarke, D.D.: Investigating the influence of organizational role on perceptions of risk in deep coal mines. J. Appl. Psychol. **88**(3), 404 (2003)
21. Bellanca, J.L., Orr, T.J., Helfrich, W., Macdonald, B., Navoyski, J., Eiter, B.: Assessing hazard identification in surface stone mines in a virtual environment. In: Duffy, V. (ed.) Advances in Applied Digital Human Modeling and Simulation. Advances in Intelligent Systems and Computing, vol. 481. Springer, Cham (2017)
22. MSHA: Section 8: Coding Manual. Part 50 Data User's Handbook. Mine Safety and Health Administration (2007). https://www.cdc.gov/niosh/mining/data/default.html
23. Wang, M.: Generalized estimating equations in longitudinal data analysis: a review and recent developments. Adv. Stat. **2014**, 1–11 (2014)
24. Hsee, C.K., Hastie, R.: Decision and experience: why don't we choose what makes us happy? Trends Cognit. Sci. **10**(1), 31–37 (2006)
25. Ross, L., Greene, D., House, P.: The "false consensus effect": an egocentric bias in social perception and attribution processes. J. Exp. Social Psychol. **13**(3), 279–301 (1977)
26. Tversky, A., Kahneman, D.: Judgement under uncertainty: heuristics and biases. Science **185**, 261–275 (1974)
27. Sicard, B., Jouve, E., Couderc, H., Blin, O.: Age and risk-taking in French naval crew. Aviat. Space Environ. Med. **72**(1), 59–61 (2001)
28. Fries, J.F.: Aging, natural death, and the compression of morbidity. N. Engl. J. Med. **303**(3), 130–135 (1980)

EXAMiner: A Case Study of the Implementation of a Hazard Recognition Safety Intervention

Brianna M. Eiter[✉] and Jonathan Hrica

Pittsburgh Mining Research Division, National Institute for Occupational Safety and Health, Centers for Disease Control and Prevention, Pittsburgh, PA, USA
{BEiter, JHrica}@cdc.gov

Abstract. Within the mining industry, the ability to recognize hazards is essential for mineworkers completing hazard avoidance tasks such as workplace examinations and pre-shift inspections. To help build the necessary skills for hazard recognition, researchers at the National Institute for Occupational Safety and Health (NIOSH) have focused their efforts on the development of practical solutions derived from research findings. One such solution is the "Search Like an EXAMiner Safety Intervention," developed by NIOSH. This program includes EXAMiner, a workplace examination simulation software designed to address critical competencies associated with hazard recognition ability, a training module, and a supervisor worksheet designed for use in conjunction with the software and training module. The purpose of this paper is to present a case study where the "Search Like an EXAMiner Safety Intervention" program was implemented, including the process used to design the intervention, implement the intervention in the field, and evaluate its impact in the field.

Keywords: Hazard recognition · Simulation · Evaluation

1 Introduction

The metal/nonmetal (M/NM) mining sector recently experienced an increase in fatal injuries. Over the course of two years, the fatal injury rate nearly doubled from 7.9 per 100,000 mineworkers in 2012 to 14.7 per 100,000 mineworkers in 2014 [1]. The Mine Safety and Health Administration (MSHA) responded to this increase with a call for an increased focus on "daily and effective workplace exams to find and fix hazards" as one of several areas where improvement was necessary to promote the health and safety of the mineworker [2]. MSHA formalized this call by updating the workplace examination regulation for M/NM mines. Under the new rule, which went into effect on June 2, 2018, mine operators are now required to conduct examinations of the working place prior to or as a mineworker begins work in an area and to promptly alert mineworkers of any adverse conditions occurring in an area before they are potentially exposed to a hazardous condition. In addition, workplace examination records must include descriptions of adverse conditions that are not immediately corrected, along with the dates of when corrective actions for those conditions are made [3]. Overall, the new

D. N. Cassenti (Ed.): AHFE 2019, AISC 958, pp. 274–286, 2020.
https://doi.org/10.1007/978-3-030-20148-7_25

rule increases regulation; however, there continues to be gaps within the new rule. For instance, the new rule does not specifically identify what qualifications and competencies are necessary to conduct a workplace examination.

Within the Code of Federal Regulation (30 CFR 56.2), workplace examinations must be performed by the "competent person." The competent person is the person designated by the mine operator as having the abilities and experiences to qualify him or her to perform the workplace examination. One of the competent person's primary responsibilities is to recognize and mitigate hazards. While the legal responsibility for finding and fixing hazards belongs to the competent person, to maintain a healthy and safe work environment, all mineworkers must recognize hazards throughout their workday. Hazard recognition is a cognitively complex process [4]. It represents a special challenge for mineworkers, including the competent person, because of the diverse worker activities that take place in a dynamic environment [5]. While hazard recognition is critical for mineworker health and safety, research indicates that mineworkers are not identifying a significant number of hazards [6, 7].

To support the M/NM mining sector as it works to comply with the new regulation related to workplace examinations, the National Institute for Occupational Safety and Health (NIOSH) has focused laboratory-based research efforts on identifying barriers mineworkers encounter when recognizing workplace hazards [6, 23]. NIOSH researchers recently created a safety intervention to address these barriers called "Search Like an EXAMiner." This safety intervention includes EXAMiner, a workplace examination simulation software designed to address critical hazard recognition competencies [8] and the "Search Like an EXAMiner" training module with an accompanying supervisor worksheet designed to address barriers to hazard recognition. The purpose of this paper is to describe the development of the "Search Like an EXAMiner" safety intervention and present a case study demonstrating the initial implementation and impact of the safety intervention at a M/NM mine site.

2 Hazard Recognition and Workplace Examinations

A hazard is defined as an act or a condition in the workplace that has the potential to cause injury, illness, or loss [9]. Hazard recognition is the realization that a condition or behavior can cause harm [7]. Hazard recognition is fundamental to every safety activity, and hazards that go unrecognized and unmanaged can potentially result in catastrophic accidents and injuries [10]. While it is critical to the health and safety of the worker, research indicates that workers in dynamic and unpredictable environments are unable to recognize a significant number of hazards [4, 11].

To identify factors that affect hazard recognition performance, researchers at NIOSH examined hazard identification through a laboratory study where subjects performed a simulated workplace exam. In this study, researchers asked experienced and inexperienced mineworkers, mine safety professionals, and mining engineering students to perform a simulated workplace examination [6]. Participants performed this hazard recognition task within a virtual environment, and were instructed to search for hazards in true-to-life-size panoramic images of typical locations at surface limestone mines—the pit, plant, roadway, and shop. After completing the hazard recognition task, all participants received feedback about their hazard recognition performance and were debriefed about the hazards they accurately identified and missed. Consistent with the findings from other studies [12], the hazard recognition task results showed that experienced and inexperienced mineworkers and mining engineering students, on average, identified only 53% of the hazards in the images. In addition and importantly, the participating mine safety and health professionals—those people at the mine site tasked with ensuring that mineworkers are trained and capable of recognizing hazards —were only able to accurately identify 61% of the hazards. These identification rates are well below the previously established 90% standard for mastery [13].

Within the mining industry, there are no regulations related to the process used to perform the workplace examination. An informal assessment of the M/NM industry indicates that mine operators typically use a checklist approach to perform pre-shift hazard inspections and workplace examinations. Inspection and examination checklists or worksheets are often constructed to focus on risks and hazards specifically called out in the CFR regulation (30 CFR 56.2). They tend to direct examiners to search for routine hazards (buildup of material on the catwalks or loose rollers on the conveyor belt), to inspect specific locations (e.g., stockpiles and berms), and to check for required personal protective equipment (PPE; wearing hearing protection where required or tying off in an appropriate location). In order to meet compliance expectations, workplace examination checklists also include space for the user to describe a hazardous situation and to affix a signature and date.

Workers in other high-risk industries (e.g., aviation and heavy construction) [14, 15] use checklists while conducting inspections. Research indicates that taking a checklist approach during inspections benefits examiners [16], particularly when checklists are used in standardized situations, when time is not critical, when the series of tasks or items is too long to commit to memory, and when the environment is conducive to carrying the list. Relying solely on checklists, however, may limit the hazards workers search for and find during workplace examinations. There is ample evidence within the literature that task instructions can cause inattentional blindness to a salient object or event [17]. In other words, directing workers to use a checklist as a guide may focus their attention to items on the checklist to such an extent that they are blind to items and events that are not included on the checklist, which could be hazardous.

3   Search Like an EXAMiner Safety Intervention Development

Recent research indicates a need for improved hazard recognition for all mineworkers, not just the competent person [6]. To address this need, NIOSH developed a safety intervention called "Search Like an EXAMiner" that targets competencies critical to hazard recognition and highlights potential barriers mineworkers encounter when identifying hazards. "Search Like an EXAMiner" was developed as a multilevel intervention (MLI) as it was designed to target more than one organizational level within a mining company (superintendent, supervisor, health and safety (H&S) professional, and mineworker) so as to encourage and sustain behavior [18]. MLIs have been shown to successfully engage supervisors and workers in identifying health and safety practices to manage workplace risks [19]. The following includes a description of each of the tools that comprise the multilevel intervention and considerations that guided development of the "Search Like an EXAMiner" intervention program.

3.1 EXAMiner Software

EXAMiner is a stand-alone, downloadable software application. It is intended for use by safety and health trainers to deliver training to mineworkers. EXAMiner can be used during mandatory (see CFR Part 46.5 and 46.8 of Title 30 Mineral Resources) and elective (e.g., pre-shift safety meetings or toolbox talks) training. EXAMiner was developed for use with groups of M/NM mineworkers in a classroom setting (see Eiter et al. 2018, for a full explanation of design specifications [8]).

The EXAMiner software was developed to improve mineworker hazard recognition ability. Because mineworkers must possess a complex set of competencies to successfully recognize hazards, EXAMiner targets four specific competencies: general hazard recognition knowledge, site-specific hazard recognition knowledge, visual search, and pattern recognition. These competencies were targeted because they were determined to be necessary for all mineworkers, and they aligned with software design specifications in that they could be easily visualized within the software.

As an instructional tool, EXAMiner is grounded in proven training strategies that incorporate information, demonstration, practice, and feedback [20]. General and site-specific hazard information is delivered through descriptions, statistics about associated accident and injury rates, and mitigation strategies. Hazard recognition is demonstrated through the workplace examination search task, which is a computer-based simulation that allows interactive and immersive activity by recreating all or part of a work experience [21] and encouraging experiential learning [22]. EXAMiner provides mineworkers the opportunity to practice by performing the workplace examination search task where mineworkers search panoramic scenes for hazards. During the debrief, mineworkers receive feedback about hazard recognition performance and descriptions of all hazards.

3.2 Search Like an EXAMiner Module and Supervisor Worksheet

"Search Like an EXAMiner" was developed in conjunction with the EXAMiner software. It is comprised of two components—a hazard recognition training module and a corresponding supervisor worksheet. The training module is appropriate for all mineworkers and can be used during mandatory (see CFR Part 46.5 and 46.8 of Title 30 Mineral Resources) and elective (e.g., pre-shift safety meetings or toolbox talks) training. The supervisor worksheet is intended for use by the competent person.

Search Like an EXAMiner Training Module. The objective of the "Search Like an EXAMiner" training module is to increase knowledge about barriers to hazard recognition in order to improve a mineworker's ability to find hazards when performing a workplace examination. To meet this objective, the training module contains three sections: introductory, instructional, and summary. The introductory section includes basic definitions related to hazards and risk and information pertinent to the workplace examination regulation. The instructional section focuses on barriers to hazard identification and is broken into three parts: barriers related to experience, barriers related to complexity in the workplace, and barriers related to change. The summary section reviews the barriers to hazard recognition and provides tips for how to overcome the barriers in the field. The sections were designed to take between 10 and 45 min to deliver. They can be delivered as a whole or as separate sections depending on time constraints and informational needs.

The instructional section addresses three barriers to hazard recognition: experience, complexity, and change. Information in support of these barriers was derived from the NIOSH hazard recognition laboratory studies [6, 23] as well as from the research literature [16]. Each barrier is addressed in the following manner: a first slide introduces and defines the barrier. A second slide or set of slides provides examples and demonstrations highlighting the barrier and its impact on hazard recognition. A third slide or set of slides then presents thought exercises for trainees to complete using information that was just presented.

Search Like an EXAMiner Supervisor Worksheet. The goal of the supervisor worksheet is threefold. The first goal is to translate training into practice. Transfer of training occurs when the knowledge learned is used on the job [24], and the effectiveness of training ultimately depends on whether learned outcomes are used in the workplace [20]. The second goal is to move beyond the checklist approach to safety inspections. While checklists are effective memory aids [16], reliance on a set list of hazards can limit what the competent person searches for during an examination [17]. The third goal is to provide the competent person with a systematic process to follow for performing a workplace examination.

Figure 1 shows the Supervisor Worksheet with an example workplace examination included. To address the transfer of training to the workplace, the content of the Supervisor Worksheet was derived from the barriers to the hazard recognition instructional section of the "Search Like an EXAMiner" training module. The three barriers—experience, complexity, and change—are listed in the left column of the worksheet along with representative examples for each barrier. The Supervisor Worksheet moves beyond a checklist as the barriers represent broader, high-level concepts instead of specific types of hazards (e.g., slip, trip, fall or housekeeping). Finally, the Supervisor Worksheet provides the competent person a systematic process to follow that is based on a workplace risk assessment methodology developed by the Health and Safety Executive of the United Kingdom [25]. Within this methodology, there are five recommended steps including: (1) Identify the hazards, (2) Decide who might be harmed and how, (3) Evaluate the risks and decide on precautions, (4) Record your significant findings, and (5) Regularly review your assessment and update if necessary. NIOSH researchers adapted and expanded these five steps to create a workplace examination process for the competent person to follow. Within the Supervisor Worksheet, the competent person follows these five steps. Step 1 is to identify the barriers to hazard recognition. Step 2 is to decide who may be injured and how, and Step 3 is to identify what precautions are already being taken and to decide what further action is necessary to mitigate the hazard. For Step 4, the competent person determines who to communicate with to mitigate a hazard, who is responsible

Location:	Date:		Overcoming Barriers to Hazard Recognition				
			Supervisor Worksheet				
Step 1:	Step 2:	Step 3:		Step 4:			Step 5:
What are the Barriers?	Who may be injured and how?	What are you already doing?	What further action is necessary?	Communicate with?	Action by whom?	Action by when?	Date Complete
							Results
Experience							
Experience with Mining	Mineworkers with less than 2 years of total mining experience may have less knowledge of site hazards and be at higher risk for injury	New mineworkers receive 24 hours of initial safety training. Annual refresher training is given each year.	Shop supervisor will give toolbox talks to increase new mineworker knowledge of hazards in the field.	Mine safety and maintenance crew members.	Shop supervisor	6/22/18	New shop mechanics are receiving weekly toolbox talks given by shop supervisor. Ongoing.
Mindset							
Experience with Location							
Complexity							
Crowded Work Environment							
Busy Location							
Multiple Hazards							
Multiple safety controls							
Change							
Environmental Factors (Weather, time of day, etc.)							
Modifications to Work Environment							
Changes to tools, equipment, and structures							
Fit for Duty							

Fig. 1. The overcoming barriers to hazard recognition supervisor worksheet with an example of a hazard identified during the workplace examination.

for performing necessary actions, when actions are to be taken, and what the results are of the actions. Finally, in Step 5 the competent person indicates the completion date or the date to review the mitigation strategy.

4   Evaluation Framework

NIOSH developed the "Search Like an EXAMiner" safety intervention in order to address identified issues related to mineworker hazard recognition. However, it is currently not known whether this safety intervention is demonstrably effective at affecting mineworker hazard recognition. Thus, it is critical to evaluate whether the "Search Like an EXAMiner" safety intervention affects the targeted hazard recognition competencies and gives mineworkers an effective process to follow to address the barriers to hazard recognition.

NIOSH researchers adapted Kirkpatrick's taxonomy [26] to create a safety intervention evaluation framework. Within Kirkpatrick's taxonomy, interventions are evaluated on four levels: according to how workers feel about the intervention (i.e.

Table 1. This table represents the objectives and evaluation questions and methods used within the "Search Like an EXAMiner" safety intervention evaluation framework.

Level of evaluation	Evaluation objective	Organizational level of evaluation	Evaluation methodology
Level 1 Reaction	Reported levels of satisfaction and reaction to the EXAMiner software and "Think Like an EXAMiner" module	What satisfaction levels do mineworkers report following use of safety intervention?	Evaluation questionnaire and semi-structured interviews
Level 2 Learning	Demonstration of competencies addressed with EXAMiner software and ability to identify hazards based on barriers within the "Think Like an EXAMiner" module	Are mineworkers [workers and supervisors] able to recognize hazards during the safety intervention?	Hazard recognition accuracy collected using EXAMiner software and analysis of Supervisor Worksheets
Level 3 Behavior	Behavior change as knowledge and skills are applied in the work setting following the delivery of the training	What hazards did supervisors identify and what mitigation strategies did they use to address those hazards?	Analysis of Supervisor Worksheets and semi-structured interviews
Level 4 Results	Organizational processes and procedures that change or improve as a result of the application of the training	What changes did management [supervisors, superintendents, and H&S professionals] make to organizational practices and processes to address hazards?	Analysis of Supervisor Worksheets and semi-structured interviews

reactions), whether they learned anything (i.e. learning), whether the learning transfers to the field (i.e., behavior), and whether the intervention achieved its objectives (i.e., results). The goal of using this evaluation framework is to show evidence of change at each of the four levels of evaluation; to accomplish this, it is necessary to take a MLI approach [18, 19]. The "Search Like an EXAMiner" safety intervention was designed as a MLI aimed at affecting change within and between different organizational levels (superintendent, supervisor, health and safety professional, and mineworker). Table 1 includes a description of the levels of evaluation, the objective of the evaluation for each level, the targeted organizational level, and evaluation methodology associated with each construct.

5   Evaluation Methodology

5.1   Case Study Approach

In order to evaluate the efficacy of the "Search Like an EXAMiner" safety intervention, a case study approach was taken. Case studies provide important information from the viewpoint of participants by using multiple sources of data [27]. Thus, this approach was appropriate because the goal was to focus on the implementation of the safety intervention at one surface sand and gravel mining operation and evaluate its effectiveness across time. The case study is descriptive in nature, and all data collected from the surface sand and gravel mining operation is qualitative in nature. A summary and synthesis of the collected data is reported within the lens of the "Search Like an EXAMiner" evaluation framework. As this is the first time the safety intervention was implemented in the field, potential changes are recommended based on identified barriers to use.

5.2   Surface Mining Operation and Evaluation Participants

The focus of this case study is a surface sand and gravel operation located in the western region of the United States. The mining company is part of a larger multi-industrial corporation. The mine site is a typical sand and gravel mine operation as it is comprised of a pit and a relatively fixed and routine crushing, washing, and stockpiling process. The organizational structure of the operation is as follows: superintendents manage overall areas of the operation (e.g., maintenance, production, etc.), supervisors manage specific areas within the operation (e.g., within production, a supervisor manages the pit while another manages the wash facility), and mineworkers perform a function within an area (e.g., haul truck operator in the pit, frontend loader operator at the wash facility). Within this company, supervisors serve as the competent person and are responsible for performing workplace examinations.

There are approximately 55 employees at the mining operation; this includes 3 superintendents, 10 supervisors, and 40 mineworkers. To address health and safety (H&S) concerns, the mine operation also employs two H&S professionals; one of the H&S professionals is primarily responsible for training workers.

Table 2. Timeline for the longitudinal, multilevel safety intervention with organizational groups, research tasks, associated research constructs, and Kirkpatrick [26] level of evaluation.

Data collection timeline	Participant group and their research task	Data collection methodology, research constructs, and evaluation level
Month 1 (Visit 1)	• Mineworkers train with EXAMiner and complete evaluation questionnaire • Supervisors train with "Search Like an EXAMiner" module and complete first supervisor worksheet • Superintendents participate in initial semi-structured interviews • H&S professionals participate in semi-structured interviews	• EXAMiner evaluation measures reaction and satisfaction [29] [Level 1] • Identify hazards, current and future mitigation strategies, who is responsible for mitigating, and results of mitigation strategy (see Fig. 1) [Level 2] • Identify current practices related to workplace examinations, hazard recognition, and mitigation. [Level 4] • EXAMiner evaluation measures reaction, satisfaction [29], and usability, as defined by the Technology Acceptance Model [30] [Level 1 and 4]
Month 2	• Supervisors complete second supervisor worksheet	• Identify hazards, current and future mitigation strategies, who is responsible for mitigating, and results of mitigation strategy (see Fig. 1) [Level 2 and 3]
Month 3	• Supervisors complete third supervisor worksheet	• Identify hazards, current and future mitigation strategies, who is responsible for mitigating, and results of mitigation strategy (see Fig. 1) [Level 2 and 3]
Month 4 (Visit 2)	• Supervisors complete fourth supervisor worksheet • Supervisors participate in semi-structured interviews • Superintendents participate in final semi-structured interviews • H&S professionals participate in semi-structured interviews	• Identify hazards, current and future mitigation strategies, who is responsible for mitigating, and results of mitigation strategy (see Fig. 1) [Level 2 and 3] • Identify perceived barriers to using the supervisor worksheet, identifying hazards, implementing mitigation strategies, and determining the impact of results [Level 1] • Identify changes to organizational processes and procedures and, determine the impact of the results. [Level 4] • Overall safety intervention assessment, identify usability concerns and barriers to use [Level 1, 3, and 4]

Recruitment of individuals for the study involved a convenience, purposive sampling strategy. After obtaining IRB approval, NIOSH researchers recruited individuals who were easily accessible and willing to participate [28]. In total, data was collected from 17 workers at the mine operation. This includes 3 superintendents, 4 supervisors, 8 mineworkers, and 2 H&S professionals.

5.3 "Search Like an EXAMiner" Evaluation Procedure

Implementation and evaluation of the "Search Like an EXAMiner" safety intervention was longitudinal in nature. Data collection was completed during two visits to the mine site and across four months. See Table 2 for a full explanation of the data collection timeline.

6 Validating the Evaluation Framework

We present results for each of the four levels to validate the evaluation framework as a means to measure efficacy of the safety intervention. These results are not exhaustive, but instead provide supporting evidence that the "Search Like an EXAMiner" safety intervention was designed to impact multiple levels of the evaluation framework.

Level 1: Reaction. Satisfaction and usability were measured through evaluation questionnaires with Likert scales and open-ended questions and further assessed through a qualitative analysis of semi-structured interviews. Mineworkers who participated in training with EXAMiner had an overall positive reaction [using a scale from 1 (completely unacceptable) to 10 (very exceptional); on average, mineworkers rated the program as an 8.6]. H&S professionals reported having no technical usability issues or concerns with the EXAMiner software; however, H&S professionals expressed a usability concern with the structure and organization of the supervisor worksheet. Specifically, they indicated that the worksheet is complex, can overwhelm users, and take too much time to complete.

Level 2: Learning. Learning was assessed in two ways: through hazard recognition accuracy that is calculated after completing the EXAMiner workplace examination search task and through analysis of the supervisor worksheets. Mineworkers were able to find approximately 80% of the hazards during the workplace examination search task, and supervisors were able to use the supervisor worksheet to find hazards at their workplace. As an example of a hazard associated with barrier change, one supervisor identified buildup of spilled material around a portable crushing plant.

Level 3: Behavior. Behavior was assessed through analysis of the supervisor worksheets. The goal of this analysis was to identify the current and future mitigation strategies to address these hazards. Regarding the Level 2 examples, the supervisor and his work crew typically clean up spilled material throughout the shift, whenever they see it. There is a lot of mobile equipment traffic into and around the portable crushing plant. Getting out of equipment, such as a frontend loader or haul truck, puts the workers at risk of being struck by mobile equipment. To mitigate this hazard, the supervisor implemented a material cleanup policy where cleanup was scheduled for the end of every shift. There are two workers dedicated to performing the task, and there is a procedure in place to communicate the start of cleanup to all incoming mobile equipment traffic.

Level 4: Results. The results of mitigation strategies were identified through analysis of semi-structured interviews with superintendents and H&S professionals. In the Level 3 example, the supervisor identified a hazard and implemented a procedure to mitigate the hazard. The superintendents supported the implementation of the scheduled cleanup, and it was determined that the scheduled cleanup not only decreased worker exposure to mobile equipment, but also to the risk of falling from equipment because workers were no longer climbing on and off several times throughout a shift.

7   Conclusion

The purpose of this paper was to describe the development of the "Search Like an EXAMiner" multilevel safety intervention, define the evaluation framework, and then present a case study where the safety intervention was implemented. Preliminary case study results suggest that the "Search Like an EXAMiner" multilevel safety intervention does affect mineworker hazard recognition and that it can be used to affect change to workplace examination processes and procedures. Future work will further validate the "Search Like an EXAMiner" multilevel safety intervention and provide recommendations for implementation of the intervention.

Disclaimer. The findings and conclusions in this paper are those of the authors and do not necessarily represent the official position of the National Institute for Occupational Safety and Health, Centers for Disease Control and Prevention. Mention of company names or products does not constitute endorsement by NIOSH.

References

1. United States Department of Labor (n.d.). https://www.msha.gov/data-reports/statistics/mine-safety-and-health-glance
2. MSHA, Mine Safety and Health Administration: MSHA announces increased education, outreach and enforcement to combat increase in NMN mining deaths (2015). http://arlweb.msha.gov/stats/review/2014/mnm-fatality-reduction-effort.asp
3. MSHA: Proposes changes to final rule on workplace examinations in metal and nonmetal mines. https://www.msha.gov/news-media/press-releases/2017/09/12/msha-proposes-changes-final-rule-workplace-examinations-metal
4. Kowalski-Trakofler, K.M., Barrett, E.A.: The concept of degraded images applied to hazard recognition training in mining for reduction of lost-time injuries. J. Saf. Res. **34**(5), 515–525 (2003)
5. Scharf, T., Vaught, C., Kidd, P., Steiner, L., Kowalski, K., Wiehagen, B., Rethi, L., Cole, H.: Toward a typology of dynamic and hazardous work environments. Hum. Ecol. Risk Assess. **7**(7), 1827–1841 (2001)

6. Eiter, B.M., Bellanca, J.L., Helfrich, W., Orr, T.J., Hrica, J., Macdonald, B., Navoyski, J.: Recognizing mine site hazards: identifying differences in hazard recognition ability for experienced and new mineworkers. In: International Conference on Applied Human Factors and Ergonomics, pp. 104–115. Springer, Cham (2017)

7. Bahn, S.: Workplace hazard identification and management: the case of an underground mining operation. Saf. Sci. **57**, 129–137 (2013)

8. Eiter, B.M., Helfrich, W., Hrica, J., Bellanca, J.L.: From the laboratory to the field: developing a portable workplace examination simulation tool. In: Advances in Intelligent Systems and Computing Advances in Human Factors in Simulation and Modeling, pp. 361–372 (2018)

9. Guidance for Hazard Determination for Compliance with the OSHA Hazard Communication Standard. https://www.osha.gov/dsg/hazcom/ghd053107.html

10. Albert, A., Hallowell, M.R., Kleiner, B., Chen, A., Golparvar-Fard, M.: Enhancing construction hazard recognition with high-fidelity augmented virtuality. J. Constr. Eng. Manag. **140**(7), 04014024 (2014)

11. Carter, G., Smith, S.D.: Safety hazard identification on construction projects. J. Constr. Eng. Manag. **132**(2), 197–205 (2006)

12. Perlman, A., Sacks, R., Barak, R.: Hazard recognition and risk perception in construction. Saf. Sci. **64**, 22–31 (2014)

13. Barrett, E., Kowalski, K.: Effective hazard recognition training using a latent-image, three-dimensional slide simulation exercise. Report of Investigation, Bureau of Mines (1995)

14. Degani, A., Wiener, E.L.: Cockpit checklists: concepts, design, and use. Hum. Fact. J. Hum. Fact. Ergon. Soc. **35**(2), 345–359 (1993)

15. Hinze, J., Godfrey, R.: An evaluation of safety performance measures for construction projects. J. Constr. Res. **04**(01), 5–15 (2003)

16. Clay-Williams, R., Colligan, L.: Back to basics: checklists in aviation and healthcare. BMJ Qual. Saf. **24**(7), 428–431 (2015)

17. Simons, D.J., Chabris, C.F.: Gorillas in our midst: sustained inattentional blindness for dynamic events. Perception **28**(9), 1059–1074 (1999)

18. Haas, E.J.: Encyclopedia of Health Communication, pp. 900–902. Sage, Thousand Oaks (2014). Multilevel Intervention

19. Haas, J., Cecala, A., Hoebbel, C.: Using dust assessment technology to leverage mine site manager-worker communication and health behavior: a longitudinal case study. J. Prog. Res. Soc. Sci. **3**, 154–167 (2016)

20. Salas, E., Cannon-Bowers, J.A.: The science of training: a decade of progress. Annu. Rev. Psychol. **52**, 471–499 (2001)

21. Maran, N.J., Glavin, R.J.: Low- to High-fidelity simulation—a continuum of medical education? Med. Educ. **37**, 22–28 (2003)

22. Kolb, D.A.: Experiential Learning: Experience as the Source of Learning and Development. Prentice-Hall, Englewood Cliffs (1984)

23. Orr, T.J., Bellanca, J.L., Eiter, B.M., Helfrich, W., Rubinstein, E.N.: The effect of hazard clustering and risk perception on hazard recognition. In: International Conference on Applied Human Factors and Ergonomics, pp. 349–360. Springer, Cham (2018)

24. Olsen, J.J.: The evaluation and enhancement of training transfer. Int. J. Train. Dev. **2**(1), 75 (1998)

25. United Kingdom Health and Safety Executive: Risk assessment: a brief guide to controlling risks in the workplace, Sudbury, Suffolk (2014)

26. Kirkpatrick, D.L.: Evaluation of training. In: Craig, R.L., Bittel, L.R. (eds.) Training and Development Handbook, pp. 87–112. McGraw Hill, New York (1967)

27. Tellis, W.M.: Introduction to case study. Qual. Rep. **3**(2), 1–14 (1997). https://nsuworks. nova.edu/tqr/vol3/iss2/4
28. Given, L.M.: 100 Questions (and Answers) About Qualitative Research, vol. 4. Sage, Thousand Oaks (2016)
29. Phillips, J., Stone, R.: How to Measure Training Results: A Practical Guide to Tracking the Six Key Indicators. McGraw Hill Professional, New York (2002)
30. Davis, F.D.: A technology acceptance model for empirically testing new end-user information systems: theory and results. Diss. Massachusetts Institute of Technology (1985)

Development of Visual Elements
for Accurate Simulation

Timothy J. Orr[(✉)], Jennica L. Bellanca, Jason Navoyski,
Brendan Macdonald, William Helfrich, and Brendan Demich

Pittsburgh Mining Research Division, National Institute for Occupational Safety
and Health (NIOSH), Centers for Disease Control and Prevention (CDC),
Pittsburgh, PA, USA
{TOrr, JBellanca, JNavoyski, BMacdonald,
WHelfrich, BDemich}@cdc.gov

Abstract. Virtual reality (VR) is a powerful tool to study human behavior and performance under the controlled conditions of a laboratory simulation. Simulation fidelity, particularly in the visual representation of the virtual environment (VE), continues to improve with ongoing advances in computer graphics technology. Because the software tools used to create these VEs are typically the same as those used by the gaming industry, the pervasiveness of high-fidelity video games raises expectations of quality, detail, and lighting in VR applications. However, high-quality models of equipment and environments are often not readily available, and developers may not be familiar with how to implement the latest techniques like physically based rendering (PBR) and photogrammetry. In addition, creating these assets in a timely manner may require several team members with different skillsets working together, and maintaining a consistent aesthetic can be a challenge. Because visual consistency is critical for maintaining immersion and controlling visual stimuli in research as well as in training, researchers from the National Institute for Occupational Safety and Health (NIOSH) created an asset development workflow. This paper describes the workflow, precision modeling techniques, and review process as well as PBR and photogrammetry techniques. It also details lessons learned throughout the process of creating numerous photorealistic VR assets.

Keywords: Virtual reality · Visual fidelity · Mineworker safety

1 Background

Virtual reality (VR) describes the simulated experience generated by an assemblage of hardware and software that is used to immerse the user. Over many years, VR developers targeted the development of virtual environments (VE) toward various purposes: entertainment, the treatment of phobias [1], rehabilitation [2], and training [3]. A large part of the VE development includes digital assets, where digital assets refer to the nonprogrammed elements, including 3D models, audio, and user interface. Because vision provides nearly 10 Mbps of data [4] and accounts for the majority of all sensory input combined [5], a focused effort on asset appearance in VEs is warranted. Research has shown that higher-resolution imagery increases search performance and

This is a U.S. government work and not under copyright protection in the U.S.;
foreign copyright protection may apply 2020
D. N. Cassenti (Ed.): AHFE 2019, AISC 958, pp. 287–299, 2020.
https://doi.org/10.1007/978-3-030-20148-7_26

training transfer, necessitating high-fidelity assets [6, 7]. In addition to being visually appealing, it is important that the assets and environment are also dimensionally accurate and consistent in order to elicit appropriate responses from the end users as they interact with and among these items [8, 9].

Researchers at the National Institute for Occupational Safety and Health (NIOSH) have developed VEs and digital assets under several research projects over the course of the last two decades [10–13]. This work provided researchers with a strong knowledge base for digital asset development, but the assets themselves have had limited reuse value. The extended timeframes and limited scope of these projects resulted in inconsistent visual quality and dimensional accuracy. Moving forward, researchers wanted to optimize time and resources and adopt new asset development tools for the latest game engine technology. Therefore, the aim of this paper is to document an asset development workflow including geometry, materials, UV layout, and shader implementation combined with review steps in order to produce high-fidelity simulation assets that are accurate with consistent characteristics, using repeatable methodologies that are well documented.

2   Introduction

2.1   Workflow

A workflow is a formalization of the steps required to move a concept from ideation to realization. Each process step encapsulates the input and output required to flow along a directed path. Steps for asset development include the primary tasks, review, and documentation. Reviews help catch errors before additional effort is expended on downstream processes. Review and documentation also help provide metrics for maintaining consistency, particularly between various artists. Additionally, formalization of a workflow affords the opportunity to identify tasks that can be parallelized to improve efficiency. Minimally, an asset development workflow for a VR simulation must include the development of basic geometry, the development of materials, the creation of UV layouts, and the configuration of the rendering engine. A description of each of these steps and their importance to the ultimate fidelity of the simulation is described in the sections below.

2.2   Geometry

The geometry of a digital asset describes the underlying mesh of polygons and vertices that makes up the basic shape and dimensions of an object. The geometric representation of an object can be realized through a number of methods such as computer-aided design (CAD), laser scanning, photogrammetry, or box modeling. While starting out with detailed digitalizations of the equipment to be modeled might seem ideal, these formats each come with challenges that can require significant effort to format, simplify, and optimize for real-time rendering in VR. CAD models often do not include small details like welds or small fasteners that might seem trivial, but will be noticeably absent to system users. Additionally, because large equipment is often customized, as-built measurements of a machine may be required, depending on the requirements of the simulation. Even in the case of laser scans or photogrammetry, verification of model dimensions requires some manual measurements of the physical object. Furthermore, laser scans and photogrammetry methods produce holes or discontinuities when one part of the geometry obscures another. Lastly, all of these methods require some form of optimization because the total number of polygons on screen affects the rate at which objects can be rendered. In order to achieve real-time rendering at high frame rates, polygons should be minimized. Building geometry from scratch—box modeling—affords the most flexibility in quantity, scale, and UV layout of mesh detail for a model of an object.

Regardless of the method of geometry creation, to further optimize rendering speeds, lower polygon models can be substituted when an object is farther away from the rendering camera. Level of detail (LOD) models can be generated using the mesh optimization tools with minimal artist manipulation. Mesh optimization software tools further refine a mesh by systematically reducing vertices while maintaining the critical features of the original model.

2.3   Material

Materials are a collection of image bitmaps—textures—that control the way light interacts with a surface. Early computational rendering work used measurements of light intensity at various angles of incidence and reflectance to produce mathematical models that describe reflection and energy conservation [14, 15]. Termed physically based rendering (PBR), these algorithms adhere to the principles that the total light energy coming off a surface cannot be greater than the incident light and that light behaves with reciprocity, adding to scene realism by accounting for light bounce from object to object. Overall, the intention of PBR is to create synthetic images that are indistinguishable from actual ones.

Practically, PBR techniques provide consistent and predictable rendered images for a given rendering engine using various lighting conditions. In this way, several artists can create content for a simulation, and by adhering to common color and reflectivity principles they can produce comparable assets. These principles are realized by standard textures applied to a material based on the prospective rendering engine. Some example textures include diffuse, normal, reflectivity, metallic, and roughness maps. Base properties for common materials are available for most of the popular game engines as libraries included in commercial software. Additionally, software tools can be integrated into some game engines like Unity and Unreal to further streamline workflows and provide nonartists the ability to tweak material characteristics.

2.4   UV Layout

The creation of a UV layout for a digital asset describes the process of applying 2D materials to a 3D geometric mesh. The U and V refer to the x and y axes in the material coordinate space to avoid confusion with the labels for x and y in the geometric coordinate space. UV denotes the plane typically used for mapping these materials, as this is the standard top-down view in most 3D modeling environments. Materials are mapped to the geometry so that the desired properties are represented at the intended location. The process of making this transformation from 2D to 3D space is referred to as UV mapping. Mapping 2D materials onto complex 3D geometry directly is challenging. Therefore, polygons comprising the 3D geometry can be flattened out into 2D space in a process called unwrapping. The artist can then resize regions of the unwrapped mesh to control how much of the material is mapped to each mesh region in order to change the local texture resolution—texel density. In doing this, artists should provide space around the mesh regions for edge padding in order to avoid mipmapping induced artifacts. These artifacts occur when background pixels of a texture bleed over the boundary as the image is rescaled for efficient rendering. The mesh can also be divided among different materials to improve texture resolution. However, artists should try to minimize the number of materials assigned to each mesh because this can reduce rendering speed. There are various software packages that include tools designed to make UV mapping easier.

2.5　Shader

Previously, PBR methods had been used to render photorealistic computer-generated imagery off-line for movies and animations [15]. However, advances in hardware graphics processing and ray tracing algorithms now allow these sophisticated techniques to improve visual realism in real-time game engines like Unity and Unreal with the use of PBR shaders.

Shaders are software components embedded in a game engine's rendering pipeline that calculate the color of each pixel based on the material and lighting. They are the last component in a real-time rendering pipeline. Because PBR shaders vary in their mathematical modeling of light interpretation, they require different material parameters and textures in order to be implemented [16]. PBR shaders vary from engine to engine, and different PBR shaders can even be used within an engine to support stylized visual effects (e.g. cartoon shading).

3　Asset Development

The following sections outline how NIOSH researchers have implemented an asset development workflow in order to develop more consistent, high-fidelity mining assets.

3.1　Workflow

The workflow begins with a customer request for a simulation asset. The team and customer create a draft specifications document at the initial customer meeting. The specifications document includes key information about the requested model, such as purpose, geometric detail, dimensional accuracy, animation, rigging, and material requirements. The customer and team will also agree upon what source material will be used for model development (e.g. CAD drawings, photographs, or physical access to the equipment) and who is responsible for providing this information. Acquiring this material can be time intensive, but critical, as the foundation of development. Lastly, the team and customer must agree upon a delivery date based on the required features and level of fidelity.

After this initial meeting, the asset team refines the request to produce an asset concept. The concept may contain marked-up sketches, photographs, or prototype 3D models and descriptive text to document the customer expectations. This touch-point is important to formalize the acceptance criteria for this asset. Scope creep and mismatches in expectations have led to serious delays in past projects. Shifting expectations often arise when a secondary customer is identified who has additional constraints or use cases. Identifying all customers at the outset can minimize such impacts. Overall, effective, frequent communication with the asset customer is key, particularly when moving through the initial steps of the workflow.

Team communication is also vital for success as much of the asset development can be done in parallel. Figure 1 depicts the generic asset development workflow, showing how geometry and material development can be completed in parallel. Depending on the project, the workflow may be more or less parallelized. While efficiency is important, it is also critical that the team has good understanding as to how the components fit together, are independent, or are interdependent. Early in the adoption of this workflow, team members tried to work ahead, rush through reviews, or skip reviews entirely. Errors discovered downstream led to greater inefficiencies. For example, if an error in the geometry was not identified until the final asset review, it could require additional changes to the materials as well as remapping of the UV layout.

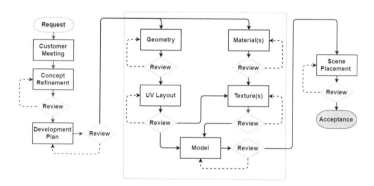

Fig. 1. Flowchart depicting NIOSH's generic asset development workflow. The blue box highlights processes that can be worked in parallel.

Similarly, the development team also discovered that prototyping solutions all the way through the workflow to see the final effect in the engine shader was helpful. Because material properties differ based on the rendering engine, rendered objects in the development software do not look the same as in the game engine. Prototyping can make the final asset development proceed more efficiently with fewer errors overall. For example, the team created prototype models to depict welds using modeled geometry, with only material normal maps, and a combination of both geometry and normal maps (Fig. 2). This allowed the customer and team to weigh the cost in time versus the benefit in fidelity for each method. This additional effort is valuable to avoid reworking the entire model. More time for experimentation and testing as a formal step in the workflow is recommended for future asset development.

Fig. 2. Detail view of heavy equipment showing seams between plates with and without welds.

Documentation also plays a pivotal role for team communication about expectations, responsibilities, and task scheduling. A team wiki was used to provide a central location for this information and proved to be a valuable resource by providing a living document for reference as assets were added to the library. Providing detail about each step of the process was important. If changes to an asset are required, it is critical to be able to go back into the early steps of the workflow and understand the design choices that were made previously, possibly by a different artist. To assist with these cases, the documentation included information regarding software tools, key settings, and process walkthroughs. The detailed documentation helped maintain a more complete shared knowledge base; this can be especially critical for project continuity if team members rotate or leave the project.

3.2 Geometry

The geometric accuracy and detail requirements of an asset are driven by the intended purpose of the model in the simulation as identified in the specifications document. For generating engineering precision models, CAD files from the product manufacturer can save a significant amount of time and effort. Dimensioned, orthographic views (top, side, and front/rear) may also be available. As depicted in Fig. 3, such drawings can be used for development or as verification of accuracy and scale. Heavy equipment brochures also often include tabular data on parameters, such as wheelbase, tire type, and minimum or maximum values, for key features that are invaluable to the artist. Online model warehouses provide a wide range of options in both cost and quality. Lastly, if access to the equipment is available, direct measurement of key dimensions is advisable. This is true even if CAD data is readily available, especially for high-cost equipment that may undergo significant customization either during manufacture or at some point during the service life of the equipment. If the goal is to simulate the equipment as it exists in the real world, an idealized version will not suffice.

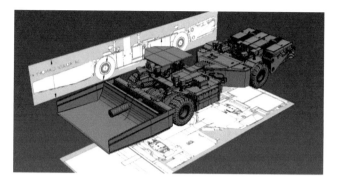

Fig. 3. A view from the geometric development software, depicting the use of orthographic drawings of the side and top views that have been scaled and positioned around the model to be used to verify the accuracy of the dimensions.

Laser scanning and photogrammetry can be useful when a high level of precision is required, but CAD files or drawings are not available and direct measurement is difficult or risky. These methods can be particularly useful for as-built applications. For example, in order to correctly calculate a point of regard on a curved 360° screen, an exact mesh was required. The screen was fabric stretched over a metal structure resulting in an irregular shape, but it was unable to be touched, and therefore could not be physically measured. Figure 4 shows the progression from the actual screen, to a wireframe, and finally to the polygon mesh developed using photogrammetric methods.

Fig. 4. The progression of the creation of a photogrammetric model of a 360° VR theater screen showing, from left to right, the actual screen, wireframe, and polygon mesh.

For most assets, the team uses standard box-modeling techniques. Creating a model from base shapes, extrusions, and other vertex and edge manipulations allows for better control and organization of polygons to ensure a clean topology, meaning there are no unintended discontinuities or overlapping polygons. A good topology aids in the ease of editing when performing overall mesh optimization, UV layout tasks, and LOD optimization as when developing LODs for tire models (Fig. 5).

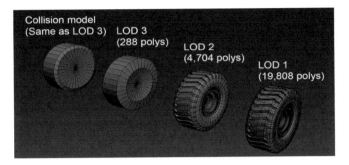

Fig. 5. Level of detail and collision meshes for a mining vehicle tire.

When developing geometry, it is also critical to examine how it all fits together in 3D. One of the main assets developed by the team was a set of tileable mine environment meshes used to generate the underground mine. Tile edges were modeled to have coincident vertices so that the tile edges would appear seamless. But, as the meshes were separated in the development software—3DS Studio Max—the vertex normal values on either side of the split were automatically recalculated, causing a visible discontinuity at the boundary in some lighting conditions in the game engine. Because this issue was not discovered during the geometry review step, it required significant additional effort to identify and correct the issue, test the solution, and then rework the geometry, materials, and UV layouts of assets.

3.3   Materials

The team uses a variety of sources and tools to develop materials including Allegorithmic's Substance suite. The software license includes access to a vast library of pre-made materials that require little modification for application in simulating industrial materials. Additional materials are also readily available at no cost from several online material libraries. Realistic objects can be built up from blending several materials and adding additional nonprocedural layer effects. For example, the general material developed for a scoop was created by layering rust and wear on a base material of painted metal as depicted in Fig. 6.

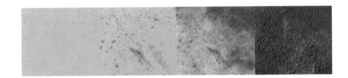

Fig. 6. Four stages of rust and wear on a painted metal surface created by the artist by applying various material layers in Substance Painter software.

In addition to material libraries, new materials can be created from base photographs. Substance's B2M (bitmap to material) software helps the artist create tileable materials from photographs. Basic parameters can be changed to generate albedo, normal, metallic, roughness, and ambient occlusion texture maps, where diffuse, normal, and specular/metallic maps are the minimum for a basic PBR material. As an example, the team used the B2M tool to create a coal material for an underground mine VE. Starting from a photograph, color values on the albedo map were brightened so that the values fell within the recommended PBR range. The other texture maps were then automatically generated from the edited image (Fig. 7).

Fig. 7. Material texture maps developed from a photograph of coal using Substance B2M software. From left to right: original albedo map, color flattened albedo map, normal map, and metallic map.

3.4   UV Layout

UV mapping should efficiently allocate the texture space to use more pixels in areas where detail is required, such as vehicle controls or gauges, and use less in areas where detail is not required, such as the canopy of the cab. However, care should be taken to not create a stark contrast in texel density between adjacent geometry, which could result in a distinct discontinuity.

Part of optimizing a UV layout includes accounting for the texture space required for padding around the edges to avoid mipmapping artifacts. As a general rule of thumb, one pixel of padding should be given for every 128 pixels of texture resolution. For an 8K texture, that means that 64 pixels of padding are needed. Figure 8 displays an albedo texture that was created for a mine vehicle, where the striped areas of the image are the padding regions.

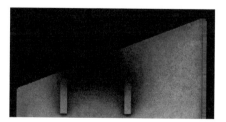

Fig. 8. Albedo texture map for a mine vehicle model depicting edge padding to eliminate artifacts introduced during mipmapping.

Mipmapping errors are also a good example of why prototyping can be important for complex or unfamiliar methods. In this case, an underground environment was being created to be viewed both as a first person player and in a zoomed-out overhead configuration, but the overhead view was not included in the review. Therefore, the padding error was not identified until the final implementation. In this case, all the geometry, materials, and UV mapping had to be redone to fix the rendering artifact.

Lastly, the UV layout of a model should be created from the highest level of detail (LOD) before lower-polygon LODs are generated. The lower LODs inherit the UV layout of the higher LOD. Furthermore, it is important to inherit the UV layout from the highest LOD because it includes the detailed geometry and smaller subcomponents such as welds, bolts on tires, and cab controls on vehicles that would not otherwise be mapped.

3.5 Shader

Shaders and the lighting pipeline of a game engine are the last direct line of control for visualizing an asset in a VE. The NIOSH development team uses Unity's standard shader for most of the developed assets. The standard shader by default includes a metallic map to control areas of the material that represent exposed metal. With the metallic setup, the color and intensity of specular highlights (direct reflections of light) are generated in a more naturalist manner based on the underlying albedo texture and reflections from the environment. For efficiency, the metallic option uses the alpha channel of the metallic map to modify the overall smoothness of the material. Another configuration of this shader is "specular option." This configuration adds the functionality to directly control the color and intensity of specular highlights (e.g., the perceived glint on objects). This configuration is still considered a PBR shader in that it respects conservation of energy principles, but because the specular highlights are manually manipulated, it can result in unrealistic material behavior.

Different materials can be created that share the same textures, but produce a unique rendered image based on parameter settings. Exemplifying some of these parameters, Fig. 9 displays the user interface for Unity's standard shader in the "standard" configuration. In this case, there is a multiplier on the normal texture map—currently set to 1.2—that can change the perceived depth of geometry that is modeled as a texture (e.g., wrinkles or folds). Overall, shaders help the developer to fine-tune the appearance of assets in the VE. Because of this, it may be tempting to use shader parameters to compensate for poorly constructed materials (e.g., increasing the intensity for a normal map). However, such errors should be returned to the artist for correction. For example, decreasing the normal map multiplier to reduce the overall surface roughness may make it difficult to use this parameter later to modulate the appearance of wear and tear in the VE. Ideally, keeping shader properties standardized across a project helps maintain a consistent visual aesthetic.

Fig. 9. A screenshot of the Unity Inspector window, illustrating how individual texture maps are assigned to the material channels, and parameters are set in the standard shader.

4   Conclusions

Virtual reality (VR) investment has increased rapidly in recent years, and the VR market could exceed 117 billion U.S. dollars by 2020 [17]. The human factors research community stands to benefit from this growth as VR solutions (hardware, content, and development tools) become widespread. Asset development methods for VR must evolve as continual improvement of VR technologies drives the fidelity of these systems closer to reality (Fig. 10). As simulations become more convincing, it is more important than ever to ensure the underlying validity of these visualizations.

Fig. 10. An example of the realism provided by physically based rendering as demonstrated by the rendering of the operator's cab of an underground mine vehicle.

Adherence to structured workflows can be helpful in achieving this goal, especially for part-time or nontechnical developers as they continue to use PBR and other advanced techniques. Focus on communication, including documentation, ensures that the development team meets the customer needs throughout the process. Attention to detail with frequent review steps provides the ability for teams to self-correct before effort is wasted. Prototyping and experimentation are valuable for team members to try new techniques, learn new skills, and manage customer expectations. Overall, a well-thought-out and executed asset development plan can significantly improve the quality and consistency of virtual environment (VE) digital assets.

Disclaimer. The findings and conclusions in this paper are those of the authors and do not necessarily represent the official position of the National Institute for Occupational Safety and Health, Centers for Disease Control and Prevention. Mention of company names or products does not constitute endorsement by NIOSH.

References

1. Parsons, T.D., Rizzo, A.A.: Affective outcomes of virtual reality exposure therapy for anxiety and specific phobias: a meta-analysis. J. Behav. Ther. Exp. Psychiatry **39**(3), 250–561 (2008)
2. Parsons, T.D., Riva, G., Parsons, S., Mantovani, F., Newbutt, N., Lin, L., Venturini, E., Hall, T.: Virtual reality in pediatric psychology. Pediatrics **140**(Suppl. 2), S86–S91 (2017)
3. Seymour, N.E., Gallagher, A.G., Roman, S.A., O'Brien, M.K., Bansal, V.K., Andersen, D. K., Satava, R.M.: Virtual reality training improves operating room performance: results of a randomized, double-blinded study. Ann Surg. **236**(4), 458 (2002)
4. Koch, K., McLean, J., Segev, R., Freed, M.A., Berry II, M.J., Balasubramanian, V., Sterling, P.: How much the eye tells the brain. Curr. Biol. **16**(14), 1428–1434 (2006)
5. Rosenblum, L.D.: See What I'm Saying: The Extraordinary Powers of Our Five Senses. WW Norton & Company, New York (2011)
6. Ni, T., Bowman, D., Chen, J.: Increased display size and resolution improve task performance in information-rich virtual environments. In: Proceedings of Graphics Interface, pp. 139–146 (2006)
7. Lee, C., Rincon, G.A., Meyer, G., Höllerer, T., Bowman, D.A.: The effects of visual realism on search tasks in mixed reality simulation. IEEE Trans. Vis. Comput. Graph. **19**(4), 547–556 (2013)
8. Sargent, R.G.: Verification and validation of simulation models. J. Simul. **7**(1), 12–24 (2013)
9. Schofield, D.: Guidelines for using game technology as educational tools. Educ. Learn. **12** (2), 224–235 (2018)
10. Orr, T.J., Filigenzi, M.T., Ruff, T.M.: Hazard recognition computer based simulation. In: Proceedings of the Thirtieth Annual Institute on Mining Health, Safety and Research, pp. 21–28 (1999)
11. Navoyski, J., Brnich Jr., M.J., Bauerle, T.: BG 4 benching training software for mine rescue teams. Coal Age **120**(12), 50–55 (2015)
12. NIOSH: Underground Coal Mine Map Reading Training (2009). https://www.cdc.gov/niosh/mining/works/coversheet1825.html. Retrieved 27 Oct 2018
13. Orr, T.J.: NIOSH Mine Emergency Escape Simulation Technology Available for Developers. NIOSH Science Blog (2016). https://blogs.cdc.gov/niosh-science-blog/2016/05/12/mine-escape-simulation/. Retrieved 27 Oct 2018
14. Greenberg, D.P., Torrance, K.E., Shirley, P., Arvo, J., Lafortune, E., Ferwerda, J.A., Foo, S. C.: A framework for realistic image synthesis. In: Proceedings of the 24th Annual Conference on Computer Graphics and Interactive Techniques, pp. 477–494 (1997)
15. Pharr, M., Jakob, W., Humphreys, G.: Physically Based Rendering: From Theory to Implementation. Morgan Kaufmann Publishers, Burlington (2016)

16. Friston, S., Fan, C., Doboš, J., Scully, T., Steed, A.: 3DRepo4Unity: dynamic loading of version-controlled 3D assets into the unity game engine. In: Proceedings of the 22nd International Conference on 3D Web Technology, p. 15 (2017)
17. MarketWatch: Virtual and Augmented Reality Market is Expected to Exceed US$ 117 Billion by 2022 (2018). https://www.marketwatch.com/press-release/virtual-and-augmented-reality-market-is-expected-to-exceed-us-117-billion-by-2022-2018-07-26. Retrieved 27 Oct 2018

Influence of Social Norms on Decision-Making Against Landslide Risks in Interactive Simulation Tools

Pratik Chaturvedi[1,2(✉)] and Varun Dutt[1]

[1] Applied Cognitive Science Laboratory, Indian Institute of Technology Mandi,
Kamand 175005, India
varun@iitmandi.ac.in
[2] Defence Terrain Research Laboratory, Defence Research and Development
Organization, Delhi 110054, India
prateek@dtrl.drdo.in

Abstract. Landslide disasters, i.e., movement of hill mass, cause significant damages to life and property. People may be educated about landslides via simulation tools, which provide simulated experiences of cause-and-effect relationships. The primary objective of this research was to test the influence of social norms on people's decisions against landslides in an interactive landslide simulator (ILS) tool. In a lab-based experiment involving ILS, social norms were varied across two between-subject conditions: social (N = 25 participants) and non-social (N = 25 participants). In social condition, participants were provided feedback about investments made by a friend against landslides in addition to their investments. In non-social condition, participants were not provided feedback about friend's investments, and they were only provided feedback about their investments. People's investments were significantly higher in the social condition compared to the non-social condition. We discuss the benefits of using the ILS tool for educating people about landslide risks.

Keywords: Human factors · Human-systems integration ·
Systems engineering · Simulation games · Social norms · Experience ·
Landslides

1 Introduction

Worldwide, landslides cause huge losses in terms of human fatalities, injuries, and infrastructure damages [1]. In fact, landslides are a major concern for disaster-prevention groups in regions with steep terrain, especially in the Himalayan mountains [2]. Due to the catastrophic effects of landslides, it is essential to make people at risk understand the causes-and-consequences of landslide disasters [3]. This understanding is likely to help vulnerable communities make informed decisions against landslide disasters. However, prior research suggests that people in developing countries (like India), who reside in landslide-prone areas, possess misconceptions about landslide risks [4–6]. One way of improving people's understanding about landslides may be via exposure to landslide simulation tools and games [7–9].

© Springer Nature Switzerland AG 2020
D. N. Cassenti (Ed.): AHFE 2019, AISC 958, pp. 300–310, 2020.
https://doi.org/10.1007/978-3-030-20148-7_27

Prior research shows that interactive simulation tools and games have been effective in providing experience and visibility of underlying system dynamics to people across a wide variety of problems [10–12]. For example, references [10, 11] developed a Dynamic Climate Change Simulator (DCCS) tool, which provided feedback to people about their decisions against climate change. This feedback enabled people to reduce their misconceptions about climate change. In a similar way, references [8, 9] developed an Interactive Landslide Simulator (ILS) tool based on the hypothesis that experience and recency of events through a simulation exercise, may be influential in improving the public's awareness and perceptions about landslide disasters [10, 11]. Results revealed that repeated feedback in ILS improved the people's decision-making against landslide risks [8, 9].

Although the use of feedback in improving learning in simulation tools has been explored across a variety of domains [6–11], there is less literature on how social norms, i.e., an acceptable behavior in a particular group, influences learning in simulation tools. Prior literature [13–15] suggests that social norms may play a significant role in shaping people's decision-making in the real-world. Also, social norms may be effective in multi-agent simulations [16, 17]. However, the applications of social norm in simulation tools for educating people about landslide disasters has not been explored in literature.

The current research addresses this literature gap by investigating the impact of social norms on people's decision-making against landslides in the ILS tool. The central hypothesis under test is that monetary contributions against landslide risks in the ILS tool (which is an indicator of improved understanding) will be larger under the influence of social norms compared to when these norms are not present. Based on results of a lab-based experiment involving human participants, we propose some benefits of the use of social norms in the ILS tool for educating people about landslide risks.

In what follows, first, we discuss background literature about use of simulation tools and social norms. Next, we report an experiment in which human participants performed in the ILS tool in the presence or absence of social norms. Finally, we report results from the experiment and highlight the potential of using social norms in simulation tools for imparting landslide awareness and education.

2 Background

Prior research has clearly established that social norms not only spur but also guide action in direct and meaningful ways [18–21]. A number of researchers have applied the concept of social norms in computer simulation of real-world phenomena [13, 22–25]. For example, reference [21] applied and examined the use of psychological model of social identity and social norms to computer models of pedestrian motion during evacuation in an emergency situation. Similarly, reference [22] compared empirical data on human behavior to simulated data on agents with values and norms in a psychological experiment involving the ultimatum game. Given this asserted power of social norms, there has been a surge of programs that have delivered normative information as a primary tool for changing socially significant behaviors, such as alcohol consumption, drug use, disordered eating, gambling, littering, and recycling [26, 27]. Such social-norms marketing campaigns have emerged as an alternative to more traditional approaches (e.g., information campaigns, moral exhortation, fear inducing messages) designed to reduce undesirable conduct [26]. In a study, reference [13] conducted a field experiment in which normative messages were used to promote household energy conservation and found that a descriptive normative message detailing average neighborhood usage produced either desirable energy savings or the undesirable boomerang effect, depending on whether households were already consuming at a low or high rate. They also found that adding an injunctive message (conveying social approval or disapproval) eliminated the boomerang effect [13]. Social norms have also been used in public goods games (PGGs) for studying how people exhibit cooperative behavior [13, 28, 29]. In PGG, participants may either contribute an amount to the public good or defect by contributing nothing. Reference [13] has shown that punishments in PGG as a societal norm can foster cooperative behavior.

Although there is extensive literature on social norms in simulation tools [13, 22–25], applications of social norms to decision-making against landslide risks in simulation tools has not been explored. References [8, 9] have used the ILS tool to understand the role of amount and availability of feedback in decision making against landslides. In this research, we use social norms in the ILS tool to improve people's decision-making and investments against landslides. We create norms in the ILS tool by exposing participants to the investments of other participants against landslides. We believe that an increase in investments by others in the ILS tool will make people contribute more against landslides and these contributions will enable people to improve their understanding about landslides. Thus, we expect participants' investments against landslides to be higher when norms are present in the ILS tool compared to when these norms are absent.

3   Interactive Landslide Simulator (ILS) Tool

The ILS tool is an interactive dynamic system for studying people's decisions against landslide risks [8, 9]. Details about the ILS tool were already discussed by [8, 9], and here we briefly cover the tool's working. Figure 1 shows the investment screen of ILS tool, where people need to invest some part of their daily income against landslide risks. Investment against landslides would be used to mitigate landslide risks via interventions like building reinforcements and planting trees.

Feedback is shown to participants in three ways (Fig. 2A, 2B): monetary information about total wealth, text messages about different losses, and imagery corresponding to losses. There is a decrease in the daily income due to an injury or fatality due to a landslide. Also, damages to property due to landslides cause a loss of property wealth. In addition, in situations involving the use of social norms, the feedback screen shows the contribution from a friend against landslides (Fig. 2A). The friend function is linear: It starts at half of the total investment (=292/2) and increases towards 292 by the last trial. A normally distributed noise is added to the linear function (normal distribution's mean = 0 and standard deviation = 5). The noise component makes it difficult to find patterns in the friend function. In situations not involving the use of social norms, the friend function is missing from the feedback screen (Fig. 2B). Positive feedback is presented to participants on the feedback screen if a landslide does not occur in a certain trial (Fig. 2C). The friend's investments against landslide across trials is shown in Fig. 2D.

Human induced landslide risks were generated using the probability equation from reference [30]. The likelihood of landslides due to physical factors is calculated considering the combination of the effects of rainfall, slope, and soil properties. It is based on the method proposed by reference [31]. The probability of a landslide due to rainfall P(R) is a random event, based on the computed value z, where

$$P(R) = \frac{1}{1 + e^{-z}} \tag{1}$$

and,

$$z = -3.817 + (DR) * 0.077 + (3DCR) * 0.058 + (30DAR) * 0.009$$

$$z : (-\infty, +\infty) \tag{2}$$

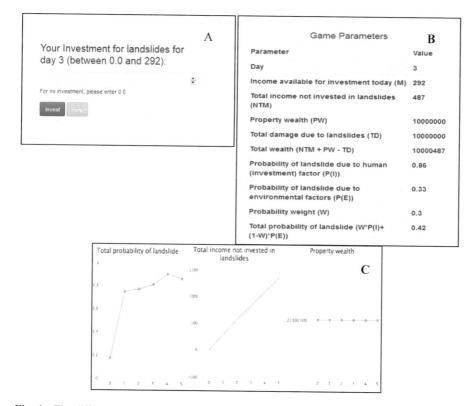

Fig. 1. The different components on the investment screen of ILS tool. (A) The text box giving choice for daily investments to reduce landslide risk. (B) Game parameters window showing values of parameters used in ILS model. (C) Dynamic plots of changing outcomes with every decision.

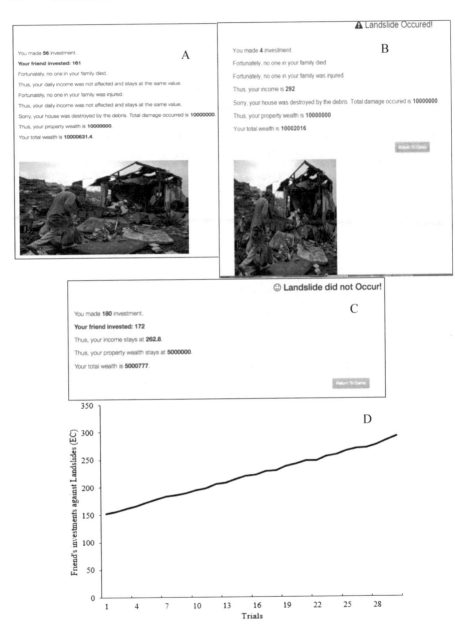

Fig. 2. Feedback screen in the ILS tool. (A) Feedback screen with information on the contribution of a friend as well as negative feedback due to a landslide's occurrence. (B) Feedback screen with no information on the contribution of a friend as well as negative feedback due to a landslide's occurrence. (C) Feedback screen with positive feedback when a landslide does not occur. (D) Friend's investments against landslide risk across trials

With the daily (DR), 3-day cumulative (3DCR) and 30-day antecedent rainfall (30DAR) as significant predictors influencing slope failure. These rainfall values were calculated using daily rain data from Indian Metrological Department (IMD). Five years of daily rain data (2010–14) was averaged to find the average probability of landslide due to rainfall P(R) over an entire year.

The slope and soil characteristics are expressed as P(S), which represent the local probability of landslides, given the geological features of the location. The determination of spatial probability of landslides, P(S) is done from Landslide Susceptibility Zonation (LSZ) map of the area [8, 9].

A landslide occurs on a certain day when an independent random number ($\sim U$ $(0, 1)$) become less than or equal to the corresponding net probability of occurrence of landslide which is a weighted sum of landslide probability due to environment (spatial and triggering factors) and human factors. The natural environmental probability of a landslide event is the product of the two probabilities, P(S) and P(R). Data from reference [32] helped set the monetary levels in the ILS tool.

4 Experiment Involving Social Norms in the ILS Tool

The ILS tool considers both environmental factors (spatial geology and rainfall) and human factors (people's investments against landslides) for calculating landslide risks. The simulation, its validation, and full functions are explained in different publications [8, 9]. In this paper, we present results of an experiment in which human participants interacted with the ILS tool in the presence or absence of social norms. We expected participants' investments against landslides to be higher when norms were present compared to when norms were absent.

Experimental Design. The study was approved by the ethics committee at the Indian Institute of Technology Mandi, India. Fifty participants from diverse fields of study participated across two feedback conditions: social (N = 25) and non-social (N = 25). In the social condition, participants were provided feedback about contributions made by a friend against landslide risks on the feedback screen. In non-social condition, feedback about contributions made by a friend against landslide risk was absent. Data were analysed for all participants in terms of their investment ratio in both the feedback conditions. The investment ratio was defined as the ratio of total investments made by participants up to a trial divided by the total investments that could have been made up to the trial. Also, we computed the ratio of participants' investment ratio to friend's investment ratio. This latter statistic allowed us to match participants' investments to friend's investments. Given the effectiveness of feedback and social norms in simulation tools [26, 33–36], we expected participant investments to be greater in the social condition compared to the non-social condition. Also, we expected the ratio of participants' investment ratio to friend's investment ratio to be higher in the social condition compared to non-social condition.

Participants. All participants were from Mandi district and adjoining areas and their ages ranged in between 18 and 26 years (Mean = 21.4 years; Standard Deviation = 2.81 years). Around 30% participants were from science, technology, engineering, and mathematics (STEM) backgrounds and the remaining were from non-STEM backgrounds. Out of a total of 50 participants, 30 cited basic understanding about landslides, 14 cited little understanding about landslides, 5 responded as knowledgeable about landslides, and 1 possessed no idea about landslides. All participants received a base payment of INR 50 upon completing their performance in the ILS tool. In addition, a performance incentive was also provided to participants in the ILS tool. To calculate the performance incentive, participants were ranked based upon the total wealth remaining at the end of their ILS play and top-10 participants were put in a lucky draw. After the entire study was over, one participant was randomly selected and awarded a cash prize of INR 500. Participants were told about the lucky draw via written instructions before they started performing in the ILS tool.

Procedure. Participants were invited to a landslide awareness study via a flyer advertisement in Mandi, India. Participants signed a consent form and participation was entirely voluntary. Participants were first provided with instructions on the task in the study. Participants started their study once they were ready to begin and did not have any questions on the study. Participants started their study by providing demographic information and knowledge about landslides. Next, participants interacted in the ILS tool for a 30-day simulated time period. Experimental sessions were about 30-min long per participant. Participants were not given any information concerning the nature of the environment or conditions in the ILS. They were told that their goal was to maximize their total wealth across repeated decisions in the ILS tool.

5   Results

We performed a repeated-measures ANOVA with condition as a between-subjects factor and investment-ratio over the 30-day period as a within-subjects factor. As per our expectation, the average investment ratio was significantly higher in the social

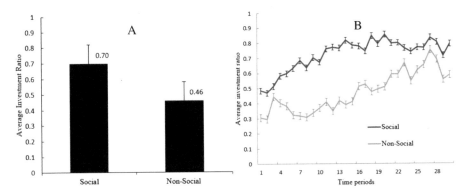

Fig. 3. (A) Average investment ratio in social and non-social conditions. The error bars show 95% CI around the point estimate. (B) Average investment ratio in social and non-social conditions across 30 trials.

condition (0.70) compared to that in the non-social condition (0.46) (F (1, 48) = 21.53, $p < 0.05$, η^2 = 0.31) (see Fig. 3A).

Furthermore, the trend of investment ratio across trials in the social condition was different from the trend of investment ratio across trials in the non-social condition (F (29, 1392) = 2.38, $p < 0.05$, η^2 = 0.05; see Fig. 3B). It can be clearly seen in Fig. 3B that there is a higher increase in investment ratio in the social condition compared to non-social condition. Overall, the results suggest that social norms feedback helped participants to increase their investments for landslide mitigation.

Next, we performed a repeated-measures ANOVA with condition as a between-subjects factor and ratio of participants' investment ratio to friend's investment ratio over the 30-day period as a within-subjects factor (see Fig. 4A). As per our expectation, the ratio of participants' investment ratio to friend's investment ratio was significantly higher in the social condition (1.13) compared to that in the non-social condition (0.73) (F (1, 48) = 59.34, $p < 0.001$, η^2 = 0.32). Furthermore, the trend of ratio of participants' investment ratio to friend's investment ratio across trials in the social condition was different from the trend of this dependent measure across trials in the non-social condition (F (29, 1392) = 2.77, $p < 0.001$, η^2 = 0.06; see Fig. 4B). It can be clearly seen in Fig. 4B that there is a higher increase in ratio of participants' investment ratio to friend's investment ratio in the social condition compared to non-social condition. Overall, the results suggest that social norms feedback helped participants' investments to match and exceed the friend's investments for landslide mitigation in the social condition compared to the non-social condition.

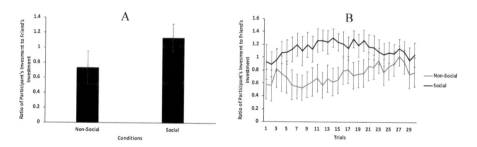

Fig. 4. (A) Ratio of participant's investment ratio to friend's investment ratio in social and non-social conditions. The error bars show 95% CI around the point estimate. (B) Ratio of participant's investment ratio to friend's investment ratio in social and non-social conditions across 30 trials.

6 Discussion and Conclusions

In this research, we investigated the use of social norms in a simulation tool, Interactive Landslide Simulator (ILS), against landslide risks. On average, people's contributions against landslides were significantly higher when social norms were present compared to when these norms were absent. Also, there was a consistent increase in people's contributions against landslides over trials in the presence of social norms compared to in the absence of social norms.

Our results are in agreement with those reported in prior literature [13–17, 21, 33–36]. In fact, reference [13, 21] found that telling people about what other people did in the neighborhood enabled people to agree with the societal norm. In the current research involving ILS, the presence of friend's contributions created a similar effect. In the current research, participants found the friend to be contributing against landslides and did not want to deviate from this norm. However, when the friend's contribution was missing, there was an absence of the norm. In the absence of the norm, people were not affected by it and they kept their contributions low.

This research has a number of implications involving the use of simulation tools. First, this research showed that social norms are impactful even in situations where there is a prevailing uncertainty about when landslides occur. Second, via a simulation, this research showed that social norms could be used to change people's decisions against landslide disasters. Overall, the presence of social norms in simulation tools like ILS is likely to be an effective method for educating people about landslide disasters. Researchers may also use tools like ILS to perform "what-if" analyses in the presence of social norms. These analyses may stretch over certain time periods and cover certain geographical locations. However, the assumptions made in the ILS tool may have to be first evaluated for the study area before it is used for policy research.

In the current research, we made certain assumptions in the ILS tool, and these assumptions may be different in the real world. For example, the model behind the ILS tool assumed that people's contributions against landslides influenced the probability of landslides linearly. Second, the damages due to landslides were fixed at the same constant value across all trials. We plan to overcome some of these assumptions as part of our future research.

First, as part of our future research, we plan to assume different models of how people's investments in the ILS tool may influence the probability of landslides. Second, we plan to try a number of models concerning damages due to landslides, where the landslide damages are different over trials. Also, as part of our future research, we plan to compare the proportion of investments against different insurance models available in the real world. Some of these ideas form the immediate next steps in our research program on landslide education and awareness.

Acknowledgments. This research was partially supported by the following grants to Varun Dutt: IITM/NDMA/VD/184 and IITM/DRDO-DTRL/VD/179. We thank Akshit Arora for developing the website for ILS. We also thank students of IIT Mandi who helped in data collection in this project.

References

1. Margottini, C., Canuti, P., Sassa, K.: Landslide science and practice. In: Proceedings of the Second World Landslide Forum, Rome, Italy, vol. 2 (2011)
2. Chaturvedi, P., et al.: Remote sensing based regional landslide risk assessment. Int. J. Emerg. Trends Electr. Electron. (IJETEE) **10**(10), 135–140 (2014). ISSN 2320–9569
3. Wagner, K.: Mental models of flash floods and landslides. Risk Anal. **27**(3), 671–682 (2007)

4. Oven, K.: Landscape, livelihoods and risk: community vulnerability to landslides in Nepal. Doctoral dissertation, Durham University (2009)
5. Wanasolo, I.: Assessing and mapping people's perceptions of vulnerability to landslides in Bududa, Uganda. Doctoral dissertation (2012)
6. Chaturvedi, P., Dutt, V.: Evaluating the public perceptions of landslide risks in the Himalayan Mandi Town. Accepted for presentation in the 2015 Human Factor & Ergonomics Society (HFES) Annual Meeting, L.A (2015)
7. Knutti, R., Joos, F., Müller, S.A., Plattner, G.K., Stocker, T.F.: Probabilistic climate change projections for CO2 stabilization profiles. Geophys. Res. Lett. **32**(20), L20707, 1–3 (2005)
8. Chaturvedi, P., Arora, A., Dutt, V.: Learning in an interactive simulation tool against landslide risks: the role of strength and availability of experiential feedback. Nat. Hazards Earth Syst. Sci. **18**(6), 1599–1616 (2018)
9. Chaturvedi, P., Arora, A., Dutt, V.: Interactive landslide simulator: a tool for landslide risk assessment and communication. In: Advances in Applied Digital Human Modeling and Simulation, pp. 231–243. Springer, Cham (2017)
10. Dutt, V., Gonzalez, C.: Human control of climate change. Clim. Change **111**(3–4), 497–518 (2012)
11. Dutt, V., Gonzalez, C.: Why do we want to delay actions on climate change? Effects of probability and timing of climate consequences. J. Behav. Decis. Mak. **25**(2), 154–164 (2012)
12. Sterman, J.D.: Learning in and about complex systems. Syst. Dyn. Rev. **10**(2–3), 291–330 (1994)
13. Schultz, P.W., Nolan, J.M., Cialdini, R.B., Goldstein, N.J., Griskevicius, V.: The constructive, destructive, and reconstructive power of social norms. Psychol. Sci. **18**(5), 429–434 (2007)
14. Neighbors, C., Larimer, M.E., Lewis, M.A.: Targeting misperceptions of descriptive drinking norms: efficacy of a computer-delivered personalized normative feedback intervention. J. Consult. Clin. Psychol. **72**(3), 434 (2004)
15. Hauert, C.: Replicator dynamics of reward & reputation in public goods games. J. Theor. Biol. **267**(1), 22–28 (2010)
16. Fehr, E., Fischbacher, U.: Social norms and human cooperation. Trends Cogn. Sci. **8**(4), 185–190 (2004)
17. Pan, X., Han, C.S., Dauber, K., Law, K.H.: A multi-agent based framework for the simulation of human and social behaviors during emergency evacuations. AI Soc. **22**(2), 113–132 (2007)
18. Aarts, H., Dijksterhuis, A.: The silence of the library: environment, situational norm, and social behavior. J. Pers. Soc. Psychol. **84**(1), 18–28 (2003). https://doi.org/10.1037/0022-3514.84.1.18
19. Cialdini, R.B., Kallgren, C.A., Reno, R.R.: A focus theory of normative conduct: a theoretical refinement and reevaluation of the role of norms in human behavior. In: Advances in Experimental Social Psychology, vol. 24, pp. 201–234. Academic Press, Boston (1991)
20. Griskevicius, V., Goldstein, N.J., Mortensen, C.R., Cialdini, R.B., Kenrick, D.T.: Going along versus going alone: when fundamental motives facilitate strategic (non) conformity. J. Pers. Soc. Psychol. **91**(2), 281 (2006)
21. von Sivers, I., Templeton, A., Künzner, F., Köster, G., Drury, J., Philippides, A., Neckel, T., Bungartz, H.J.: Modelling social identification and helping in evacuation simulation. Saf. Sci. **89**, 288–300 (2016)
22. Mercuur, R., Dignum, V., Jonker, C.: The value of values and norms in social simulation. J. Artif. Soc. Soc. Simul. **22**(1), 1–9 (2019)

23. Savarimuthu, B.T.R., Purvis, M., Purvis, M., Cranefield, S.: Social norm emergence in virtual agent societies. In: International Workshop on Declarative Agent Languages and Technologies, pp. 18–28. Springer, Heidelberg, May 2008
24. Axelrod, R.: An evolutionary approach to norms. Am. Polit. Sci. Rev. **80**(4), 1095–1111 (1986)
25. Gilbert, N.: Agent-based social simulation: dealing with complexity. Complex Syst. Netw. Excellence **9**(25), 1–14 (2004)
26. von Sivers, I., Templeton, A., Köster, G., Drury, J., Philippides, A.: Humans do not always act selfishly: social identity and helping in emergency evacuation simulation. Transp. Res. Procedia **2**, 585–593 (2014)
27. Donaldson, S.I., Graham, J.W., Hansen, W.B.: Testing the generalizability of intervening mechanism theories: understanding the effects of adolescent drug use prevention interventions. J. Behav. Med. **17**(2), 195–216 (1994)
28. Chu, M.L., Parigi, P., Law, K., Latombe, J.C.: SAFEgress: flexible platform to study the effect of human and social behaviors on egress performance. In: Proceedings of the Symposium on Simulation for Architecture & Urban Design, p. 4. Society for Computer Simulation International, April 2014
29. Hasson, R., Löfgren, Å., Visser, M.: Climate change in a public goods game: investment decision in mitigation versus adaptation. Ecol. Econ. **70**(2), 331–338 (2010)
30. Mathew, J., Babu, D.G., Kundu, S., Kumar, K.V., Pant, C.C.: Integrating intensity–duration-based rainfall threshold and antecedent rainfall-based probability estimate towards generating early warning for rainfall-induced landslides in parts of the Garhwal Himalaya. India. Landslides **11**(4), 575–588 (2014)
31. Parkash, S.: Historical records of socio-economically significant landslides in India. J. South Asia Disaster Stud. **4**(2), 177–204 (2011)
32. Roos, P., Gelfand, M., Nau, D., Lun, J.: Societal threat and cultural variation in the strength of social norms: an evolutionary basis. Organ. Behav. Hum. Decis. Process. **129**, 14–23 (2015)
33. Sterman, J.D.: Risk communication on climate: mental models and mass balance. Science **377**, 532–533 (2008)
34. Cialdini, R.B.: Crafting normative messages to protect the environment. Curr. Dir. Psychol. Sci. **12**, 105–109 (2003)
35. Dutt, V.: Social influence encourages conservation behaviours (2012). Financial Chronicle: http://www.mydigitalfc.com/knowledge/social-influence-encourages-conservation-behaviours-011
36. Kumar, M., Aggarwal, K., Dutt, V.: Modeling decisions in collective risk social dilemma games for climate change using reinforcement learning. In: IEEE Conference on Cognitive and Computational Aspects of Situation Management (CogSIMA), Las Vegas, Nevada (2019, in press, accepted)

Author Index

© Springer Nature Switzerland AG 2020
D. N. Cassenti (Ed.): AHFE 2019, AISC 958, pp. 311–312, 2020.
https://doi.org/10.1007/978-3-030-20148-7